SALESFORCE CONFIGURATION WITH CHATGPT

CONVERSATION TO ACCELERATE YOUR CLICKS (AND CODE)

ANIRUDH GANESH BHARADWAJ

ANDY FORBES

Roadmap Press
Hillsboro, Virginia, United States

ISBN: 978-1-969489-14-3 (paperback)
ISBN: 978-1-969489-04-4 (ebook)

Cover design by: Roadmap Press

This book is dedicated to our colleagues and collaborators, whose curiosity, patience, and practical judgment turn complex systems into tools that actually work. We are grateful to the teams at **Salesforce** *and* **OpenAI**, *whose platforms and ideas are shaping a new generation of human-computer collaboration, and to the many coworkers, partners, customers, architects, administrators, developers, and analysts who put these tools to use every day in the real world. More broadly, we thank everyone who has contributed to building, advancing, teaching, and using computers over the decades. This work rests on the accumulated effort of countless people who believed that knowledge, when shared and applied with care, could make work better, systems more humane, and progress possible.*

Anirudh Ganesh Bharadwaj & Andy Forbes

Perfection is achieved not when there is nothing more to add, but when there is nothing left to take away.

— ANTOINE DE SAINT-EXUPÉRY

CONTENTS

PREFACE

This book starts from a simple observation: configuring a modern SaaS platform is no longer just a matter of knowing where the checkboxes are. Platforms like **Salesforce** have grown into living systems, shaped as much by judgment, tradeoffs, and context as by features and code. At the same time, tools like ChatGPT have introduced a new way of thinking through complexity. Not by replacing expertise, but by giving experienced practitioners a thinking partner that can accelerate exploration, clarify options, and surface patterns that might otherwise take hours or days to assemble.

That combination changes how good work gets done. Used well, ChatGPT does not "configure Salesforce for you." Instead, it helps you reason more clearly about requirements, constraints, data models, security boundaries, automation strategies, and long-term maintainability. It helps you pressure-test assumptions, explore alternative designs, and explain your thinking in language that both technical and non-technical stakeholders can understand. Used poorly, it can just as easily amplify confusion, propagate bad patterns, or produce answers that sound confident but collapse under scrutiny. The difference is not the tool. It is the practitioner.

This book is written for people who already take responsibility for real outcomes. Architects, administrators, developers, business analysts, product owners, and delivery leaders who live with the consequences of configuration decisions long after the sprint ends. It assumes you understand that "working" is not the same as "working well," and that every declarative setting, flow, data model, and integration carries downstream implications. ChatGPT is treated here as an accelerator for that kind of professional judgment, not a shortcut around it.

You will not find step-by-step click paths for every Salesforce feature, nor a catalog of prompts meant to be copied blindly. Instead, you will learn how to *think with* ChatGPT. How to frame questions so context matters. How to challenge its answers. How to use it to compare approaches, document intent, uncover edge cases, and translate between business goals and technical design. In other words, how to integrate it into the way experienced teams already work, without surrendering accountability or rigor.

This book also reflects a broader moment in computing. For decades, progress has come from better abstractions layered on top of prior ones. ChatGPT, built by teams at **OpenAI** and others across the industry, represents another such layer. Powerful, flexible, and still imperfect. Like every major shift before it, its value will ultimately be determined not by what it can generate, but by how thoughtfully people choose to use it.

If this book does its job, you will finish it with a clearer mental model for using ChatGPT as part of your Salesforce practice. Not as an authority, not as automation magic, but as a disciplined collaborator that helps you do better work, faster, with fewer blind spots. The responsibility remains yours. This book is here to help you carry it well.

1

GETTING STARTED

Welcome to Salesforce Configuration with ChatGPT! In an era where businesses demand rapid innovation and agility, Salesforce's "clicks, not code" philosophy has emerged as a transformational approach. Imagine being able to deliver a full-featured CRM enhancement or a custom app for your company in days or weeks – without writing a single line of code. It might sound like magic, but it's exactly what Salesforce's declarative tools empower you to do.

This chapter sets the stage by highlighting the power of Salesforce's declarative development toolkit and how it can fundamentally change the way your organization builds solutions. If you're a **new Salesforce admin**, you'll discover that you don't need a computer science degree to create robust business processes and dynamic user experiences. If you're an **experienced developer**, you'll gain a renewed appreciation for what can be achieved with clicks alone – and how reserving code for only truly complex scenarios can save time and reduce technical debt. For the **business analyst or product owner**, this introduction reveals how you can directly shape system capabilities to match business needs, translating requirements into working features with unprecedented speed. In short, whether you're technical or not, mastering declarative configuration unlocks a more collaborative and efficient way to build on the Salesforce platform.

Salesforce's platform was built from the ground up with a configuration-first mindset. Over the years, the suite of "point-and-click" tools has grown from simple field setups to incredibly powerful builders – think **Flow Builder** for automation, **Lightning App Builder** for user interfaces, and **Dynamic Forms** for responsive page layouts. The philosophy of "clicks, not code" isn't about oversimplifying development; it's about **maximizing impact while minimizing complexity.** Declarative does not mean "basic" – in fact, it often means **sophisticated solutions without the bloat** of custom code. A well-crafted Flow or a cleverly designed declarative approval process can handle scenarios that once required dozens of lines of Apex code, all while being easier to maintain and update. Salesforce continuously enriches these tools (for example, Dynamic Forms was a long-awaited enhancement that finally arrived after 13 years of anticipation, instantly giving admins far more flexibility with page layouts). The result is that the gap between what can be done declaratively versus programmatically keeps narrowing.

In the spirit of innovation, this book also introduces a modern sidekick into the mix: **ChatGPT** and generative AI. Think of ChatGPT as your intelligent co-pilot as you navigate Salesforce configuration. Throughout the chapters, we'll show how AI tools can dramatically accelerate and improve every aspect of declarative development – from brainstorming automation ideas, to documenting configuration for governance, to troubleshooting issues. You'll see examples of interacting with ChatGPT by typing or even using voice, of uploading Salesforce export files for analysis, and even leveraging advanced features like the ChatGPT **Operator** for web browsing and **Sora** for generating illustrative videos. These aren't gimmicks; they represent a new wave of productivity. For instance, you might draft a complex validation rule with ChatGPT's help, or quickly get a summary of your org's security settings by feeding them to the AI. By weaving ChatGPT into our Salesforce toolkit, we combine human creativity and intuition with AI's speed and breadth of knowledge.

Before we dive in, let's map out where we're headed in this opening chapter. We'll start by clearly defining what **declarative development** means in the Salesforce context and how it contrasts with traditional coding. Understanding this foundation will make it evident why Salesforce invested so heavily in "low-code" tools. From there, we'll explore the **business and technical benefits** that make declarative configuration so appealing – from faster time-to-market to lower maintenance overhead. Next, we will discuss adopting a **Salesforce Well-Architected mindset**, a set of guiding principles (Trusted, Easy, Adaptable) that help ensure your solutions are built for long-term success. With that mindset, we'll look at the importance of **governance and proactive org management** – because even clicks-not-code can get chaotic without a plan. We'll then examine how to **design declarative solutions for scalability**, ensuring your apps and automations can grow with your business. No introduction would be complete without touching on **security and compliance**, as Salesforce offers powerful declarative means to keep data safe and processes auditable. Finally, we'll wrap up the chapter with strategies for **performance optimization and user adoption**, so that your declarative solutions run smoothly and actually get used by people. Each section ends with a quick narrative bridge to the next, guiding you through this story of declarative power.

By the end of this chapter, you should feel both informed *and* inspired. Salesforce declarative configuration isn't just a set of admin tasks – it's a strategic advantage when approached with the right knowledge and mindset. And with ChatGPT's help, you have a new ally to push the boundaries of what's possible, faster than ever. Let's embark on this journey to unlock the full potential of "clicks, not code," and see how it can make you a hero in your organization.

UNDERSTANDING SALESFORCE DECLARATIVE DEVELOPMENT

Salesforce **declarative development** means building applications and automations using configuration and clicks instead of writing code. In practice, this involves using Salesforce's point-and-click tools – creating objects and fields in Object Manager, designing pages with Lightning App Builder, writing formulas, setting up validation rules, building flows for automation, and so on. It's the opposite of **programmatic development** (where you would write Apex code, Visualforce, or Lightning Web Components). Declarative development is often summed up by the mantra **"clicks, not code."** This doesn't imply that no code exists at all – under the hood, Salesforce's platform handles the code for you. It means you, as the builder, can focus on *what* the solution should do, and Salesforce handles *how* to execute it. The result is that you can deliver functionality much faster and usually with fewer errors, because you're assembling tested, pre-built components rather than crafting everything from scratch.

One way to appreciate the power of declarative development is to compare it directly with coding on a few key dimensions:

Speed of Delivery (Cost and Time): Building with clicks is typically much faster than writing and deploying custom code. You don't need to set up development environments or write extensive test classes just to deploy. A process that might take a developer several days to code and unit test could potentially be configured in a few hours with declarative tools. This translates to lower development costs and faster time-to-market for business solutions.

Maintenance: Declarative solutions tend to be easier to maintain. There are no classes or scripts to debug line-by-line; instead, you adjust settings or tweak a flow. Salesforce updates its platform three times a year, and your declarative customizations automatically benefit from improvements in those releases. (If Salesforce enhances a declarative feature, your org can usually start using that enhancement with minimal effort on your part.) In contrast, code requires ongoing care – developers must ensure their custom code still works with each Salesforce release and incorporate new platform features manually. Fewer lines of custom code also mean fewer opportunities for bugs and less worry about hitting low-level Salesforce governor limits.

Complexity and Scalability: It might sound counterintuitive, but you can build very complex logic without code by using tools like Flow Builder. Declarative doesn't mean "unsophisticated." In fact, because declarative features operate within Salesforce's framework, they inherently handle many scalability concerns for you. For example, a well-designed flow or roll-up summary field will respect bulk operations and large data volumes in ways that custom code (if written poorly) might not. There are of course limits to be mindful of in declarative development (e.g. a total number of fields on an object, or a limit to how many flows can run on an object per hour), but these are more about design best practices than about technical failure points. When you approach those limits, it's usually a sign to refactor your configuration, not necessarily a need to resort to code.

To make this concrete, let's look at a simple scenario. Suppose your sales team wants that when an Opportunity is marked "Closed Won," a renewal Opportunity should be created for six months in the future, with some fields copied over. A traditional developer might think, "I'll write an Apex trigger to do that." An admin or declarative developer, however, can implement this in a few clicks using Flow Builder: a record-triggered flow on Opportunity update can check the stage and create the follow-up record – no Apex needed. Similarly, if the requirement is to prevent a record from being saved unless certain fields are filled out, you don't write a bunch of conditional code; you create a Validation Rule in Salesforce's setup, which is essentially a formula that throws an error message if the condition is not met. Need to calculate a field value, like Total Order Amount = Sum of Line Items? Before reaching for code, you can likely use a Roll-Up Summary field or a Formula field to do the job. In the past, even something as complex as orchestrating a multi-step approval process (routing a record through several people for sign-off) could be done declaratively with Approval Processes. Today, Flow Builder has become the go-to tool for most automation, surpassing the older Workflow Rules and Process Builder. The key takeaway is that Salesforce has invested in making declarative tools cover the majority of use cases. By knowing these tools well, you can often accomplish 80-100% of a given business requirement without custom code.

ChatGPT Prompt (Text): "We need to automatically create a renewal Opportunity 6 months after one is closed. Should I use a declarative solution (like a Flow) or write an Apex trigger?"

When you present a question like this to ChatGPT, it will analyze the scenario and likely advise using a declarative approach (in this case, a Flow), given Salesforce's current capabilities. This can be a great way to double-check your approach. If you're ever in doubt whether something can be achieved with clicks, asking ChatGPT (or using Salesforce's documentation) can save you from writing code unnecessarily. ChatGPT might even outline the steps: e.g. "Use a record-triggered flow on Opportunity that fires on update of the stage to Closed Won, then add a Create Records element to make the renewal Opportunity..." – giving you a blueprint to follow.

Now, does this mean code is never the answer? Not at all. There will always be edge cases or extremely custom requirements where writing Apex or building a Lightning Web Component is the right (or only) solution. Programmatic development allows for ultimate flexibility – if you can imagine it, you can probably code it. But it comes at a cost of increased complexity. Salesforce's best practice (and a theme of this book) is to **use declarative tools first** whenever possible, and resort to code only for the requirements that **cannot be met** with out-of-the-box features. This approach keeps your org simpler and more maintainable over time. Many orgs find that a declarative-first strategy significantly reduces their backlog and accelerates delivery, because admins and analysts can implement many changes without waiting for a development cycle.

Pro Tip: Even for experienced developers, a declarative-first approach is a win. Try building the skeleton of a solution with declarative tools; you might be surprised how far you get. You can always extend it with a bit of code at integration points or for complex calculations. By minimizing custom code, you also make it easier for other admins or consultants to understand your org. Remember that any code you do write in Salesforce must have accompanying test classes and will need to be debugged, deployed, and documented. If a Flow or formula can achieve the same outcome, it often leads to a more **agile and resilient** solution.

By understanding what declarative development entails and its relationship with programmatic solutions, you've taken the first step. In the next section, we'll dive deeper into **why** this approach matters so much. Beyond just ease of development, there are profound business and technical benefits to leveraging Salesforce's clicks-not-code philosophy.

WHY DECLARATIVE MATTERS — BUSINESS AND TECHNICAL BENEFITS

Why insist on clicks over code? The advantages of Salesforce's declarative approach aren't just theoretical – they translate into real benefits for the business and for the technology teams maintaining the system. In this section, we'll break down those benefits into two categories: **business benefits** (what this means for your company, your users, and your bottom line) and **technical benefits** (what this means for the health and evolution of your Salesforce org).

From a **business perspective**, declarative development accelerates value delivery. Because admins and solution designers can build features rapidly without a lengthy development cycle, organizations can respond to needs almost in real-time. Imagine a sales manager comes to the Salesforce team with a request: a new field to track a competitor on deals, with an alert if an opportunity is marked lost due to that competitor. With traditional development, this might become a mini-project with requirements documents, developer assignments, and a deployment scheduled weeks out. With declarative tools, an admin could add the field, create a quick automation (perhaps a Flow or even just a reporting dashboard with a notification), and deploy it within a day or two. Faster delivery means faster feedback and the ability to iterate quickly. In today's agile business environment, that responsiveness is a huge competitive advantage. Companies can pilot new processes or features and refine them without a heavy upfront investment.

Another business benefit is **cost efficiency**. Not every company has a large team of developers on standby, and even if you do, their time is expensive and often backlogged with critical projects. Declarative configuration allows a broader pool of talent (like business admins or analysts) to contribute to building solutions. This reduces the reliance on hard-to-hire technical specialists for every little change. It also means that when your superstar developer is on vacation or leaves the company, your CRM doesn't grind to a halt – many configurations can be handled by others on the team. Lower dependence on custom code can also reduce your long-term **operational costs**; you won't need as much custom development work for upgrades or bug fixes when you've leaned on Salesforce's built-in capabilities that Salesforce itself maintains and improves.

Declarative solutions also tend to be **more transparent to stakeholders.** Because they are often configured through visual tools (like the Flow Builder's flow diagrams or the simple logic of a validation rule), it's easier for non-developers to understand what's been built. A product owner can literally look at a flowchart of an automation in Flow Builder and verify if it matches the intended business process. This fosters better collaboration between business and IT – everyone can speak the language of the platform, to some extent. When business stakeholders can see and even experience a prototype of their request quickly, it builds trust and alignment. They feel empowered because the conversation shifts from "Can we do this?" to "We've done this – try it out and give feedback."

Now let's consider the **technical benefits** of a declarative-first approach. One major advantage is **reduced technical debt.** Every custom code snippet you add to an org is something you have to maintain over the long run. Code might need to be refactored as requirements change, or debugged when issues arise. In contrast, configurations either work within Salesforce's managed framework or they don't exist (you typically can't save a broken formula or inactive flow – it must meet Salesforce's standards to be activated). It's still possible to build a messy org with clicks (we'll address governance in the next sections), but generally speaking, leveraging standard features means you're not reinventing the wheel. An insightful statistic some Salesforce architects cite is that *for every $1 spent on developing custom code, an estimated $6 can be spent later fixing or adjusting that code as technical debt.* Declarative development helps avoid racking up that debt in the first place, because there's simply less bespoke code that can become outdated or require rework.

Another technical benefit is **upgradability and resilience.** Salesforce's thrice-yearly releases often include performance improvements and new capabilities for declarative features. If you've built an approval process or a flow, it's likely to just get better over time as Salesforce optimizes the platform. Contrast this with a custom-coded solution: while it might not "break" when Salesforce updates the platform, it also doesn't automatically get faster or richer – it's static until a developer enhances it. We already touched on how declarative customizations ride along with platform upgrades. This also means fewer **regression issues** to worry about. When Salesforce adds a new feature (say, a new type of component for Lightning pages or an enhancement to Flow), you can often adopt it seamlessly. If you had custom code providing similar functionality, you might face a tough decision to stick with your code or refactor to use the new feature. With declarative, you embrace the new goodness with much less friction.

Declarative configurations, by virtue of being built-in, also adhere to Salesforce's **security and governance guardrails** by default. For example, a record visibility rule you set up with clicks (using Sharing Settings or Roles) will uniformly enforce across all standard components like list views and reports. If a developer were to write a custom UI or custom queries, they would have to remember to enforce those same rules in code. Similarly, a Flow running in user context will respect field-level security and object permissions automatically. This consistency means you're less likely to inadvertently introduce a security hole or inconsistency – a boon for system architects who care about system integrity.

ChatGPT Prompt (Voice): 🎤 "What are the biggest business advantages of using declarative configuration in Salesforce instead of custom code?"

By using a voice query like the above, you can quickly get an articulate summary from ChatGPT that you might share with an executive sponsor or in a team meeting. ChatGPT might respond with points about faster delivery, cost savings, and flexibility – exactly the talking points you need. You could follow up with, "And what about technical advantages?" to have it enumerate things like easier maintenance, automatic upgrades, and lower risk. This is a great way to prepare your own narrative when advocating for a declarative-first strategy.

Pro Tip: To maximize these benefits, treat declarative development as a disciplined practice. Just because something is easy to configure doesn't mean you should do it without a plan. Experienced Salesforce professionals measure the impact of their declarative changes. For example, if you replace a manual process with a flow, track how much time is saved or how much data quality improves. These metrics can help you demonstrate the ROI of declarative solutions to management. Also, remember that "clicks not code" isn't an all-or-nothing mandate; it's about making smart choices. Sometimes a small bit of code (a single Apex class or a Lightning Web Component) combined with mostly declarative setup yields the best result. The goal is to use the right tool for the job, leaning towards the tools that deliver results with less complexity.

The business and technical benefits establish *why* declarative development is so powerful. However, to truly capitalize on these benefits, you need to approach Salesforce configuration with the right mindset and guidelines. Next, we'll look at adopting a **Salesforce Well-Architected mindset** – a set of principles that will help ensure your declarative solutions are not only fast and cost-effective, but also trusted, easy, and adaptable as your organization grows.

ADOPTING A SALESFORCE WELL-ARCHITECTED MINDSET

Having the tools and knowing the benefits is crucial – but equally important is *how* you approach designing solutions. Salesforce encourages architects and admins to follow a **Well-Architected** mindset, distilled into three core principles: **Trusted, Easy, Adaptable**. While the TEA model offers helpful guardrails, real-world design often involves tension between these principles—especially when "Trusted" and "Adaptable" pull in different directions. For example, adding flexibility for power users might complicate auditability or data control. These principles, originally outlined in Salesforce's architecture guidance, provide a compass for making design decisions that lead to long-term success. Let's unpack each principle and see what it means for declarative configuration:

Trusted: A trusted Salesforce solution is *secure, compliant, and reliable*. In other words, it protects the business and its stakeholders. For declarative development, this starts with using Salesforce's robust security model correctly. Are you setting up the right **profiles, permission sets, and role hierarchies** so that users only see what they should? Are you using **validation rules** to enforce data quality and compliance (for example, ensuring a field is filled in to meet a regulatory requirement)? Trust also means reliability: if you build a complex flow, have you considered how to handle exceptions or errors so it doesn't fail unexpectedly? A real-world example of the Trusted principle might be designing a case management process where sensitive cases are automatically restricted (using declarative sharing rules) and ensuring every critical step is captured in an audit log (perhaps by utilizing Field History Tracking or a custom logging object with a flow). The goal is that users and auditors can trust the system to behave correctly and secure the data.

Easy: An easy Salesforce solution *delivers value fast* and is *intentional, automated, and engaging*. "Easy" in this context has two facets: easy to build/maintain and easy for end-users. Declarative configuration shines here because it enables rapid development (as we discussed in the last section) – that's delivering value fast. But being *intentional* means you're not just throwing together clicks haphazardly; you're making design choices that solve today's problem *and* set up the platform for future value. For example, you might use **Lightning App Builder** to create a custom sales dashboard app, assembling standard components. It's quick to do (maybe a few hours of configuration), and it's intentional in that you're putting the most important information and shortcuts in front of sales reps. The solution is *automated* where possible – perhaps you include a Flow-driven widget that automates a task like updating all related records with one click. And it's *engaging* in that it simplifies the user's experience (maybe you used **Dynamic Forms** to only show relevant fields on a page, or added helpful text on a screen flow to guide users). An easy solution tends to have higher **user adoption** because it feels tailored and convenient. Declaratively, this might mean using point-and-click tools to streamline a process that used to require several manual steps or emails. When users find a solution intuitive and helpful, they embrace it, which in turn means the business realizes the value of what was built.

Adaptable: An adaptable solution *evolves with the business*; it is *resilient and composable* in the face of change. This principle is a reminder to design with future changes in mind. In Salesforce declarative terms, being adaptable might involve choosing configuration options that allow easy adjustments. For instance, suppose you set up a set of **email templates and automations for customer onboarding.** If it's adaptable, you might use **Custom Metadata Types or Custom Settings** (configured declaratively) to store the content of the emails or the schedule, rather than hard-coding those details in many different flows. That way, when the business decides to tweak the process (say, add a new email or change timing), an admin can update a record in a custom setting instead of editing multiple flow elements. Resilient solutions handle not only planned changes but also unplanned ones – for example, if a flow fails due to an integration outage, perhaps a well-architected solution has a built-in retry or sends an alert (achieved via declarative error handling in Flow). Composable means you build components that can be re-used or re-purposed. In clicks-not-code, this could translate to **modular flow subflows** or **invocable actions** that can be plugged into different processes as needed. Think of it like Lego blocks: you might have one subflow that handles sending an email, and many parent processes can call that subflow whenever an email needs to be sent. If the email sending logic needs to change, you update one subflow and all processes benefit. That's adaptability in action.

Adopting these principles while you design declaratively will guide you toward better outcomes. You might ask yourself at each step: *Am I building something that users will trust with their data and processes? Is this solution easy to use and delivering quick wins? If tomorrow the business changes course, will this design bend or break?* It can be helpful to even create a checklist. For example:

- **For Trusted:** Security review passed? Compliance considerations (like data retention or PII encryption) met? Proper error handling in place?
- **For Easy:** Minimal clicks for end users? Performance acceptable (no long page loads)? Clear UI and labels so users aren't confused?
- **For Adaptable:** Using declarative tools that are configurable (e.g., custom metadata for business rules)? Modular design (no monolithic 300-step flow that's hard to modify)? Future needs anticipated (if expansion to another region is likely, did we avoid hardcoding region-specific logic)?

ChatGPT Prompt (Design Guidance): "I'm designing a Salesforce solution for a new hiring process. How can I ensure the design is Trusted, Easy, and Adaptable?"

ChatGPT Prompt: "In a scenario where marketing wants to allow custom campaign fields per region, but audit controls need consistent structure, how can I balance 'Trusted' vs. 'Adaptable' in my solution design? What questions should I ask myself or the team?"

If you feed a high-level description of your design to ChatGPT, you can ask it to evaluate or suggest improvements based on these Well-Architected principles. For instance, ChatGPT might respond: "To make it **Trusted**, ensure only HR personnel can see candidate records (using role hierarchy and sharing rules) and log any changes to a candidate's status. To make it **Easy**, use a Screen Flow for hiring managers that surfaces just the needed fields and automates approvals, ensuring a smooth UX. For **Adaptable**, consider storing hiring criteria in Custom Metadata so they can be changed without altering the flow, allowing the process to easily evolve with policy changes." This kind of AI-assisted review can act like a second pair of eyes, helping experienced architects double-check their approach or giving newer admins a structured way to think through design improvements.

Pro Tip: Seasoned Salesforce architects often create architecture **guardrails** aligned with Trusted, Easy, Adaptable. For example, they might mandate that every automation be documented with its purpose and owner (to maintain trust and reliability), or that no single flow should handle completely unrelated processes (ensuring solutions remain easy to understand and maintain). There are sometimes trade-offs between the principles – for instance, adding rigorous error handling (Trusted) can add complexity that slightly reduces ease of build. Part of the Well-Architected mindset is knowing how to balance these. Don't let the pursuit of one (say, making something ultra adaptable via complex abstraction) undermine the others. Over time, you'll develop an intuition for building solutions that hit the sweet spot of all three principles.

Armed with the Trusted, Easy, and Adaptable mindset, you're better prepared to make declarative choices that stand the test of time. The next step is to put this mindset into practice through strong **governance and proactive org management**. After all, even well-designed features can become chaotic in a Salesforce org without proper oversight. Let's explore how governance frameworks help keep your declarative kingdom in order.

GOVERNANCE AND PROACTIVE ORG MANAGEMENT

As your Salesforce org grows with more declarative components – custom objects, fields, flows, and so on – it's essential to have a strong governance framework. **Governance** in the Salesforce context means having the policies, processes, and oversight to manage changes and maintain your org's health. Without governance, the very ease of declarative development can lead to chaos: multiple people building similar automations that conflict, a proliferation of slightly different fields that all serve the same purpose, or inconsistent user experiences across departments. The mantra "with great power comes great responsibility" applies here. Clicks are powerful, and governance ensures that power is used wisely and consistently.

A key aspect of governance is preventing unnecessary **duplication** and **ensuring consistency**. For example, imagine at an international company, one admin creates a custom field called "Client ID" on the Account object, and another admin elsewhere creates "Customer Number" on the Account object not realizing a similar field already exists. Now you have two fields holding similar data – a recipe for confusion. Good governance would have a process to check if a requirement can be met with existing functionality before creating something new. This could be as simple as maintaining a data dictionary or an inventory of metadata. Many organizations establish a **Center of Excellence (CoE)** or an architecture review board that evaluates significant new configurations. That doesn't mean every new field needs a committee approval, but major items like "we need a new custom object" or "we're introducing a new automation on Case records" might go through a design review to ensure it aligns with overall standards and doesn't conflict with anything.

Governance also encompasses **naming conventions and documentation.** A best practice is to come up with a consistent naming scheme for your metadata. For instance, you might decide that all flows for Sales processes start with "SalesFlow_" in their name, or all custom fields related to finance have a "Fin_" prefix. Consistent naming makes it easier to scan through the setup and understand what things are for. Even more importantly, always use the description fields Salesforce provides for custom objects, fields, flows, etc. It may seem tedious when you're in the heat of building, but six months later you (or someone inheriting your work) will be extremely grateful for a note explaining "This flow updates renewal opportunity dates and must run after Opportunity is closed-won." Documentation can also extend outside of Salesforce Setup – some teams maintain spreadsheets or use specialized tools to track all customizations, their purpose, and owner. Modern DevOps tools for Salesforce (like DevOps Center or third-party solutions) can help by treating config changes similar to code, tracking who changed what, when, and why in a change log.

Another pillar of governance is **proactive org management.** Salesforce provides an **Optimizer** tool that analyzes your org and highlights potential issues (like too many fields on an object, unused workflows, slow queries in reports, etc.). Beyond one-time diagnostics, org governance works best as a rhythm. Teams can use Optimizer data—and even summaries from past flow runs or deployment patterns—to create a lightweight monthly or quarterly review. ChatGPT can assist by helping set this cadence, flag key focus areas, and even prep meeting agendas based on system signals. Running Optimizer periodically – say, every few months – and reviewing its recommendations is a great governance habit. Likewise, you should periodically audit your profiles and permission sets to ensure people's access is up-to-date (for example, remove permissions for users who no longer need them). For data integrity, you might implement validation rules not just to enforce requirements, but also to gently police data consistency (e.g., a rule to ensure every Opportunity has a close date in the current fiscal year, to avoid lazy data entry like 12/31/9999 as a placeholder). Proactive management means you're not just reacting to problems after they occur; you're anticipating and preventing them. This could include establishing a **regular release cycle** for changes (even declarative ones) –

perhaps you bundle changes into bi-weekly or monthly releases, which gives time for proper testing in a sandbox and documentation, rather than clicking changes directly in production on the fly.

ChatGPT Prompt (File Upload): "I've uploaded our Salesforce Optimizer PDF report. Please summarize the top three issues identified and suggest actions to address each."

ChatGPT Prompt : "We want to establish a recurring governance review every month to monitor system health. Based on data from Optimizer, Flows, and field usage reports, what should we include in our review? Can you suggest a 5-point checklist or agenda?"

In this example, you've taken a file generated by a Salesforce tool and let ChatGPT digest it for you. Always be cautious and **remove or anonymize any customer data** before sharing files with an AI. (The Optimizer report is mostly metadata-focused, so it's a good candidate.) ChatGPT might come back with something like: "1. **Issue:** Many fields on the Account object are unused – *Recommendation:* review and delete or hide redundant fields to simplify the data model. 2. **Issue:** A workflow rule and a process builder are both updating the Contact address – *Recommendation:* consolidate this logic into a single Flow to avoid conflicts. 3. **Issue:** Several reports are using unindexed fields, causing slow performance – *Recommendation:* add indexes or adjust filters on those reports for faster load times." By offloading the initial analysis to ChatGPT, you can quickly create a to-do list for org cleanup and governance improvements. You could even ask ChatGPT to draft an email to your team summarizing these findings, turning a dense report into an actionable summary.

Pro Tip: Treat your Salesforce configuration like a living codebase that needs version control and refactoring. Implement a source control strategy for your metadata (yes, even things like flows and profiles can be tracked in Git or another system). This way, you have a history of changes and can even roll back if a declarative change has unintended consequences. Also, schedule regular "spring cleaning" sessions for your org. For instance, an experienced admin might run a report of fields that haven't been populated in the last 6 months and bring a proposal to the governance board to eliminate or repurpose them. By keeping the org tidy and lean, you maintain agility – it's easier to make changes in a well-organized org. Governance might sound bureaucratic, but good governance actually *speeds you up* in the long run by avoiding pitfalls and rework. Think of it as sharpening the axe before cutting wood.

With governance practices in place, your declarative development efforts will be more sustainable and controlled. You've set the rules of the road; now it's time to design solutions that not only follow those rules but also scale as your business grows. In the next section, we'll look at designing declarative solutions for **scalability**, ensuring that what you build today will still perform and deliver value as demand increases and requirements evolve.

DESIGNING DECLARATIVE SOLUTIONS FOR SCALABILITY

Scalability is all about making sure that the solutions you build can handle growth – more users, more data, and more complex requirements – without a drop in performance or maintainability. When we talk about designing declarative solutions for scalability, we're considering both **technical scale** (e.g. can this handle thousands or millions of records? Will it perform well as data volume grows?) and **functional scale** (e.g. can this process or framework accommodate new features or variations without a complete redo?). Fortunately, Salesforce's platform, combined with good design practices, can achieve quite a bit of scale purely with declarative configuration.

One of the foundational elements of scalable design is a sound **data model**. Declaratively, this means taking advantage of Salesforce's schema capabilities in a thoughtful way. For example, if you find yourself needing to track something like multiple addresses for a customer, resist the temptation to add fields "Address 1," "Address 2," up to "Address 5" on an object. A more scalable design would be to create a related **Address object** with a lookup or master-detail relationship to the customer (Account or Contact). That way, there's no arbitrary limit to the number of addresses, and each address record can have its own lifecycle. Using master-detail relationships when you need tight coupling (and roll-up summaries) or lookup relationships for looser coupling is key. Also consider **picklist vs. custom object**: if you have a fixed list of values that might grow slowly over time (like product categories), a picklist is simple and scalable; but if those values have attributes of their own or change frequently by users, making them a custom object (managed by data records) might scale better long-term. The mantra is **normalize your data model** where appropriate – it not only avoids hitting limits (Salesforce has limits on number of fields per object, etc.), but also makes future changes easier (adding a new record type is easier than adding 10 new fields).

Another aspect of scalability is how you design your **automation processes.** Salesforce Flow is extremely powerful, but you should design flows with growth in mind. This might mean using **subflows** to break up a large process into manageable, reusable pieces. For instance, if several processes in your org need to send an email to a client, instead of each flow having its own separate "Send Email" logic, you could build one subflow "Send Client Email" that takes in parameters (like recipient and email template) and have all other flows invoke that. This way, if the email sending process needs to be updated (say the template changes), you update one place. That's functional scalability by modularity. It also makes testing easier – you can test that subflow independently.

Scalability also means considering **bulk operations and performance.** Declarative tools abstract a lot of the heavy lifting, but you should still be mindful. For example, when building a record-triggered flow that runs "after save," think about how it will behave if someone triggers it on 200 records at once via a data import. Salesforce will batch those, but if your flow, for instance, does a SOQL query or a loop inside it, you want to ensure it's built to handle collections of records efficiently (using collection processing elements in Flow rather than one-at-a-time logic). Similarly, if you use formulas or roll-up summary fields extensively, consider the performance: a formula that references 10 other fields with complex logic recalculates on every record view – make sure that's necessary and test load time with a large dataset. Salesforce has some features like **indexed fields** and deferred calculations for roll-ups that kick in at scale, but you should design as if you might hit thresholds. For instance, to scale a roll-up summary beyond the standard limits (if you have extremely high child record counts), you might choose to use a declarative tool like **Declarative Lookup Rollup Summaries (DLRS)**, which uses a blend of clicks and a tiny Apex trigger under the hood to handle huge volumes in batches.

When it comes to UI, designing for scalability means ensuring the user interface remains snappy and user-friendly even as complexity grows. This might involve using **Lightning App Builder** features like tabs and accordion components to avoid showing everything to everyone all at once. If one Lightning page needs to display a lot of related lists or data, consider spreading that out over several tabs, so that only the needed info loads initially (Lightning pages load tab content on demand, which improves initial load times). With **Dynamic Forms**, you can even show or hide sections of a page based on criteria – for example, only show the heavy "Related Orders" section if the account has a high-value status, or show certain fields only when relevant. This keeps the default experience lightweight. Salesforce has an "Analyze" button in the Lightning App Builder that will flag if your page is getting too component-heavy and might hurt performance. Paying attention to those warnings is part of scalable design.

Lightning App Builder interface showing a Dynamic Forms configuration. Here, an admin is adding a "Field Section" component (highlighted in red) to a candidate record page, demonstrating how fields can be grouped and made conditionally visible with clicks.

Dynamic Forms and the Lightning App Builder enable you to tailor page layouts heavily without code. In the image above, note how fields are being placed into a new section and that there's a "Filters" option to set component visibility – all through the setup UI. This visual approach to building UIs is not only convenient but also ensures consistency and easier maintenance compared to custom-coded pages.

ChatGPT Prompt (Diagram Request): "Please generate a simple diagram of my data model: We have an Account object, which can have many Projects, and each Project can have many related Tasks."

Even for design, ChatGPT can assist in visualizing your architecture. If you ask for a diagram or an ERD (Entity-Relationship Diagram), ChatGPT (especially with an image generation capability enabled) might produce a basic image showing Account → Project → Task with the correct one-to-many relationships. It might not be a beautifully polished Visio diagram, but it can give you a quick visualization to confirm you're thinking about the relationships correctly. Some users have even gotten creative and used the Markdown or ASCII art capabilities of ChatGPT to sketch out relationships or flowcharts. The takeaway: don't hesitate to leverage AI to **prototype your architecture visuals.** This can be handy when explaining your design to others – a picture is worth a thousand words, even if it's a simple one generated on the fly.

Pro Tip: As you design for scale, always consider the **Salesforce limits and quotas** as guardrails. For example, there's a cap on how many automation rules or flows can fire on an object in a single transaction, how much CPU time a transaction can consume, etc. While Salesforce is generous with limits, a complex org can edge closer to them over time. An experienced architect will do a capacity planning exercise: "What happens if our record count doubles? Triples? Will this solution still work?" Sometimes the answer is to adjust the design (perhaps use a scheduled job to process records in batches rather than all at once, which can be done with declarative tools like scheduled flows). Also, if your org is going to grow big, start thinking about data lifecycle – can some records be archived (using Big Objects or external storage) after a few years to keep active data volumes manageable? These considerations ensure that your declarative solution doesn't just work for the first 100 records, but for the first 100,000 and beyond.

By building with scalability in mind, you create Salesforce solutions that are not only powerful now but remain robust as usage grows. Next, we'll shift focus to another critical aspect that goes hand-in-hand with scale: **security and compliance.** As you scale up your system and handle more data, keeping that data secure and meeting compliance requirements becomes even more paramount.

SECURITY AND COMPLIANCE IN DECLARATIVE CONFIGURATION

In any system, especially one managing customer and business data, **security and compliance** are non-negotiable. Salesforce provides a robust security framework, and the good news is that it's largely declarative – meaning you can configure a highly secure org without writing a single line of code. The platform is built with a "defense-in-depth" approach, offering multiple layers of security that you, as an admin or architect, can leverage.

At the core of Salesforce security are **profiles and permission sets.** These control what users can do (create records, edit certain fields, run reports, etc.) and see at the object and field level. Designing a secure declarative solution means carefully assigning these permissions based on roles. For example, you might have a custom object for "HR Cases" that only HR users should access. By creating a profile or permission set for HR and granting access to that object (while others are denied), you've declaratively enforced a data silo. Even without code, you're implementing the principle of least privilege – users get exactly the access they need, and nothing more. Over time, permission sets have become the preferred way to grant incremental access (because you can layer them for users with multiple responsibilities), while profiles serve as the baseline. All of this is done through Setup menus and clicks.

Then there's **record-level security**: who can see which individual records. Salesforce uses a combination of **Organization-Wide Defaults (OWD)**, **role hierarchy**, and **sharing rules** to handle this declaratively. Suppose your organization's default is that no one can see each other's Opportunities (OWD set to Private). You could then set up a role hierarchy so that Sales Managers can see the Opportunities owned by their team members (role-based access). Additionally, you might create a sharing rule: e.g. if an Opportunity's "Strategic Account" checkbox is true, share that record with a special Strategic Sales team, even if they're not in the same hierarchy. All of this is clicks configuration. No Apex needed to enforce these sharing rules – Salesforce's engine takes care of it once you define the criteria. The result is a flexible yet powerful way to ensure the right people see the right records.

Beyond internal access control, Salesforce declarative config also supports compliance-oriented features like **Field History Tracking** and **Encrypted Fields.** With a few clicks, you can turn on history tracking for sensitive fields (say, on a Contact's phone number and email, to record changes) which is great for audit trails. If you have data that needs encryption at rest, you can mark fields as encrypted (using Platform Encryption – once enabled, this is managed through point-and-click in Setup). This means even if someone were to access the raw database, those fields are gibberish without the encryption keys – yet users inside Salesforce with proper access see the data normally. Another powerful declarative tool is **Salesforce Health Check**. This utility gives you a security score based on your org's settings (like password policies, session timeouts, etc.) compared against Salesforce's recommended baseline or against industry standards you choose. It will highlight, for instance, if your session timeout is too long or if you haven't enabled multi-factor authentication, so you can adjust those settings in a couple of clicks to improve your security posture.

Speaking of compliance, many industries have regulations like GDPR (privacy) or HIPAA (health data security). While complying with these often goes beyond just Salesforce settings, using declarative features helps a lot. For example, GDPR might require that you can erase a person's data upon request – you could implement a **declarative retention policy** by using a checkbox "Data Deletion Requested" and a Flow that anonymizes or deletes certain fields when it's checked, rather than writing a script. Or use **Data Classification** tags (a feature in Salesforce where you label fields as PII, Confidential, etc.) to keep track of which fields hold sensitive data – all done via config in the Schema Builder or Object Manager.

ChatGPT Prompt (Operator): "Search Salesforce Help for 'Security Health Check' and tell me the steps to run a health check on my org."

Imagine you're not sure how to use a particular security feature – you can ask ChatGPT to find and summarize Salesforce's documentation. By enabling the browsing/Operator mode, ChatGPT can fetch the official help article and boil down the instructions. It might return: "To run a Health Check, navigate to **Setup** -> **Security** -> **Health Check**. There you'll see your current score and a list of settings with risk categories. You can edit the settings directly from that page to improve your score. The steps are: 1) Go to Health Check in Setup, 2) Review the score and identified risks, 3) For each risk item, click on it and use the provided link or button to change the setting to the recommended value, 4) Save changes and refresh Health Check to see your new score." Using ChatGPT in this way is like having a smart assistant that can instantly reference Salesforce's manuals for you. This is particularly handy when preparing for compliance audits – you can quickly get info on how to tighten certain settings without wading through multiple web pages yourself.

Pro Tip: For a scalable security model, leverage **permission set groups** and a role-based access strategy. Rather than creating dozens of profiles for every combination of job function, create permission sets for discrete access needs (e.g., "Can Approve Contracts" or "Read HR Cases") and combine them into permission set groups that correspond to roles or departments ("Sales Manager Access", "HR Team Access"). This modular approach (entirely declarative) makes it much easier to onboard a new user or adjust access when someone's role changes – you just assign or remove a permission set or group, rather than cloning or editing profiles. Also, periodically perform access reviews. Even though profiles and permission sets are declarative, it's easy for cruft to build up (e.g., users who changed departments but still have old permissions). Run a report of users and their permission sets every quarter and have data owners confirm who should have what. Keeping your org's access clean ensures that all the security configuration you've set up remains effective and compliant over time.

With strong security configurations in place, you've ensured trust in your Salesforce solutions – both in terms of data protection and meeting regulations. The final piece of the puzzle is making sure your org and solutions not only scale and stay secure, but also **perform well and are loved by users.** In the next section, we'll focus on performance tuning and user adoption, which ultimately determine the success of your declarative innovations.

ENSURING PERFORMANCE AND USER ADOPTION

Finally, even the most well-designed, secure, and scalable solution can falter if it doesn't perform well or if users don't embrace it. In this section, we'll look at best practices to keep your declarative solutions **performant** and strategies to drive **user adoption** – both are critical for long-term success.

Performance Optimization: Salesforce is generally optimized out-of-the-box, but as you add customizations, you need to be mindful of their impact. One common area to watch is page load times in Lightning Experience. If users complain that a record page takes too long to load, consider how much is being displayed and when. We've mentioned splitting complex pages into tabs (so data loads only when a tab is clicked) and using dynamic visibility to hide components until needed. These simple tweaks can drastically cut down initial load times. Also, be cautious with **related lists** on a page – loading 10 related lists, each showing 50 records, can add up. If some related information is rarely used, move those related lists to a secondary tab or use an on-demand loading strategy. Salesforce's Lightning Usage App can show you page performance metrics for different Lightning pages in your org, which is great for identifying slow spots.

For declarative automations (Flows, etc.), performance is about doing the minimal work necessary for a given transaction. If you have multiple record-triggered flows on the same object, see if they can be consolidated or at least set to run in specific orders (Salesforce now allows you to configure flow trigger order). Sometimes a poorly placed automation can cause a save to take extra seconds. For example, if on every save of an Opportunity you're triggering a flow that queries a bunch of related records and does heavy calculations, a user editing that Opportunity might experience a delay. Could that be changed to run asynchronously (e.g., using a "Scheduled Path" in Flow to execute 5 minutes later, or an Apex-defined action to handle it in a batch)? You can often design declaratively such that the user's immediate action isn't bogged down by non-critical updates. Also consider validation rules: while generally not a performance issue, too many complex validation rule formulas on one object might add overhead – ensure each rule is necessary and as simple as possible.

Reports and dashboards are another area of user experience. A report that has to churn through millions of rows of data will naturally be slow. Encourage the use of report filters (especially on indexed fields like Record Type or Created Date) to narrow data sets. If you have very large data volumes, consider partitioning data by time (e.g., report on current fiscal year data vs. all years) or archiving old records out of the main object (which can be done with declarative means like scheduled export and delete, or using Big Objects for archival storage). The key is to be intentional about performance. Don't assume "clicks not code" means "set it and forget it." Test your automations and pages with a realistic amount of data. If you're rolling out a new flow that will mass-update records, try it on a batch of 200 in a sandbox.

User Adoption: Now onto user adoption – essentially, getting people to actually use the shiny new declarative solution you built. One advantage you have is that declarative tools often produce more **user-friendly** results by default (since they leverage Salesforce's standard UI and patterns). But you still need to consider change management to drive adoption. Firstly, involve users early. If you're designing a new screen flow for case intake, talk to a few support agents while designing it (or even better, demo a prototype) to gather feedback. Because declarative development is fast, you can iterate based on their input without much delay. This inclusivity makes users feel heard and increases the likelihood they'll like the final product.

Next, ensure solid training and onboarding for new features. The good news: because your solution is declarative and aligned with Salesforce's standard behavior, users might find it intuitive. Still, you should utilize tools like **Salesforce In-App Guidance** to create pop-up tips or walkthroughs. For example, after deploying a new feature, you could have a little guided prompt appear the first time a user encounters it, highlighting a new button or explaining a new field. This is all point-and-click to set up. Additionally, consider short videos or live demos. Many admins host quick "lunch and learn" sessions or record a 2-minute demo of a new process and share it. Users are often more likely to watch a short video than read a long email.

ChatGPT Prompt (Wireframe): "Generate a simple wireframe for a Salesforce home page tailored to a sales rep, including a dashboard chart, recent opportunities, and a 'New Lead' quick action."

To improve adoption, it helps to design with the user in mind, and sometimes you need to experiment with layouts. ChatGPT's image generation capabilities can assist by creating rough wireframes from a description. By providing a prompt like the above, you might get back an autogenerated sketch of a possible home page. This can spark ideas – maybe it shows a two-column layout with a chart on the left and tasks on the right, and you realize that would indeed help your users focus on key info. You can then configure the Lightning page accordingly. It's like having a digital brainstorming partner for UI/UX design. A well-thought-out user experience is a huge factor in adoption, and tools like ChatGPT can help you envision the user's perspective before you build.

Once you've deployed changes, gather metrics and feedback. Salesforce provides adoption dashboards (there's a free package on the AppExchange) that show things like login rates, record creation trends, and feature usage. For instance, if you rolled out a new "Deal Registration" form for partners declaratively, check a month later: how many deal regs came in? If numbers are low, maybe partners aren't aware of it or find it cumbersome – that's a cue to provide additional training or simplify the process. You can even use ChatGPT to help craft a friendly announcement or an FAQ document about the new feature, saving you time in writing it from scratch.

ChatGPT Prompt: "I'm using ChatGPT to draft internal Salesforce FAQs and release notes. What kinds of content should I avoid auto-generating in compliance-sensitive industries like healthcare or finance? Can you list red flags to look out for?"

ChatGPT Prompt (Sora Video): "Create a 15-second tutorial video showing how a sales rep can log a call in Salesforce, with captions."

A novel way to drive adoption is through bite-sized videos or animations, which is where ChatGPT's Sora video generator comes in. By providing a prompt like the one above, ChatGPT (with Sora enabled) can produce a short, narrated screencast-style video. It might show a simplified Salesforce interface with a cursor clicking "Log a Call," filling out details, and hitting save, accompanied by text captions. Such a video can be shared in your company's chat or email as a quick how-to. While the technology is still evolving (the visuals might be somewhat generic), it's a fun, engaging way to reinforce training. Always review the output to ensure it matches your actual process, but imagine the time you save versus filming a tutorial yourself. This kind of AI-generated assist can make ongoing education of users much more scalable.

Pro Tip: Driving adoption is an ongoing effort. Seasoned admins know the job isn't done when the feature is deployed – it's done when the feature is being used effectively. Continuously solicit feedback: set up a Chatter group or Teams/Slack channel where users can post their questions and suggestions. Use what you learn to refine the app. Maybe a field label isn't clear, or a screen flow could use an extra help text – these are quick declarative tweaks. Also, identify and support **power users** or champions in each department. They can help colleagues and advocate for the tools, creating a positive feedback loop. From a performance standpoint, keep monitoring after launch. If you see a flow failing occasionally or a page element loading slowly, investigate and optimize it. Incremental improvements (like reducing a screen flow by one step, or making a report load 2 seconds faster) can significantly boost user satisfaction over time. In summary, a fast system that people love to use is the ultimate goal of all your declarative efforts.

SUMMARY

In this opening chapter, we explored the transformative power of Salesforce declarative configuration and how embracing "clicks, not code" can lead to faster, smarter solutions. We began with an introduction to what declarative development means, contrasting it with traditional coding and seeing how Salesforce empowers us to build complex functionality with point-and-click tools. We then looked at the myriad benefits – from the business agility gained by rapid, cost-effective delivery, to the technical advantages of reduced maintenance and alignment with platform upgrades. Adopting a Salesforce Well-Architected mindset (Trusted, Easy, Adaptable) provided guiding stars to ensure our solutions are secure, user-friendly, and future-ready.

With principles in mind, we discussed governance and proactive org management, emphasizing that even declarative development needs oversight and cleanliness to prevent chaos. Strong governance sets the stage for sustainable growth of your org. Building on that, we looked at designing for scalability – making choices in data modeling and automation that will stand up as data and requirements grow. We saw that security and compliance are not hurdles but integral parts of declarative design, easily handled with Salesforce's built-in features when used diligently. Finally, we rounded out the discussion with performance and user adoption, because a solution only truly succeeds if it runs well *and* people use it.

Along the way, we demonstrated how ChatGPT and generative AI can be a powerful companion in this journey – whether it's brainstorming solutions, analyzing org metadata, fetching documentation, generating diagrams and videos, or assisting with training.

As you proceed to the subsequent chapters of this book, you'll dive deeper into specific tools and tactics – from a comprehensive overview of the declarative toolkit to advanced techniques and real-world case studies. But you now have a solid foundation: a strategic understanding of why declarative development matters and how to approach it wisely. With this knowledge (and ChatGPT by your side), you're well-equipped to build Salesforce solutions that are not just clicks-and-configurations, but truly enterprise-grade innovations. Here's to becoming a highly effective Salesforce maker in the era of AI-assisted development – happy building!

2

THE SALESFORCE PLATFORM & DECLARATIVE TOOLKIT OVERVIEW

Salesforce's true power lies in its rich **declarative toolkit** – the point-and-click features that let you build applications without writing code. In this chapter, we take a friendly guided tour through Salesforce's native declarative tools across Sales Cloud, Service Cloud, and the Lightning Platform. You'll discover how to configure data models, automate processes, tailor the user interface, build reports, secure data, optimize performance, and maintain your org – all with clicks, not code. By the end, you'll have a solid grasp of what tools are available and when to use them, arming you with confidence to solve business challenges declaratively.

This journey begins with the **fundamentals of objects, fields, and relationships.** A good data model is the foundation of any Salesforce solution – get it right and everything built on top (automation, UI, reporting) becomes smoother. We'll cover how Salesforce organizes data and how you can extend the data model to fit your business needs using custom objects and relationships. Understanding these core platform essentials will set the stage for more advanced configurations.

Next, we'll dive into **declarative automation**. Salesforce provides powerful tools like Flow Builder, Approval Processes, and Validation Rules to automate business processes and enforce data quality. You'll learn what each of these tools does, where they shine, and how the modern Flow Builder has evolved to become the go-to solution for most automation needs. We'll include practical examples – from simple data validations to multi-step approvals – to illustrate how you can save time and reduce errors with automation.

With data and process automation in hand, we then explore **user interface configuration and dynamic experiences.** This is where you learn to create intuitive apps and page layouts using Lightning App Builder, Dynamic Forms, and other UI tools. A well-designed UI can dramatically improve user adoption. We'll show how to tailor pages for different users, display the right information at the right time with Dynamic Forms, and even how to leverage new features to declutter screens for better performance. Real-world tips will help you make the Salesforce interface both efficient and delightful.

No Salesforce overview would be complete without covering **analytics – reports and dashboards.** In this chapter, you'll also learn to declaratively turn raw data into actionable insights. We'll discuss best practices for building meaningful reports and visual dashboards that executives and teams can rely on for decision-making. Topics include choosing the right report type, effective chart visualization, and dashboard filters. You'll see how non-technical users can create reports on the fly, and how a few design principles ensure your analytics truly support your business goals.

Rounding out our toolkit, we dive into **declarative security configuration.** Salesforce's "clicks not code" philosophy extends to robust security features: profiles, permission sets, roles, sharing rules, and more. We'll explain how to use these to enforce the principle of least privilege (giving users only the access they need). By mastering declarative security, you can protect sensitive data, comply with audits, and still empower users with the access required for their job. We'll keep it practical with examples of setting up a flexible, secure model without writing a single line of code.

Finally, we address two crucial aspects of enterprise-scale Salesforce solutions: **performance optimization** and **maintainability**. Even declarative apps can suffer if not designed for scale, so you'll learn techniques to ensure your configurations run fast and efficiently – from indexing data for quicker queries, to streamlining automation logic. And because Salesforce orgs evolve, we'll share strategies for maintaining and scaling your declarative solutions over time. This includes using tools for documentation, version control, and org health checks to avoid clutter and technical debt.

Throughout each section, you'll find **ChatGPT prompt callouts** demonstrating how you can leverage ChatGPT as your personal Salesforce assistant. Whether via typed prompts, voice commands, the Operator browsing tool, file uploads, SORA (text-to-video generation), or image creation, we'll show examples of ChatGPT in action. These callouts illustrate how to brainstorm solutions, generate examples, and even create visual aids – all using ChatGPT. *(Remember to review your privacy settings before uploading any real Salesforce metadata or files to an AI service!)*

So, let's get started. In the next section, we begin with the building blocks of the Salesforce platform: objects, fields, and relationships. Understanding these core elements is essential before tackling automation, UI, and more. Ready to explore Salesforce from the ground up? Let's go!

SALESFORCE CORE PLATFORM ESSENTIALS

Every Salesforce solution starts with **data modeling** – defining how information is structured and related in your org. Salesforce provides a robust, declarative way to model data using **objects, fields, and relationships.** In this section, we'll explore standard vs. custom objects, different field types, and how to establish relationships like lookup and master-detail. Mastering these essentials ensures your foundation is solid for any app you build.

Standard Objects vs. Custom Objects: Salesforce comes with many standard objects out-of-the-box to support common business needs. For example, Sales Cloud includes objects like Account, Contact, Lead, Opportunity, and Case; Service Cloud adds objects like Case (for support tickets), Knowledge, and Service Appointment. These standard objects have pre-defined fields and functionalities tailored to typical CRM processes. However, every business is unique. That's where

custom objects come in – you can create your own objects to store information specific to your organization. For instance, a property management company might create a custom object Property__c, or a university might have Student__c and Course__c. Custom objects are created with clicks in **Object Manager**, where you define the object's name and behavior (like whether it tracks activities or has notes). They behave much like standard objects but are fully configurable by you.

Fields and Data Types: Objects (standard or custom) have fields that capture individual pieces of data. Salesforce offers a wide range of field types to accommodate different data formats. Common field types include Text (for names, descriptions), Number and Currency (for financials or scoring), Date/Time, Picklist (drop-down values), Checkbox (yes/no), and more. Some fields are **formula fields**, which are special because their values are calculated based on other fields or conditions you define (declaratively writing a formula). Another powerful type is **roll-up summary fields**, which automatically aggregate data from a related detail (child) object, for example summing up all Invoice Amounts related to an Account. (Roll-up summary fields are only available on the master side of a master-detail relationship – more on that soon.) By combining these field types, you can enforce data formats and even business logic (with formulas) at the data layer, without any code.

It's worth noting that Salesforce also provides **Schema Builder**, a drag-and-drop interface to visualize and manage objects and fields. This can be helpful for getting a bird's-eye view of your data model and how objects relate. However, many admins simply use Object Manager and the Setup menu for precision control. As you add fields, always consider naming conventions and descriptions. A **Pro Tip:** *Always fill in the description for custom fields and objects to document their purpose. This small step greatly aids future admins and developers in understanding your data model.*

Relationships: Lookup and Master-Detail: In a relational database (which Salesforce is at its core), relationships link records of one object to another. Salesforce supports a few types of relationships declaratively. The most common are **Lookup relationships** and **Master-Detail relationships.** A *lookup relationship* essentially creates a loose link between two objects. For example, a custom Property__c object might have a lookup field to Account (to indicate which customer owns that property). Lookup relationships can be optional, and the two objects remain mostly independent – if you delete the parent record, the child just loses that reference (it could even optionally be configured to clear or prevent delete). In contrast, a *master-detail relationship* is a tighter coupling: the "detail" record exists as a part of the "master" record. A classic example is an Invoice Line Item (detail) under an Invoice (master). With master-detail, if you delete the master record, all its detail records are automatically deleted (cascading delete). Detail records inherit some attributes from their master (like the owner, and sharing/security settings in many cases). Master-detail also enables that powerful roll-up summary field we mentioned – you can only create roll-ups on a master object, rolling up values from its details. Each custom object can have up to **two master-detail relationships** (meaning it can be detail to two different masters), but can have many lookups for more loosely related information. This design enforces a hierarchy where a detail can't itself have details ad infinitum; it ensures data integrity.

When modeling relationships, think about **data integrity and requirements.** Use master-detail when: the child should not exist without the parent (strict containment), you want aggregated roll-up calculations, or you need the child's security to follow the parent. Use lookups when: the relationship is optional or more flexible, or when it's a connection between two independent objects. For example, a "Project" might have a lookup to "Client (Account)" – the project can exist even if you remove the client link or reassign it, so lookup is appropriate. On the flip side, "Project Task" might be a detail under "Project" if a task is meaningless without an associated project.

Junction Objects (Many-to-Many): Salesforce doesn't natively allow direct many-to-many relationships with a single field (since a lookup or master-detail field always sits on one object referencing another as a one-to-many). However, you can achieve many-to-many by introducing a **junction object.** A junction object is simply a custom object that has two master-detail fields, each linking to one of the two objects you want to relate. For instance, suppose you have a custom object Student__c and another Course__c, and you want to track enrollments (each student can enroll in many courses, and each course has many students). You would create an Enrollment__c object with a master-detail to Student and another master-detail to Course. Each Enrollment__c record represents a junction (or link) between one student and one course. The junction object approach allows many-to-many connections declaratively. Keep in mind, a custom object can have at most two master-detail relationships, which is perfect because a junction needs two (one to each side). A *Pro Tip: Give junction objects clear names that indicate the two things they connect (e.g., "StudentCourseEnrollment" or similar) to make the data model easy to understand at a glance.* Also remember that master-detail fields require the user to have access to the master object to create the detail, so plan your security roles accordingly when using junctions.

Reference & External Relationships: Beyond lookup and master-detail, Salesforce also offers a special lookup called **External Lookup** or **Indirect Lookup** (used in connecting to external data via Salesforce Connect) and **Hierarchical relationships** (a special lookup on the User object to relate one user to another, often for org charts). These are more advanced and less common but good to know exist. Most Salesforce architects spend 90% of their time with lookup and master-detail relationships.

As you design your data model, keep an eye on **limits and best practices.** Salesforce has limits on how many fields per object (often hundreds, varying by edition) and how many lookups you can have (usually up to 40 relationships per object in total, with up to 2 being master-detail). In practice, if you find one object has dozens of relationships or hundreds of fields, it might indicate an overly complex design – consider refactoring by splitting into multiple objects or re-evaluating if all those fields are truly needed. Also, for master-detail, Salesforce recommends not exceeding 10,000 detail records under one master record for both performance and usability reasons. While the platform can handle more, extremely large child record counts can slow down UI displays or roll-up calculations.

ChatGPT Prompt – Typing: "I have two objects, Job__c and Contractor__c. I want to allow each contractor to work on multiple jobs, and each job to have multiple contractors. How can I set that up declaratively?"

ChatGPT: Great question! To model a many-to-many relationship between Job and Contractor, you would use a junction object. For example, create a new custom object called JobAssignment__c with a master-detail field to Job__c and another master-detail to Contractor__c. Each JobAssignment record represents one contractor working on one job. This way, a contractor can be linked to many JobAssignment records (each with a different Job), and a Job can have many contractors via many JobAssignment records. This declarative setup avoids any coding and uses Salesforce's standard relationship features.

The above example shows how ChatGPT can reinforce your understanding of relationships in a conversational way. You could also use ChatGPT to brainstorm naming conventions or validate your data model decisions.

Advanced Field Capabilities: Beyond basic fields, Salesforce's platform offers declarative enhancements like **Picklist value sets** (reusable picklist definitions across objects), **Dependent picklists** (where one picklist's options filter based on another's value), and **Record Types**. Record Types deserve special mention: they allow you to define different "subsets" of records for the same object that might have different fields or processes. For example, a "Customer Account" vs "Partner Account" could be two record types on the Account object, showing different page layouts or tracking slightly different info. Record types are a declarative way to handle variation within a single object, and they pair with page layouts and picklists to tailor the UI and data to each scenario. We'll see record types in action in the UI configuration section, but keep them in mind as part of your data model strategy as well.

Quality and Consistency: A solid data model isn't just about what objects and fields you create, but also about controlling data quality from the start. Leveraging **validation rules** (which we'll cover in the next section) at the field level can prevent bad data from ever being saved (for example, making sure a date of birth isn't in the future, or an email address contains "@"). You can also use default values on fields to populate common entries, or formula default values to auto-calculate something upon record creation. All of these are declarative configurations done on the object/field definitions and ensure your data remains reliable and useful.

Before wrapping up core platform basics, remember that designing an optimal data model often requires thinking ahead. Consider how data will be used in reports, how often it will be updated, and how large it might grow. It's much easier to adjust and add complexity early in the design than to correct an overly convoluted schema later. Start simple, meet the requirements, but try to anticipate future needs. When in doubt, iterate – Salesforce makes it easy to add fields or relationships later as business needs evolve.

ChatGPT Prompt – File Upload: *A Salesforce admin exports a list of all custom fields in the Account object (as a CSV) and uploads it to ChatGPT.* **Prompt:** "Review the list of 120 custom fields on Account and suggest any that might be redundant or could be split into a related object."

ChatGPT: *(analyzes the CSV and responds)* "I notice several sets of fields that seem related to each other. For instance, there are 10 custom fields prefixed with Project_ (Project Start Date, Project End Date, Project Budget, etc.). If multiple projects are tracked on an Account, it might be better to use a related custom object for Projects instead of many fields on Account. Likewise, fields Contact1__c, Contact2__c... Contact5__c suggest an attempt to list multiple contacts on an Account record; using the standard Contact object (related to Account) would handle multiple contacts more elegantly. Consider reviewing these for potential normalization."

In this example, ChatGPT used the uploaded metadata to intelligently pinpoint possible design improvements. This kind of review can help maintain a clean, scalable data model. (Always ensure no sensitive customer data is in such exports – and consider field names only.)

With the core platform building blocks covered, you now have a firm foundation. We've seen how to declaratively create the structure of an application: the objects and fields that store your data, and the relationships that connect it. **In the next section, we'll build on this foundation by introducing Salesforce's declarative automation tools.** These tools will bring your data to life by automating processes and enforcing rules, all without writing code. Let's explore how to make Salesforce work for you through clicks!

DECLARATIVE AUTOMATION TOOLS

Now that your data model is set up, the next step is to automate business processes and ensure data quality using Salesforce's **declarative automation tools.** Automation is the secret sauce that transforms a static database into a dynamic application that actively supports your business workflows. Salesforce historically offered multiple automation tools – Workflow Rules, Process Builder, and Flow – but in recent years, these have converged into a single powerhouse: **Flow Builder.** Alongside Flow, we also have **Validation Rules** for data quality and **Approval Processes** for orchestrating multi-step approvals. In this section, we'll cover each of these tools, illustrating when and why to use them. We'll also point out the evolution of automation on the platform, so you understand the current best practices.

Flow Builder – The Automation Powerhouse: Salesforce Flow is now the go-to declarative tool for building business logic. With Flow Builder's modern interface, you can create everything from simple field updates to complex, multi-step processes with branching logic. There are a few types of flows you should know: **Record-Triggered Flows**, which fire automatically when records are created/updated (or deleted) – these are effectively the next-gen version of Workflow Rules/Process Builder, and can run *before save* (very fast, good for validations or auto-populating fields) or *after save* (for actions like sending emails or creating related records). Then there are **Scheduled Flows**, which run at specified times (for example, a daily job to close inactive support cases). **Screen Flows** are flows that have a user interface – they can be launched by a button or embedded on a Lightning page to guide users through a step-by-step process (like a wizard). Additionally, **Autolaunched Flows** can be invoked from other flows or even from Apex code, essentially letting you create reusable logic (think of them like subroutines). The rich functionality of Flow includes the ability to query and update records, loop over sets of records, make decisions (if/then branches), call reusable subflows, interact with external systems (through HTTP callouts or platform events), and present dynamic screens to users. This breadth means **most** automation requirements can be fulfilled with Flow.

In fact, Salesforce is actively consolidating automation into Flow. **Workflow Rules** and **Process Builder** (the older tools) are being retired – as of Winter '23 you couldn't create new workflow rules, and as of Summer '23 new Process Builders are also disabled. Salesforce's guidance is to migrate all those automations into Flows. Why the push to Flow? Because Flow offers better performance and more capability than its predecessors. For example, Flow can handle *before-save updates* which are super-efficient for certain field updates (they don't even trigger a save again), and Flow supports things like screen interactions and richer logic that were impossible in Workflow Rules. If you're a new Salesforce admin today, you can focus on Flow and not worry about learning Workflow Rules at all, except for maintaining legacy orgs. For those with older orgs, plan a strategy to consolidate automations – it often makes sense to combine multiple old rules into a single flow for clarity and efficiency.

A **Pro Tip** for Flow: *Design with bulkification in mind.* Even though you're not writing code, flows have to operate within Salesforce's governor limits like anything else. This means you should avoid scenarios like performing a data lookup inside a loop for each record (instead, collect all IDs and do one lookup outside the loop). Salesforce has improved Flow's ability to handle large volumes by adding things like collection processing elements, but it's still possible to create an inefficient flow if you're not careful. One common best practice is the "**One Flow per Object per Trigger**" rule – i.e., have a single record-triggered flow for each object for a given context (one for before-save, one for after-save). This way, you avoid multiple automations competing or running in unpredictable order. You can use subflows within that one flow to keep it organized. This practice simplifies maintenance and troubleshooting.

Validation Rules – Guardrails for Data Quality: While Flow can manipulate data, **Validation Rules** act as gatekeepers *when users are entering data*. A validation rule is a declarative formula that evaluates to true or false, combined with an error message. If the formula returns true (meaning the validation condition is met), the record save is blocked and the error message is shown to the user. Essentially, "true means there's an error." For example, a simple validation rule might ensure an Opportunity's Close Date is not before today (to prevent entering past dates for new deals), or require that if Stage = "Closed Won", then a custom "Won Reason" field must be filled out. Validation rules are critical for maintaining data integrity because they enforce business rules consistently. They are defined on an object and can reference multiple fields of that object (and even fields of parent records via cross-object formulas, e.g., check that a Case's Account is of a certain type when a Case field has a value). When designing validations, aim to make the error messages clear and actionable for users ("Please enter X before saving" rather than a technical description). A strategic use of validation rules ensures that your beautifully modeled data doesn't get polluted by mistakes or missing values. One thing to note: validations only fire when a record is being created or edited by the user (or sometimes by a non-system process). They can be bypassed by certain automated processes or data loader operations if not careful. But generally, they are a straightforward way to uphold business constraints.

Approval Processes – Structured Approvals Made Easy: Many businesses have scenarios where certain records need sign-off or approval by a manager or a specific role. Salesforce offers **Approval Processes** as a declarative mechanism to handle this. An approval process is a sequence of steps that a record goes through, from an initial submission to final approval (or rejection). With an approval process, you specify the entry criteria (e.g., "If an Opportunity's discount is greater than 20%, it requires approval"), who the approver(s) are at each step (maybe first the sales manager, then if over 30% discount, also the finance director in a second step), and what actions happen on approval or rejection (perhaps an email notification, or updating a status field). The beauty of the approval process is that it's largely point-and-click to configure. Salesforce handles the routing of approval requests, the UI for approvers (they get a nice "Approve/Reject" button or can even respond via email), and tracking the history. It's worth noting that some newer Salesforce automation capabilities overlap with approvals; for example, you can build a custom approval-like process using a Screen Flow and some record fields. However, the standard Approval Process is still a reliable and quick way to implement structured approvals (and is available in most Salesforce editions). Use it when you have a well-defined, mostly linear approval scenario. If your approval needs are very dynamic or complex (lots of conditional branching), that might be a sign to use Flow for more flexibility. But for classic use cases like "manager approval for big discounts or high-value deals," the declarative approval process is hard to beat for simplicity.

Other Declarative Automation Tools: A few other tools deserve mention: **Email Templates and Alerts** – while not "automation" by themselves, you configure Email Alert actions (used in flows, workflow, approvals) with declarative templates to send standardized emails. **Outbound Messages** – an older workflow feature that can send a SOAP message to an external system on a trigger (less common now, but still declarative integration). **Escalation Rules** – specific to Case object in Service Cloud; they automatically escalate a case to a higher support tier if it remains open after a certain time. And **Entitlement Processes** – a declarative timeline for support cases that ensures SLA milestones are met. Each of these tools is a bit more niche, but it's good to know they exist in the declarative toolbox for specific needs (especially in Service Cloud scenarios).

A Look at an Example: Let's say we want to automate a common sales process: when an Opportunity is marked Closed Won, create a related "Customer Onboarding" Task and notify the onboarding team. In the past, we might use a Process Builder for this. Today, you'd create a Record-Triggered Flow on Opportunity, set it to run *after save* when Opportunity Stage equals "Closed Won". In the flow, you'd add an Action to create a Task record (setting the fields like Assigned To the onboarding queue, due date = today + 1, subject "New Customer Onboarding for {!Opportunity.Name}"). You'd also add an Email Alert action (or use a Send Email action in Flow) to notify the team. This entire automation can be done in a matter of minutes declaratively. No Apex code needed, no waiting for a developer – clicks not code at its finest. And if the process logic changes (maybe now there are two tasks, or a different team), an admin can adjust the flow quickly.

ChatGPT Prompt – Voice: (Imagine speaking to ChatGPT using the voice feature) "Hi ChatGPT, I need to automate a process. When a support Case is marked as 'Resolved', if the Case type is 'Problem', I want to email the product team with the case details and also create a follow-up task for the support rep in one week to check in with the customer. How can I do that without code?"

ChatGPT (voice response): "You can absolutely do that with Salesforce's declarative tools! You would likely use a record-triggered flow on the Case object. Set the trigger to fire after save, when the Case Status changes to 'Resolved' and the Type = 'Problem'. In the flow, use an Email Alert action or a Send Email element to send an email to the product team – you'll need an email template that includes the case details (you can create a custom HTML or text email template for this). Next, add a Create Record element to make a Task. Set the Task's WhoId/WhatId to the appropriate contacts or records (maybe WhoId = Case Contact, WhatId = Case ID for reference), Assigned To the case owner (the support rep), Subject like 'Follow up with customer', and due date = Today + 7 days. That's it! Activate the flow, and now every time a Problem case is resolved, the product team gets notified and the support rep has a reminder task a week later."

By describing your requirements via voice, ChatGPT can outline a solution. It's almost like talking through a design with a colleague, helping you ensure you don't miss any steps. This voice interaction can make brainstorming automation ideas feel very natural and quick.

Historical Perspective: For context, if you step into a long-standing Salesforce org, you might encounter **Workflow Rules** and **Process Builder** flows. Workflow Rules could do four things: field update, send email, send outbound message, or create task – all single-step. Process Builder improved on that by allowing multiple criteria and actions, including creating related records. But both had limitations and are now superseded by Flow. Salesforce's retirement timeline for these older tools extends into 2025 (with full support ending by Winter '25 or so), giving everyone time to migrate. The takeaway for you is: focus your learning on Flow, and use it for new automations. If you have to maintain an older workflow, consider converting it. Salesforce even provided a Migration Tool that converts workflow rules into flows; though interestingly, they caution not to just convert everything blindly one-to-one, as that could lead to hundreds of flows, but instead to consolidate logic where it makes sense. This highlights an important point: technology aside, **take time to design your automation architecture.** Think about all the processes in your system and try to implement them in a structured, organized way. It'll pay off in easier maintenance.

When Declarative Isn't Enough: Occasionally, you might wonder – can everything be done in Flow? What about Apex (code)? There are a few edge cases where code might still be needed (perhaps complex looping logic, heavy integrations, or transactions that need fine-grained control). But Salesforce keeps expanding Flow's capabilities (e.g., allowing screen flows on public sites, letting flows call subflows, even adding the ability for flows to invoke Apex code if needed). A general guideline: **if you can do it in Flow, you probably should**, to benefit from ease of maintenance and admin ownership. We'll discuss in Chapter 3 about how to decide between declarative vs. programmatic, but know that Salesforce's intent is to let you achieve most requirements with these declarative tools.

Before we move to the next section, let's touch on one more automation-related feature: **Formula Fields and Default Values** as a form of automation. While not "automation" in the sense of triggered actions, formula fields (which we mentioned earlier) and default field values do perform automatic calculations and data population. They're calculated in real-time and don't require any user intervention to update, so in a way they automate the consistency of certain data (for example, an "Account Score" formula that always computes based on latest activities). It's worth considering formulas as part of your solution before jumping to a record-triggered flow that just sets a value – sometimes a simple formula field is cleaner and uses no processing time during saves.

ChatGPT Prompt – Operator (Browsing): "Find Salesforce documentation on writing effective validation rule formulas for phone number format."

ChatGPT (using Operator): *ChatGPT opens Salesforce Help documentation or a community article on validation rules.* "According to Salesforce documentation, a validation rule formula for phone numbers can ensure the field contains only digits and expected symbols. For example, a formula might use REGEX to enforce a pattern like REGEX(Phone, \"^(\\\\d{3}-\\\\d{3}-\\\\d{4})?$\") for a US phone format (123-456-7890). The key is to use functions like ISBLANK, REGEX, etc., in combination to check the format and length. The documentation provides several samples and best practices."

In this example, ChatGPT's Operator tool fetched the relevant documentation and provided a summarized answer with a snippet. This can save you time when constructing more complex validation logic by hand – you get guidance without scouring the web yourself. (Always test validation formulas in Salesforce to ensure they work with your specific field and scenario.)

With declarative automation, you can streamline processes that once required extensive custom code. You've seen how flows can handle a wide array of needs, how validation rules enforce standards, and how approval processes manage sign-offs. **Next, we'll turn our attention to the user experience.** After all, we can have a brilliant data model and powerful automation, but if the end users don't have an intuitive interface to engage with the system, adoption and effectiveness suffer. In the following section, we explore Salesforce's UI configuration tools – App Builder, page layouts, and the game-changing Dynamic Forms – to create a delightful and efficient user experience.

USER INTERFACE CONFIGURATION AND DYNAMIC EXPERIENCES

An effective Salesforce solution isn't just about data and automation behind the scenes – it's also about delivering a **user interface (UI)** that is intuitive, responsive, and tailored to the user's needs. Fortunately, Salesforce's declarative toolkit shines in this area as well. With **Lightning App Builder**, **Page Layouts**, and **Dynamic Forms** (among other features), you can create custom UIs and experiences without writing a single line of HTML or JavaScript. In this section, we'll discuss how to configure record pages, apps, and other UI components declaratively. We'll pay special attention to Dynamic Forms, one of the more revolutionary recent enhancements for building flexible page layouts. Along the way, we'll consider best practices for user-centric design and how to leverage these tools for better adoption and productivity.

Lightning App Builder – Building Pages with Components: The Lightning App Builder is your WYSIWYG editor for Salesforce UI. It allows you to create and edit **Lightning pages** – these include Home pages, Record pages, and even App pages – by dragging and dropping components onto a canvas. A Lightning page is essentially a collection of components arranged in regions (usually a header, main column, sidebar, etc., depending on the template you choose). Components can be standard (provided by Salesforce, like Related Lists, Record Details, Activities timeline, Reports, etc.), or custom (built by developers using Lightning Web Components or Aura, or installed from the AppExchange). The App Builder lets you visually compose a page that surfaces the information and actions your users need most. For example, for an Account record page, you might place a highlights panel at top (key fields), then two columns: one column with related lists (contacts, opportunities) and another with details and perhaps a custom component showing a Google map of the address. All of this can be done without code.

A key concept in Lightning pages is **flexibility per user profile** or **app**: you can assign different page variations to different user profiles (maybe sales reps see a different layout than support agents on the same object) or by the app context (the Sales app's Contact page might differ from the Service app's Contact page if those departments focus on different info). This flexibility means you should plan your Lightning pages with your user personas in mind. Consider what each type of user needs at a glance, and use the App Builder to create optimized layouts for each.

Classic Page Layouts and Dynamic Forms: Before Lightning App Builder was around, Salesforce had (and still has) the concept of **Page Layouts.** Page Layouts are the traditional mechanism to arrange fields, sections, and related lists on a record detail page (in classic UI and Lightning's "Details" tab). They still play an important role: they determine the default layout of fields and sections, and which related lists appear. They also control which quick actions show up on page (if Dynamic Actions aren't used) and a few other legacy UI settings. Page Layouts are assigned to profiles via object **Page Layout Assignments**, often in conjunction with Record Types (e.g., for Record Type A, profile X uses Layout A, profile Y uses Layout B).

However, **Dynamic Forms** have taken UI flexibility to another level. Dynamic Forms, introduced for Lightning Experience, allow you to break free from the fixed "Details" panel of fields that page layouts offer. With Dynamic Forms, you can treat individual fields and sections as components in the Lightning App Builder. This means you can place specific fields anywhere on the page (not just in the pre-set details area) and, importantly, you can apply **visibility rules** to fields and sections. Visibility rules let fields or entire sections show or hide based on criteria you define (like the field's value, the user's role, the record type, etc.). The goal of Dynamic Forms is to create user-centric, **conditional page layouts that display the right information at the right time.**

For example, you might have an "Urgent Issue Details" section on a Case page that should only be visible if the Case's Priority = High. With Dynamic Forms, you can do that easily – the user only sees that section when needed, reducing clutter when it's not relevant. This is much more efficient than the old method which would have required multiple page layouts and assigning them out. In fact, one of the big benefits is **reducing the number of page layouts** you need. Instead of creating separate layouts for every user role or scenario, one dynamic Lightning page can serve multiple audiences by showing/hiding pieces dynamically. Initially, Dynamic Forms launched for custom objects and a few standard objects, but Salesforce has extended support to **most standard objects** now (with some exceptions like Activities or objects not yet upgraded to the new framework).

Enabling Dynamic Forms on a page is straightforward: in the Lightning App Builder for a record page, if Dynamic Forms is available for that object, you'll see a button or prompt to "Upgrade to Dynamic Forms". This will convert the fields and sections from the page layout into individual components on the canvas. You can then move them around or set visibility filters. It's worth mentioning that while fields become flexible, *Related Lists* are still handled via a separate component (Dynamic Related Lists is another new feature allowing filterable related list components). Also, behind the scenes, the underlying page layout still exists and is what users falling back to Classic UI or to PDF printouts would see, so you typically keep it somewhat sensible.

Dynamic Actions: Alongside Dynamic Forms, Salesforce also introduced **Dynamic Actions**, which let you manage which action buttons show up in the Highlights Panel using visibility rules (instead of via the old page layout assignments). This is another part of creating dynamic experiences – for instance, maybe a "Approve Loan" action button should only show for finance managers *and* only when a loan record is in "Pending Approval" status. Dynamic Actions can handle that declaratively.

Custom Apps and Navigation: While record pages are a big part of UI, don't forget that you can also declaratively create **apps** in Salesforce. An app in Lightning Experience is basically a collection of tabs (which can link to objects, reports, dashboards, or external web pages) with a name, logo, and branding. Using the App Manager in Setup, you can create custom Lightning apps to tailor sets of tabs for different functions or teams. For example, you might have a "Contracts App" that surfaces Contracts, related documents, and a dashboard tab, separate from the main Sales app. This is a declarative way to give users focused workspaces.

Additionally, for console-heavy users (like call center agents), Salesforce offers the **Service Console** configuration, where you can set up things like a split view, utility bar (with phone integration, notepads, etc.), and keyboard shortcuts, all through configuration. Much of that is out of the box but can be tweaked declaratively to fit user needs.

Best Practices for UI Configuration: When designing UIs, always think from the perspective of the end-user. What information do they need most frequently? What actions do they need to take, and how can we make those one-click away? Use components like **Related Record** (which can show a summarized view of a parent record on the child record page) to save users from extra navigation. Leverage **Tabs** component to break a page into logical sections (e.g., Details tab, Related tab, maybe a custom Visualforce or LWC component tab). Don't overload a single page with every possible field – even though Salesforce can handle a lot on one page, it can be overwhelming for users and can slightly affect load times. Dynamic Forms help mitigate that by showing only relevant fields, but still be mindful of overall page complexity.

Another tip: use the **Activation** settings in App Builder wisely. You can assign pages by app, record type, and profile. So if, for example, your salespeople and service agents use the same object but in different ways, consider creating two page layouts (or now one dynamic page with conditional sections) and assign appropriately. This way each group feels like the system is tailored for them – which it is!

User Testing: Because we're doing all this declaratively, it's easy to iterate. Engage a few actual users to try out the pages and give feedback. If they say "I always scroll to find X field", that might be a candidate to surface in the highlights panel or top of page. If some information is rarely used, maybe hide it behind a tab or in a collapsible section.

ChatGPT Prompt – Image Generation: "Generate a simple wireframe of a Salesforce Account page with two columns: left column with 'Account Details' fields, right column with a 'Key Contacts' list and a 'Map' showing address."

ChatGPT (Image Generation): (ChatGPT produces an image - a rough sketch style - showing a page layout: a header with Account Name, two columns, left labeled Account Details with some field blocks, right showing a Contacts list and a map image.)

Using ChatGPT's image generation, you can brainstorm UI layouts. While it might not exactly match Salesforce's look, it can help communicate a design idea to a team or stakeholder. Just remember that any actual implementation will be done in App Builder; these AI-generated visuals are for planning or discussion.

ChatGPT Prompt – Typing: "I have a custom object 'Project__c' with a field 'Status__c'. I want the Project record page to only show the section 'Deployment Details' when Status = 'Deployed'. How do I do that?"

ChatGPT: You can achieve this with Dynamic Forms visibility rules. First, ensure your Project record page is using Dynamic Forms (if not, upgrade it in the Lightning App Builder). Once you have individual sections and fields manageable, click on the 'Deployment Details' section component on the page. In the properties pane, you'll see a "Set Component Visibility" option. Add a filter condition: Status__c EQUALS "Deployed". Save and activate the page. Now, users will only see the Deployment Details section when the project's status is Deployed. If the status changes or for projects not yet deployed, that section stays hidden. This is a declarative way of conditionally showing UI components based on record data. Pretty neat!

As you polish your UI, keep accessibility in mind too (Salesforce's core components are generally accessible, but ensure your layouts don't hide important info in a way that might be missed by screen readers, etc.). Also consider localized labels if you have multi-language users – use Translation Workbench for custom text where needed.

Finally, remember that a great UI not only makes current users happy but also eases onboarding of new users. If someone can log in and the screen makes logical sense, with clearly labeled fields and intuitive layout, they can start being productive faster without fumbling around. Declarative configuration allows you to continuously refine the experience – if you observe users frequently clicking several times to get related info, maybe you bring that info onto the main page. There's always room to iterate and improve.

With a solid understanding of UI tools, you can tailor the Salesforce experience to match your business processes and user expectations. We've seen how to build pages and forms that surface the right info at the right time, improving efficiency and user satisfaction. **Next, we'll focus on turning all that data in your system into meaningful insights through Salesforce's reporting and dashboard tools.** After all, collecting data is only half the battle – making sense of it is where the real value lies.

REPORTING, DASHBOARDS, AND ANALYTICS

Data in Salesforce is valuable, but it's true power comes out when you can **report on it and glean insights.** Salesforce's declarative analytics tools – **Reports and Dashboards** – enable admins, business analysts, and even end-users to turn data into actionable information without needing a data analyst or writing SQL queries. In this section, we'll cover how to create effective reports and dashboards, best practices for using these tools, and how to leverage them to drive decision-making. We'll also touch on some analytics features and considerations to ensure your reports remain performant and meaningful.

Reports: The Basics and Beyond: A Salesforce **Report** is a list of records (rows) that meet certain criteria, displayed with whatever columns (fields) you choose, and can include grouping and summary formulas. Salesforce provides a **Report Builder** interface where you choose a **report type** (which determines what objects and relationships you can report on), then define filters (e.g., "Show me all Opportunities where Stage = Closed Won and Close Date = This Month"), and select the columns of data to display. You can also group the data (e.g., group Opportunities by Owner or by Region to see subtotals) and add summary fields (sum of Amount, record count, etc.). Reports can be displayed in different formats: Tabular (simple list), Summary (grouped with subtotals), Matrix (grouped by rows and columns, like a pivot table), and Joined (multiple different report blocks shown together, useful for comparing across objects or views).

Creating a report is entirely declarative: drag fields in, set filters via pick-lists and text inputs, and Salesforce will generate the results in real-time for you to preview. Once you're happy, you save the report into a folder (with access controls) so others can run it. Reports are live – whenever someone runs or refreshes it, they get the latest data from the database (subject to their security access). Users can also subscribe to reports to get them emailed on a schedule, or export them to Excel for further analysis if needed.

Report Types: One concept to understand is the **report type.** Salesforce comes with a lot of standard report types (like "Accounts with Contacts", "Opportunities with Products", etc.), which cover common object join scenarios. If you need a report that isn't covered, you can create a **Custom Report Type** declaratively. For example, you might create a report type "Projects with Tasks and Issues" if you have those custom objects and want a single report showing related data. Custom report types let you pick up to 4 objects (primary, plus 3 related via lookup/master-detail) and define how the joins work (like "with" or "without" related records). Doing this with clicks saves you from the world of writing database queries – it's all abstracted by Salesforce.

Best Practices for Meaningful Reports: Building a report is easy; building a *useful* report requires some thought. It helps to start with the question: *What decision or action should this report inform?* For instance, a report called "Open Cases Needing Attention" implies that support managers should act on the cases listed. That report might filter to high-priority cases open more than 48 hours, grouped by owner to see workload. This is much more actionable than a raw dump of all cases. Work closely with users to identify the key metrics they care about. Often, reviewing existing spreadsheets or legacy reports can reveal what they find important. If users are exporting data to Excel and doing manual pivots, try to reproduce that inside Salesforce so they don't have to – maybe a matrix report can replace a lot of that manual effort.

Another tip is to **clean up and consolidate reports.** Many orgs accumulate hundreds of reports over time, often with overlapping purpose. Regularly review and prune reports that aren't used. Salesforce has a feature to see last run date, or you can use an API tool to find unused reports. Keeping the report repository lean ensures people go to a few trusted reports rather than sifting through clutter. Encourage a practice where new reports are created thoughtfully, and existing ones are reused when possible.

Dashboards: Visualizing the Data: A **Dashboard** in Salesforce is a visual display of data from one or more reports, typically arranged as a set of charts, tables, or metrics on a single page. Each dashboard consists of **components** – each component is tied to a source report and displays data from that report in a certain visualization (bar chart, pie chart, line, gauge, etc., or even a plain table or metric number). Dashboards are great for at-a-glance monitoring of key performance indicators (KPIs). For example, you might have a Sales Dashboard with components showing "Total Sales This Month" (perhaps a big number from a report), "Top 5 Opportunities" (a table from a report of open opps sorted by size), "Opportunities by Stage" (a bar chart summarizing counts by stage), and "Pipeline by Region" (maybe a pie or donut chart). Users, especially managers and executives, love dashboards because it's a one-stop shop.

When building dashboards, choose the right visualization for the data. Time trends work well as line charts. Parts of a whole work nicely as a pie or donut (though too many slices become hard to read, so consider bar charts as well). If exact numbers matter more than visuals, a table component might be better. Also, use **dashboard filters** when appropriate – these allow the viewer to filter all components by certain criteria (e.g., a "Region" filter could apply to all charts, so the user can toggle between EMEA, Americas, APAC, etc., without needing separate dashboards). This can increase the usability of one dashboard for multiple contexts.

Dashboards can be set to run as a specific user or the viewer. If you want everyone to see the same totals (e.g., a company-wide total pipeline), you might run it as a certain user who has access to all data. If you want each person to see only what they have access to, you run it as the viewer (but then someone with less access might see smaller totals – which is fine if that's desired). There's also an option for dynamic dashboards which count against limits but allow per-user perspective without creating multiple copies.

Analytics Best Practices: Much like reports, dashboards benefit from curation. Don't throw every metric on one dashboard – focus it on a theme (sales performance, support team efficiency, etc.). Keep the layout clean and logical (maybe most important KPIs at top-left). Use descriptive titles and add short descriptions where useful to guide the viewer. Also, keep performance in mind: a dashboard is only as fast as the underlying reports. A report that has to crunch tens of thousands of records with heavy formulas might slow down the dashboard load. Sometimes aggregating data with a reporting snapshot (scheduled report storing results to an object) can help for very large data sets, but that's an edge case.

It's also a good practice to educate and empower power users to create their own reports and dashboards (within reason). Many sales managers, for example, might want to tweak a report for their team specifically. With folder permissions, you can allow them to save reports in certain folders. This reduces the backlog on the admin and encourages user engagement. However, balance this by maintaining some governance: you might have a few "official" dashboards that are IT-maintained and always accurate (like board-level metrics), while allowing exploratory or personal reports for day-to-day stuff.

Trends and Advanced Analytics: Salesforce has expanded analytics over time with products like **CRM Analytics (Tableau CRM, formerly Einstein Analytics)** which is a separate platform for very advanced dashboards, data exploration, and AI-driven insights. Using that requires more skills and is outside pure "declarative config" (it's more akin to a specialized tool). For our scope, standard Reports and Dashboards are the core, and they serve most needs well if used properly. That said, even standard reports have some advanced features: you can create **row-level formulas** and **summary formulas** in reports to do calculations on the fly (e.g., percentage of something, or a formula field that wasn't worth making on the object but needed in a report context). You can also bucket fields in reports (grouping values into categories). All of these are done through the report builder UI.

Performance of Reports: Salesforce reports run on Salesforce's servers and generally handle quite large data volumes, but certain things can slow them down. If a report is "unbounded" (e.g., no date filter and millions of rows to scan), it might hit limits or timeouts. It's usually wise to include at least one filter that limits scope (like records from the last 5 years, or Region = X, etc.) especially if data volume is huge. Also, filtering on fields that are indexed (like record owner, record type, etc.) can speed up reports. Most of this is taken care of by Salesforce under the hood, but as an admin, just be mindful of not creating a situation where a report tries to pull everything and the kitchen sink.

ChatGPT Prompt – File Upload/Data Analysis: *Suppose you export a Salesforce report as CSV – "Cases Closed vs. Escalated Q1.csv" – and then upload it to ChatGPT.* **Prompt:** "Analyze this support cases report and summarize the key insights. The report has columns: Agent, Cases Closed, Cases Escalated."

ChatGPT (analysis): *(It reads the CSV and processes the data.)* "After analyzing the data, it appears that Agent A closed 120 cases with 5 escalations (4.2% escalation rate), Agent B closed 90 cases with 2 escalations (2.2%), and Agent C closed 110 cases with 10 escalations (9.1%). Key insight: Agent C has a significantly higher escalation rate than others, suggesting potential training or support needed for Agent C on handling tough cases. Overall case closure numbers are relatively even across agents. The team's average escalation rate is ~5.2%."

This example shows how ChatGPT can help interpret report data. While Salesforce dashboards can show numbers, an AI can add context or highlight anomalies (like one agent's higher escalation rate). It's like having a data analyst on call. Always ensure that any data you export doesn't contain sensitive info unless you're comfortable with the analysis platform's privacy; aggregate metrics like above are generally fine.

Driving Adoption of Reports & Dashboards: One challenge organizations face is getting users to rely on Salesforce reports instead of exporting to Excel or using other tools. A strategy to increase adoption is to ensure the reports/dashboards meet their needs (so they don't *need* to go elsewhere), and to integrate them into regular meetings or workflows. For example, sales teams might start their weekly meeting by looking at a Salesforce dashboard of pipeline. If someone is doing extra manual work outside the system, find out why and see if Salesforce can be configured to do that for them. Many times, empowering users via declarative features goes a long way: e.g., show them how to use **dashboard filters** to get the views they want , or how to create a custom report if they have the access. A **Pro Tip:** *Establish a "single source of truth" for key metrics. If the leadership trusts the Salesforce dashboard, everyone will use it. That might involve data cleanup, but once achieved, it curbs the habit of "I'll just download and make my own spreadsheet."*

Pro Tip: When deploying dashboards to a wide audience, consider using the **"Subscribe" feature** to proactively send updates. Executives might not log in daily to check a dashboard, but if they get a Monday 8am email with the latest dashboard in their inbox, it keeps Salesforce data front-and-center in their decision making. Subscriptions can be set for yourself or for others (if you have permission), and you can include a PDF snapshot. This is a great way to drive engagement with the data on a regular cadence.

By leveraging reports and dashboards, you turn Salesforce from a system of record into a system of insight. Declaratively, you can slice and dice data, present it in compelling formats, and ensure each stakeholder has the information they need at their fingertips. The best configurations also consider performance and governance to keep analytics running smoothly as data grows.

Now that we've covered how to get insights *out* of Salesforce, let's pivot to an equally important topic: **security** – specifically, how to *keep data in Salesforce secure and properly accessible* using declarative tools. In the next section, we'll explore Salesforce's point-and-click security model, ensuring you can confidently share data with the right people and protect it from the wrong ones, all without writing code.

SECURITY MANAGEMENT THROUGH DECLARATIVE CONFIGURATION

Security is a paramount concern for any enterprise system, and Salesforce provides a comprehensive declarative security model to protect data while allowing collaboration. It's often said that **security isn't an afterthought in Salesforce – it's built in from the ground up**, and as an admin or architect you have a rich set of "clicks not code" tools to enforce who can see and do what. In this section, we'll cover the core elements of Salesforce security configuration: **Profiles, Permission Sets, Roles, Sharing Rules, and more.** We'll see how these pieces fit together to implement the principle of least privilege and ensure data is shared appropriately. We'll also touch on declarative tools for compliance and monitoring, like Health Check.

Profiles – The Base User Permissions: Every user in Salesforce is assigned a **Profile.** A profile is essentially a collection of settings and permissions that determine what a user can do in the org. This includes which objects they can access (read/create/edit/delete), which fields they can see or edit (field-level security), which Apex classes or Visualforce pages they can use (for more dev-y things), login hour and IP restrictions, and so on. Out of the box, Salesforce provides some standard profiles (like System Administrator, Standard User, Read Only, etc.), and you can create custom profiles for more tailored needs (in most editions, standard profiles can't be edited much, so you make clones if needed). Profiles are often aligned with a user's job role in a broad sense (e.g., a "Sales User" profile vs "Support User" profile) because they set baseline capabilities. One critical design principle is to avoid giving profiles too much power. In the past, many orgs would create profiles for each department with all needed perms, but as orgs scale, that becomes inflexible and can violate least privilege if not careful. Salesforce now provides a special low-permission profile called **Minimum Access - Salesforce** (in many orgs) which basically has the bare essentials to log in. A modern approach is to give most users a very restrictive profile and then grant additional permissions via permission sets (we'll get to those next). This way, you reduce the risk of over-provisioning access by default.

Permission Sets (and Permission Set Groups) – Granular and Flexible Access: Permission Sets are like "add-ons" to profiles. A permission set can grant the same types of permissions as profiles (object CRUD, field access, etc.) but is assigned to users individually (or via groups). The big advantage is that a user can have only one profile, but *many* permission sets. This allows you to compose user access from building blocks. For example, instead of having a profile "Support Manager" that has everything, you could have a base profile "Support User" and then a permission set "Manager Access" that gives the extra rights like editing user records or seeing all cases. Then you assign that permission set to the actual managers. If another profile (say Sales) also needs that same special access for their managers, you can give the same permission set to those individuals. This avoids duplicating profile configurations and makes it easier to manage changes. Salesforce strongly encourages using a **permission set-led security model** for this reason. In fact, in some newer Salesforce features, certain permissions are only available via permission set (e.g., some high-level system permissions).

Permission Set Groups further simplify administration by allowing you to bundle multiple permission sets together and assign the group to a user, rather than many individual sets. This way, if a role (like "Field Sales Rep") maps to 5 permission sets, you group them and assign the one group to the user. It's still more flexible than profiles because you can layer multiple groups or sets as needed.

One key thing: permission sets (and profiles) can only **grant** access, not remove. So you always start from a base of no access and add what's needed (hence the term least privilege). For example, if a profile gives edit access to an object, you can't use a permission set to take that away – you'd have to remove it from the profile and then give it back selectively via sets if needed.

Roles and Sharing – Record-Level Access: While profiles and permission sets govern what types of data and actions a user *can* access (object-level and field-level), the **Role hierarchy and sharing rules** govern which **specific records** a user can see. Salesforce's record access model starts with **Organization-Wide Defaults (OWD)**, which you set per object to say the baseline visibility: for example, Accounts could be set to Private (meaning by default users only see their own accounts unless shared) or Public Read-Only (everyone can see all accounts, but maybe only edit their own if that's the rule), etc. Once the OWD is set, you use **Roles** to open up visibility. Roles are arranged in a hierarchy (often mirroring the managerial org chart or functional hierarchy). Users at higher roles can (if you choose) see the records owned by people in roles below them (this is automatic for standard objects like Opportunities, Cases when using role hierarchy – there's a setting per object "grant access via hierarchies"). Roles are not about permissions like edit or delete – that's still profile – roles are purely about *which records you can see (and edit if you already have edit rights on that object).*

For example, if Alice is in the "US Sales Rep" role and Bob is her manager in the "US Sales Manager" role (above her in hierarchy), and if account sharing is private, Bob will nonetheless see Alice's accounts by virtue of role hierarchy. This satisfies common needs where managers oversee their team's records.

But roles alone might not cover more complex sharing needs – that's where **Sharing Rules** come in. Sharing rules are declarative rules that say "share records of Object X meeting Y criteria with Z users/groups/roles". For instance, you could have a sharing rule: "Share all Opportunities where Industry = 'Healthcare' with the Healthcare Sales Team role." This could give read-only or read/write access as needed. Sharing rules allow you to break the default silos and extend access horizontally or to specific groups. The nice part is, like everything else, you configure them with point-and-click in Setup (specifying criteria or ownership based sharing).

Salesforce has other sharing mechanisms too: **Manual Sharing** (a user manually shares a single record to someone, if they have rights to do so), **Territory Management** (an account-sharing model often used as an alternative/complement to roles for sales teams, based on geography or product), and newer stuff like **Salesforce Teams** (Account Teams, Opportunity Teams – which let you define a team of users on a record and share accordingly). These are all declarative or point-and-click features to handle specific scenarios of record access without coding.

Field-Level Security and Sensitive Data: On top of objects and records, Salesforce lets you control at the field level who can see or edit a particular field. This is configured in Profiles/Permission Sets as field permissions. For example, you might want everyone to be able to see Contact phone numbers, but only HR to see Social Security Number field on a custom object – so you'd mark that field as hidden for all profiles except HR's (or give only HR permission via a set). This is often used for sensitive personal data or confidential info. Another related feature is **Encrypted Fields** (which are a type of field that is stored encrypted at rest and has limited viewing/editing options). Encrypted fields are still configured declaratively, though if you want more advanced encryption, Salesforce Shield (a paid add-on) allows encryption of standard fields and broader encryption at the platform level.

Principle of Least Privilege: The guiding principle for security is to give users *exactly* the access they need, nothing more. Salesforce makes it possible to really fine-tune this. For example, using the **Minimum Access Profile + Permission Sets** approach, you could start every user with basically no access, then add on only the objects and actions they require. This ensures that if a user's role changes or if an account is compromised, the exposure is limited. It's also good for audit purposes – you can demonstrate that, say, only the Finance team's permission set grants access to the Salary fields, etc.

To implement least privilege effectively, regularly conduct a **permission audit.** Salesforce's Setup offers a **Profile and Permission Set** list view where you can see who has what, or you can use the **Permission Set Assignment Export** in the Data Loader to review. Look out for things like "View All Data" or "Modify All Data" – these are extremely powerful and usually should only be on System Admin or very special users. Same for object-specific "View All" or "Modify All" on objects; those bypass sharing rules, giving broad access to that object's records. Use those sparingly.

ChatGPT Prompt – File Upload (Profile XML): *Imagine you export a profile's metadata (XML) or a permission report and give it to ChatGPT.* **Prompt:** "Review this profile's permissions and highlight any risks based on least privilege."

ChatGPT: "Looking at the provided profile, I see it has Modify All on the Account object – that means users with this profile can view and edit all Account records, regardless of sharing rules. If that level of access isn't absolutely necessary, it's risky because it bypasses normal data sharing controls. I also see that 'Export Reports' permission is enabled – if these users don't need to export data, disabling that would reduce the chance of large data exfiltration. Additionally, the profile has the 'Author Apex' permission, which is typically only needed for developers; granting it broadly can be a risk. Overall, I'd recommend principle of least privilege adjustments: remove Modify All on Account if possible, and create a separate permission set for those who truly need that level (like sales ops), remove the Apex auth permission for regular users, and double-check if any other admin-level permissions are included."

This shows how ChatGPT can assist in a security review. By parsing a profile's meta or a permission spreadsheet, it can call out things an admin might overlook in a long list. Always be cautious with sharing full org exports, but reviewing a single profile's perms is likely fine if sanitized. The insight accelerates tightening security.

Sharing Architecture Example: Suppose your company uses Salesforce to manage consulting projects. You have a custom object Project__c. You want project team members to see their project, their managers to see all projects in their department, but projects should otherwise be confidential (not visible to other departments). How to do this declaratively? You'd set Project__c OWD to Private. Then define roles like "Consultant" under "Consulting Manager" under "Head of Consulting" etc. Role hierarchy will ensure managers see their team's records. To add team members who are from other departments, you might use a Project Share custom object or manual sharing – but better, you could also use a sharing rule: e.g., each Project has a lookup to an Account (client) and maybe you share all Projects for Account X with the Account Team (if using Account Teams). Another approach: use Apex sharing if very dynamic, but since we focus on declarative, you'd try to model it with those features. The point is, a combination of OWD, roles, and sharing rules usually gets you 95% of the way for record access configurations.

Compliance Tools – Shield, Health Check, etc.: Salesforce provides some declarative tools to help with compliance. **Health Check** (in

Setup) gives you a score and highlights security settings relative to a baseline (for example, are your password policies strict enough? Is session timeout appropriately short?). It's not directly about sharing or perms, but overall org security. If you have Salesforce Shield, you get things like Event Monitoring (to track who's viewing what data) and Platform Encryption which you can turn on for fields with a few clicks (though planning encryption is a big task itself, no code needed but careful thought required).

Also, the **"View All Users"** and **"Login As"** features: as an admin profile, you can typically see all users and even login as them (if enabled). Those are powerful but in many companies, they control who is allowed to use them for privacy reasons. For compliance, maybe only Tier 2 support can login as others, etc. These are profile permissions that can be adjusted declaratively to meet policies.

Communicating Security to Users: Often end-users might not understand why they can't see something ("I can't see that account my colleague mentioned"). Part of declarative security management is documenting and communicating the sharing model. A simple way is to create a matrix or a few example personas – "Sales Rep can see: their own opportunities, their team's accounts, etc.; cannot see: opportunities of other teams" – and publish that internally. Also, when leveraging permission sets, name them clearly (e.g., "Case Management – Edit" vs "Case Management – Read" sets) so it's easier to know what each does. Salesforce's naming can get confusing (especially with things like permission set groups), so clear naming conventions are an admin's friend in maintaining sanity.

Pro Tip: Use the **"User Access Verification" (Who Sees What)** series Salesforce provides (or create your own version) to double-check your sharing. There's a Salesforce Admins blog and video series called "Who Sees What" that's very helpful. It walks through how different sharing settings work together. For your own org, one trick is to create a dummy user for each profile/role combination and use the "Login As" feature to view exactly what they see. This way, you can validate that your configuration is working as intended. If something shows up that shouldn't, adjust your sharing rules or field security accordingly.

As you configure security declaratively, you'll find it's a mix of art and science – art in designing a model that fits business needs elegantly, and science in mapping those needs to Salesforce features precisely. The outcome, though, is worth it: a secure Salesforce environment where everyone has the access they need and no more. It builds trust with your users (they won't stumble into data they shouldn't see) and with management (that data is properly safeguarded).

With security best practices under our belt, we now turn to another behind-the-scenes aspect of a healthy Salesforce org: **performance.** In the next section, we will discuss how to manage and optimize performance declaratively – ensuring that all the wonderful configurations we've built scale smoothly and respond quickly as usage grows.

MANAGING AND OPTIMIZING PERFORMANCE DECLARATIVELY

Even the most feature-rich Salesforce implementation can falter if it doesn't perform well. Slow page loads, sluggish reports, or hitting governor limits can frustrate users and undermine confidence. The good news is that with mindful declarative design, you can avoid most performance pitfalls. Salesforce, being a cloud platform, abstracts a lot of the heavy lifting, but as an admin or architect you still have to design for scale and efficiency. In this section, we'll cover performance considerations and best practices for declarative configurations: how to ensure your data model and automations are optimized, using tools like indexing and selective queries, and how to keep your UI snappy. We'll also touch on monitoring tools to detect and address performance issues.

Data Model and Query Performance: Performance often starts with the data model. A well-structured data model not only makes building features easier, it also makes queries (like reports, list views, and automation logic) more efficient. Here are some tips:

Indexes: Salesforce automatically indexes certain fields (like record IDs, foreign keys like lookup or master-detail fields, audit dates, and external IDs, plus some custom fields you mark as External ID or Unique). Indexed fields make queries filter faster. If you have filters on reports or list views that are slow, check if you're filtering on a field that's not indexed. For very large data volumes, you might request a custom index via Salesforce support (this is a big hammer, but it's doable for key fields). The general rule of thumb is to make your filters selective – e.g., filtering on "Status = Closed" might be fine if only 10% of records are closed, but filtering on "Status != Closed" might be not selective if 90% are open. Using **Skinny Tables** is another advanced concept SF can do behind scenes for performance (but that's not something you configure directly, it's Salesforce internal if needed). The key takeaway: design your objects and fields such that common queries can be selective (e.g., perhaps add a checkbox like "High Priority" that gets checked via automation for certain records, so listing high priority items is a quick query by an indexed checkbox rather than a complex formula filter). According to best practices for large data volumes, "reducing the number of records that the system must process" is a primary strategy. Proper indexing is one way to achieve that.

Avoid Unnecessary Complexity in Filters: A filter that uses a formula field or a subquery might be slower than a simple field filter. If you notice you're always filtering on a formula (like "Show me all Contacts where Age > 21" and Age is a formula calculating from birthdate), it might be faster to denormalize that or have a checkbox "Adult__c" set via workflow/flow. This is that balance of normalization vs performance.

Archiving Data: Over time, some records may become irrelevant to daily operations (e.g., opportunities from 10 years ago). Consider using built-in archiving (like Salesforce auto-archives activities after a certain period) or building archive logic (export and delete old records or move them to a big object or external system) for data that isn't needed live. Less data = faster queries generally, though Salesforce can handle millions of records if well-indexed.

Automation and Transaction Performance: Declarative automations (Flows, Process Builders if any left, etc.) run as part of saving records or other events. Poorly designed flows can slow down those operations:

One Trigger, One Flow: As mentioned earlier, try to consolidate automations so that only one flow (per context) runs for an object. If you have 5 record-triggered flows all on "Opportunity update", they all run and it's harder to control order or avoid redundant work. One flow can do the logic of five, and it's easier to optimize in one place. Each additional automation adds overhead.

Use Before-Save Flows: If your flow's purpose is just to update some fields on the record being saved (e.g., set a default value, or do a minor calculation), using a **before-save flow** is much faster than an after-save flow doing an update (which causes an extra DML). Salesforce says before-save updates can be ~10x more efficient than old workflow field updates. It's basically because they execute in memory and don't perform an extra save.

Limit Loops and SOQL: If a flow needs to process a batch of records, use collection processing instead of per-record loops. For example, if you have to find related records for each opportunity in a batch, don't query inside a loop. Instead, collect all Opportunity Ids in a collection and do one query outside the loop to get all related records in one go (this is "bulkification" in declarative form). Similarly, avoid very large loops in Flow when possible; sometimes it may be better to break into chunks or use a scheduled path to batch things.

Governor Limits: Salesforce enforces limits (like max 50,000 records in a query, max CPU time, etc.) to ensure fairness in multi-tenant environment. Declarative processes can hit these too. If you find flows hitting CPU timeouts or SOQL limits, it's a sign to simplify logic or move some of it to asynchronous methods (like schedule it or use a platform event to handle later). In flows, you can use the "Scheduled Paths" feature to offload certain actions to run a bit later (which effectively runs in a separate transaction).

Parallel Execution Considerations: If you have multiple automation triggers that might run at the same time (like a bulk data load or integrations hitting Salesforce), try to design idempotent and independent processes. For example, if two different flows update the same field in different ways, a parallel update might cause contention or record locking. That level of conflict is rare in declarative-only setups, but be aware if integrating with external systems.

User Interface Performance: Users are most sensitive to how fast pages load and respond. Salesforce Lightning has made great strides in performance, but there are things you can do:

Optimize Lightning Pages: Remove unnecessary components or use component visibility to only load them when needed. If you have a homepage with 10 report charts, each of those is a query. Do all 10 need to show by default? Maybe show top 3 and let user click a tab for more. For record pages, consider using the **Accordion** or **Tabs** component to section content – components not in the active tab don't load until clicked (this is lazy loading). Also utilize **Dynamic Forms** to not render fields that aren't needed until criteria met (which reduces front-end rendering time).

Use the Lightning App Performance checker (there is a "Analyze" button in Lightning App Builder that can flag issues like too many components or components that take long to load). Also, Salesforce has a "Lightning Usage" app which, among other stats, can show slowest pages.

Client vs Server: Some performance is client-side (the browser rendering). Large related lists (1000 items) might be slow mostly because of data volume transferring. Try to use filtered related lists or limit what's shown by default (users can click "View All" if needed). Also note that certain advanced components like interactive charts might load extra libraries; use them only if necessary.

Monitoring Tools: Salesforce provides a few tools to help monitor performance:

Lightning Usage App: This built-in app (find it in App Launcher) shows metrics like daily active users, but also slowest Lightning components and pages, load times, etc. If you see that a particular page is consistently slow (maybe average load 5 seconds whereas others 2 seconds), you can investigate what on that page might cause it – maybe a heavy Visualforce component or an unbounded related list.

Salesforce Optimizer: This is a tool you can run from Setup which scans your org and provides a report on various best practices, including things that can affect performance (like too many fields on a page layout, or too many similar triggers). It gives a nice PDF output with recommendations.

Event Monitoring (if available): With Salesforce Shield or by purchasing event monitoring, you get log data on performance of queries and transactions. Even without it, you can use debug logs on specific user actions to see where time is spent.

Browser Dev Tools: Sometimes the issue is not Salesforce server but the client, e.g., older browsers or network latency. Checking in multiple environments or using the Chrome DevTools performance tab can identify if it's front-end slowness (rendering, JS) versus waiting on server.

Design for Scale: When building declaratively, always ask: "What if this object has 10 million records? Will this still work?" Often the answer is yes, if you've kept things simple and indexed. Salesforce can handle large data if configured right. However, some features have soft limits; for instance, list views that return too many results might be capped or not allowed unless filter by something indexed. Multi-select picklist filters are notoriously non-selective because of how they store data (so try not to filter reports on multi-selects if performance matters). Knowing these quirks can inform design (maybe use a junction object instead of a multi-select for certain data relationships, because then you can query it more effectively).

When Code or Apps Can Help: Sometimes, a bit of code or an AppExchange tool can help performance, ironically to supplement declarative. E.g., if you have a very complex nightly calculation, an Apex batch job might pre-compute and store results in a custom object that your dashboard then just reads quickly. But before resorting to that, see if declarative options like a scheduled flow or a reporting snapshot could do it. There is also the option of **Big Objects** for archival (which can store billions of records, but with very limited querying ability – mostly good for logging data).

ChatGPT Prompt – Operator (Knowledge Search): "Search for common causes of slow record save times in Salesforce."

ChatGPT: *After browsing knowledge base:* "According to Salesforce documentation and community discussions, common causes of slow record save times include: having many process builders or flows on the same object (which each consume processing time), triggers or processes that execute SOQL queries in loops (which can hit limits), and extensive formula fields or roll-up summary calculations that need to recalc on save. Additionally, very large numbers of related records (over 10k) on a master-detail can introduce a slight delay due to roll-up recalculation. To improve save performance, one should consolidate automation (reduce number of flow executions), ensure any queries are bulkified (perhaps using collections in flows), and consider moving non-urgent processes to asynchronous paths."

Using ChatGPT's browsing here brings in community wisdom and Salesforce's own best practice notes. It reaffirms some points we've made and gives you confidence you're covering typical issues. You might use such a summary as a checklist when diagnosing a slow transaction in your org.

Pro Tip: Remember that performance tuning is an ongoing activity. As the org grows and new features are added, periodically revisit your configurations. A flow that worked fine with 1,000 records might start straining at 1,000,000 records. Make use of Sandbox testing with larger data sets if possible to foresee issues. Also, stay updated on Salesforce releases – they often optimize things behind the scenes or introduce new features (for example, the recent introduction of **Restriction Rules** to replace some sharing rules can improve performance by reducing the number of sharing calculations for large data sets in certain cases).

By thoughtfully designing and monitoring, you can achieve **fast, efficient declarative solutions.** Salesforce's cloud infrastructure is robust, but giving it a helping hand through good design separates a decent admin from a great one. Users might not notice when things are well-tuned (because stuff just works quickly), but they'll definitely notice if it's slow. So our job is to almost be invisible – performance shouldn't even be a topic of conversation for end users; it should just meet expectations.

As we conclude this section, we have one more major topic in this chapter: **maintenance and scaling** your declarative solutions over time. After building all these features, how do we keep our Salesforce org clean, organized, and adaptable to change? In the next section, we'll discuss strategies for maintaining what you've built, governing change, and ensuring your declarative configurations continue to serve the business well into the future.

MAINTAINING AND SCALING DECLARATIVE SOLUTIONS

Congratulations – you've built a robust Salesforce solution using declarative tools! But the journey doesn't end once the features are live. One of the hallmarks of an enterprise-ready system is how well it can be **maintained and scaled** as requirements evolve. In this final section of the chapter, we'll explore best practices for sustaining your declarative configurations. This includes documentation, using sandboxes and source control for changes, monitoring org health, and strategies for cleanups and optimizations over time. By treating declarative development with the same rigor as programmatic development, you ensure your Salesforce org remains agile, efficient, and free of unnecessary clutter.

Documentation of Configuration: Good documentation is often the difference between a one-off solution and a scalable platform. Salesforce provides places to add descriptions – every field, object, flow, etc., has a Description or Notes field. **Always fill these in** when you create something new, explaining not just what it is but why it exists. For example, a validation rule description might say "Ensures end date is after start date to maintain data integrity – implemented after user reports of incorrect inputs." These help someone later understand the intent. Beyond in-app descriptions, maintain some central documentation. This could be a simple Google Doc or spreadsheet listing custom objects and their purpose, major automation and what they do, etc. Some teams use **architecture diagrams** for complex processes (tools like Lucidchart or Salesforce's own architect diagram kits can help). The goal is that if a new admin joined your team, they could get up to speed quickly by reading your docs.

Another useful artifact is a **release notes or change log.** Each time you make a declarative change, record it: "Jan 2025: Added new field X to support Y requirement, by Z admin". Over time, this becomes a valuable history. It can be as informal as a Chatter group where you post changes, or a dedicated Confluence page, etc. Many orgs are moving to treat "config as code" – meaning they track changes in version control (we'll touch that next) – which inherently provides a log of changes.

ChatGPT Prompt – Typing: "Suggest a template for documenting a Salesforce Flow so future admins can understand it."

ChatGPT: "Sure! Here's a simple template:

Flow Name & API Name: [Full name of the Flow]

Purpose: A brief description of what this flow accomplishes (e.g., 'Closes opportunity and creates renewal opportunity when an opportunity is won.').

Type & Trigger: Record-triggered (before/after), Screen Flow, etc., and when it fires (object and criteria).

Inputs & Outputs: If it's an autolaunched flow called by others, note any input variables or output variables.

Key Elements: List of important steps (Decisions, Creates, Updates, etc.) with a note on each. For example: Decision "Large Deal?" branches the logic based on Opportunity amount > $100k.

Subflows or Invocations: Mention if it calls other subflows or Apex or sends emails, etc.

Exceptions & Faults: Note any error handling implemented or potential points of failure (e.g., "If email send fails, flow continues without alert currently.").

Created By & Date: Who built it and when, plus any significant later modifications by others.

Related Automation: If related to other flows or workflows, list them (for context).

This template ensures that anyone reviewing the flow documentation can quickly grasp its function and design. Attach a screenshot of the flow diagram if possible for visual context."

Using this kind of structured documentation (which ChatGPT can help generate or format) will save time down the road. You can create similar templates for documenting objects or security settings.

Source Control and Environments: It's a best practice in Salesforce to use **multiple environments** (sandboxes) for making changes and then deploy to production. Even though we're clicking around and not writing code, tools like Salesforce DX have made it possible to track declarative changes in **version control (Git)**. If you have a team of admins or developers working in parallel, consider adopting a source control approach: you extract metadata (which includes all your declarative stuff like object definitions, flows, etc.) into files and commit to a Git repository. This allows versioning, rollback if needed, and easier deployments through scripts. It might sound technical, but even admins can contribute by learning a few commands or using a tool like Gearset or Copado (DevOps tools for Salesforce) which provide UI to move changes and track them. The payoff is huge when maintaining an org long-term: you get audit trails of who changed what and when, you can do code reviews for declarative changes (yes, even flows can be reviewed by looking at screenshots or XML differences to ensure quality), and you can quickly deploy to another org or revert a bad change.

If setting up full version control is too heavy initially, at least adopt a solid **sandbox strategy**. Have a development sandbox (or several) where changes are built and tested, a UAT sandbox for users to accept changes, and then a planned deployment to prod. Avoid making untracked changes directly in production, because that leads to undocumented differences and possible mistakes (and you can't easily undo something done in prod without backup).

Org Health and Cleanup: Over time, Salesforce orgs can accrue what we call "**metadata clutter**" – fields that aren't used, old workflows left inactive, reports no one runs, etc. Periodic clean-up is healthy. Salesforce Optimizer (mentioned earlier) will point out things like "these fields have not been populated on any records recently" or "you have 10 inactive flows, consider deleting or archiving them". A manual approach is also fine: run reports on "Field Utilization" (there are apps like Field Trip that analyze field fill rates). If a field is always blank and no one remembers why it exists, maybe it's time to remove it (in a sandbox first, to be safe, and see if anything breaks). Clean folder structures for reports/dashboards: maybe archive old ones into an Archive folder or delete if truly obsolete.

Scaling Considerations: As your org grows, some things that were not an issue might become one. We talked about performance, but scaling is also about maintainability with more users and more admins. Implement **naming conventions** for everything (flows, fields, picklist values, etc.). It's much easier to find "Flow - Case - Escalation Handler" than a flow named "CaseFlow1". Use prefixes or suffixes to group things (some orgs prefix all fields of a certain project with PjX_ etc. – do what makes sense but be consistent). If you haven't, define a **DevOps process** for how changes are requested, reviewed, and implemented. For example, a Jira ticket for every change, reviewed by a lead admin or architect if it's complex. This process ensures thoughtful design and avoids ad-hoc tweaks that could conflict.

Staying Educated: Salesforce releases updates three times a year. Each release can bring new declarative features, or changes to limits. Keep up with release notes – especially sections relevant to admins (there's usually a summary of "Flows: new feature X" or "Security: new setting Y"). Incorporate new improvements; for instance, if Salesforce introduces a better way to do something you previously did with a workaround, consider migrating. For example, if you had a custom roll-up flow and Salesforce later provides a declarative roll-up summary on that relationship, you might switch to the native feature for simplicity. Also, Salesforce might deprecate things (like Workflow/Process builder retirement we discussed). Be prepared to adjust your org accordingly. Plan a **review of your automations and features at least annually** to see if they align with current best practices.

User Feedback Loop: Maintenance also involves the end-users. They might discover edge cases or have suggestions for improvement. Establish an easy way for users to provide feedback – maybe a Chatter group "Salesforce Ideas" or periodic meetings with reps from each department. As you get feedback, you can improve the declarative solutions or fix issues. This iterative approach is essentially agile development applied to configuration. It helps ensure your org doesn't become static; it evolves with the business. However, balance rapid changes with stability

– bundle minor tweaks into regular releases rather than ad-hoc daily changes, so you have time to test and users can adapt to changes in a predictable way.

Backup and Recovery: While Salesforce is reliable, you should have a strategy for backup of your metadata and data. For metadata, if using source control, that's a backup. If not, consider using the weekly export of metadata or a tool to retrieve and store it. For data, Salesforce offers Weekly Data Export (which can be automated to an extent) and now a Data Backup service (paid). At minimum, if you are doing a destructive change (like deleting fields), back up the data from those fields first. There's nothing worse than realizing a year later that you need that data and it's gone. Some companies utilize third-party backup solutions which is extra insurance.

ChatGPT Prompt – SORA (Video): *Imagine using SORA to create a short video tutorial.* **Prompt:** "Generate a brief video script highlighting how our team handles Salesforce changes using declarative configuration and version control, to onboard new team members."

ChatGPT (SORA video response outline): *The AI might produce a script or even a simple animated video (if integrated):* "Scene 1: An admin receives a request (shown via a chat bubble). Narration: 'At Acme Corp, any Salesforce change starts with a request – maybe a new field or automation.' Scene 2: The admin moves the request into a planning board, Narration: 'Our admins review and plan the change, designing it using Salesforce declarative tools like Flow Builder.' Scene 3: A split screen of two sandbox orgs and Git, Narration: 'We use a sandbox to build and test. Our changes are saved to a source control repository, ensuring every adjustment is tracked.' Scene 4: A quick demo of a deployment, Narration: 'Once tested, the change is deployed to production during a scheduled release window. Users are notified of the update with release notes.' Scene 5: Team celebration, Narration: 'By following this process, we keep our Salesforce org scalable, documented, and reliable – and our team members, like you, can easily collaborate and contribute!'."

This might be a rough output, but it shows the concept. A video can engage new team members, and SORA or similar AI could help produce quick internal explainers from your prompts. It's a futuristic productivity boost – documenting processes not just in writing, but in multimedia.

Pro Tip: Consider establishing a **Center of Excellence (CoE)** for Salesforce in your organization. This is a cross-functional team or at least a recurring meeting where admins, power users, maybe a developer or architect, and someone from management discuss the Salesforce roadmap, governance, and standards. It's a forum to ensure everyone is on the same page about best practices, to review major upcoming changes, and to share knowledge. A CoE drives continuous improvement and keeps the platform aligned with business strategy. Even if your company is small, the concept of regularly reviewing the state of your Salesforce org with stakeholders is very valuable.

As you maintain your org, remember the old saying: *"Measure twice, cut once."* In Salesforce terms, plan and review changes carefully (especially anything that could impact a lot of users or data). Leverage sandbox testing and involve users in UAT for significant changes. It's much easier to adjust before going live than to fix something after.

We've now covered a broad spectrum of Salesforce declarative tools and best practices – from data modeling all the way through to ongoing maintenance. This holistic understanding empowers you to not only build effective solutions but also keep them running smoothly in the long haul.

SUMMARY

In Chapter 2, we embarked on a comprehensive tour of the Salesforce Platform's declarative toolkit. We began by laying the foundation with **core platform essentials** – understanding standard and custom objects, fields, and relationships. You learned how master-detail and lookup relationships structure your data, and how solid data modeling underpins all other configurations. Building on that, we explored **declarative automation tools** like Flow Builder, Validation Rules, and Approval Processes, highlighting Salesforce's shift to a Flow-centric approach for implementing business logic with clicks not code. We then shifted focus to the **user interface**, using Lightning App Builder and Dynamic Forms to create tailored, dynamic experiences that improve usability and adoption.

From capturing data and automating processes, we moved to **analytics**, demonstrating how reports and dashboards can be crafted declaratively to turn data into actionable insights for your organization. You saw the importance of meaningful metrics and keeping analytics user-friendly and relevant. Security was another crucial stop on our journey – we looked at **declarative security management**, using profiles, permission sets, roles, and sharing rules to enforce least privilege and protect data. We emphasized designing a security model that is both robust and flexible without writing code, aligning with Salesforce's built-in sharing capabilities and compliance tools.

The chapter also addressed **performance optimization**, outlining how thoughtful declarative design (efficient queries, indexed fields, streamlined automation) can ensure your Salesforce org performs well as data and users grow. We looked at ways to monitor and fine-tune performance issues before they impact users. Finally, we covered **maintaining and scaling declarative solutions**: the practices that keep your org clean, well-documented, and adaptable to change. This included leveraging modern tools like version control for metadata, instituting governance processes, and regularly reviewing configurations so that your declarative solutions remain sustainable over time.

Throughout each section, we included ChatGPT prompt callouts demonstrating how an AI assistant can augment your Salesforce declarative work – from explaining concepts and generating examples, to analyzing files and even producing visuals or video scripts. These examples illustrated that tools like ChatGPT can be a creative partner in solutioning, troubleshooting, and learning, whether through text, voice, or multimedia interactions.

By mastering the content of this chapter, you've armed yourself with a solid understanding of Salesforce's declarative toolkit. You should now feel confident in choosing the right declarative tool for a given task and understand the end-to-end lifecycle: design, build, secure, optimize, and maintain. The knowledge here sets the stage for deeper dives into architectural thinking and strategic design, which we will explore in subsequent chapters.

In the next chapter, we will build upon this toolkit and focus on **design principles and solution architecture** for Salesforce. We'll learn how to apply an architectural lens to everything we configure – ensuring solutions are not only functional, but also scalable, maintainable, and aligned with business strategy. With your firm grasp of "what" tools are available from this chapter, you'll be ready to tackle the "how" and "why" of putting them together in Chapter 3. Keep the declarative best practices we've discussed in mind – they will be the backbone of the well-architected approaches to come. Let's continue our journey into Salesforce enterprise thinking!

3

DESIGN PRINCIPLES AND SOLUTION ARCHITECTURE

In the world of Salesforce, it's easy to dive in and start building with clicks, but seasoned architects know that a strong foundation of design principles separates quick fixes from long-term success. This chapter emphasizes how applying disciplined architectural thinking to declarative configuration leads to maintainable, scalable solutions closely aligned with your business objectives. We champion a *"Clicks, but with Intent"* philosophy: harnessing Salesforce's powerful point-and-click tools deliberately and thoughtfully, rather than haphazardly. Declarative doesn't mean unplanned – every click should be purposeful and backed by an understanding of the business need.

We begin by mastering how to gather and analyze business requirements effectively. Through stakeholder interviews, workshops, and clear user stories, you'll learn to pinpoint what the business truly needs from Salesforce. This careful requirements analysis ensures that every design decision you make is grounded in real-world needs and not just technical whims.

With solid requirements in hand, the next challenge is choosing the right tool for the job. Salesforce offers a rich array of declarative tools – Flows, validation rules, formula fields, approval processes, and more. We'll discuss how to evaluate these options against your requirements and constraints, developing a framework to decide when clicks (declarative configuration) suffice or when you might need code. Through examples and comparisons, you'll see how the right choice makes all the difference in balancing functionality with maintainability.

Beyond individual tools, great solutions adhere to overarching design principles. We introduce the **Salesforce Well-Architected Framework**, focusing on its three pillars: **Trusted, Easy**, and **Adaptable**. These principles – security and reliability, simplicity and usability, flexibility and scalability – serve as guiding stars for any declarative design. You'll see how to weave these values into your configurations so that they not only meet today's needs but also inspire confidence and handle tomorrow's demands.

Even with good intentions, it's easy to fall into common pitfalls. This chapter will shine a light on typical **anti-patterns** in declarative design – from tangled, overly complex automations to misuse of relationships or formula fields that create more problems than they solve. By learning to recognize these red flags early, you can avoid them or refactor them into cleaner solutions.

Of course, we won't stop at what to avoid. We also explore **proven declarative design patterns** that exemplify elegance and efficiency, such as modular Flow structures, re-usable components, and using custom metadata to drive logic. These patterns provide templates for solving problems in ways that maximize scalability and ease of maintenance.

As important as designing well is **documenting those designs and decisions.** We cover strategies for capturing your declarative architecture in writing – whether through design specs, decision logs, or visual diagrams. Effective documentation ensures your rationale is understood by stakeholders and that your solutions remain intelligible and governable as your team and org evolve.

Finally, we'll emphasize that good design is an ongoing process, not a one-time task. Adopting an **iterative mindset** means continuously improving your configuration based on feedback and changing business needs. We'll discuss how to conduct regular reviews, learn from successes and missteps, and gradually refine your Salesforce setup to stay aligned with business strategy and operational excellence.

For any successful Salesforce project, everything begins with understanding what the business truly needs. Let's start our deep dive there – by ensuring we know how to gather and analyze requirements with intent and clarity.

GATHERING AND ANALYZING BUSINESS REQUIREMENTS

Building any Salesforce solution should start with a **clear understanding of the business requirements.** Even though Salesforce declarative configuration lets you rapidly create features with clicks, you can't afford to skip the planning. Taking time to gather and analyze requirements ensures you're solving the right problems and setting yourself up to make intentional design choices. It's easy to be tempted by the "just build it now" mindset, but seasoned architects know that *a few hours spent understanding the business need can save days of rework* down the line. This is the essence of *Clicks, but with Intent* – before you click anything in Salesforce Setup, make sure you know *why* you're doing it.

Engage stakeholders early and often. Start by conducting stakeholder interviews with everyone who has a stake in the process or system you're improving. Talk to end users on the front lines (sales reps, support agents, etc.), their managers, and any process owners or IT colleagues. Each will have unique perspectives on pain points and goals. In these conversations, focus on listening more than solutioning. Encourage stakeholders to describe *what* they need and *why* it matters, rather than jumping straight to how they think Salesforce should do it. Often a stakeholder might say, "We need a checkbox for X," but your job is to dig deeper: what outcome are they trying to achieve with that checkbox? By understanding their true intent, you might find a better way to deliver the result.

Workshops can be particularly effective for **cross-functional requirements.** Bring groups from different teams together for a brainstorming session or process walkthrough. A common technique is to map out the current process and highlight pain points or inefficiencies on a whiteboard (virtual or physical). For example, if Sales hand off deals to Service, outline that handoff process and let participants point out where things go wrong or data falls through the cracks. Collaborative workshops often uncover hidden requirements or dependencies that individual interviews might miss. They also help build a shared understanding among stakeholders – a critical foundation for getting buy-in on the solution later.

Once you've gathered raw input, **document the requirements clearly**. A best practice is to write user stories – short, simple descriptions of a feature told from the perspective of a user or role. For instance: "*As a support agent, I want to receive an alert when a high-priority case is open for more than 2 hours so that I can proactively reach out to the customer.*" This format forces you to capture the *who*, *what*, and *why* of the requirement. Along with each user story, document acceptance criteria: the specific conditions that will confirm the requirement is met. In our example, acceptance criteria might include, "*Given a high-priority case that has been open for 2 hours, when the system checks for overdue cases, then a notification is sent to the assigned support agent.*" These criteria serve as tests you'll later use to verify your solution's success. Writing these details down not only guides development but also helps avoid misunderstandings – everyone can agree on paper what "done" looks like for each need.

Now comes the *analysis* part: step back and examine the collected requirements holistically. Look for overlapping needs or even conflicting requests among different stakeholders. You might find that two departments essentially asked for the same capability in different words – an opportunity to create one solution that serves both. Alternatively, you might discover that satisfying one group's request could inadvertently cause an issue for another (for example, auto-closing a case might satisfy support managers but frustrate sales reps who lose visibility). Identifying these intersections early allows you to design a balanced solution or negotiate trade-offs before anything is built.

It's also important to **prioritize requirements**. Not everything can be delivered at once, and trying to do so often leads to half-baked features. Work with business sponsors to rank requirements by importance and urgency. A practical approach is to categorize each item as "Must have," "Should have," or "Nice to have." Tackle the must-haves first – these will deliver the core business value. Prioritization ties directly into design decisions: for example, if a requirement is a lower priority, you might choose a simpler solution for now and plan to enhance it later, rather than over-engineering upfront. Keeping complexity low for non-critical features is a hallmark of good architectural thinking.

During analysis, start thinking about Salesforce capabilities at a high level. Without jumping into the solution yet, you can begin mapping needs to Salesforce's declarative tools conceptually. Is a requirement about automating a process? Make a note that you'll likely consider Flow Builder. Is it about ensuring data quality? Perhaps validation rules or required fields. At this stage, you're not deciding the exact solution, but this mapping ensures that for each need, you have some idea of what types of tools could fulfill it. It's a sanity check – if you find a requirement with no obvious way to implement it declaratively (or at all), that's a red flag to address now (maybe by refining the requirement or planning for a custom solution).

After you've written and reviewed the requirements, **validate your understanding** with the stakeholders. Play back what you've documented: "Here's what I heard you need... did I get that right?" This not only catches any misunderstandings early but also reassures the business that you're aligned with them. It's much cheaper to adjust a requirement document than to rewrite a Flow or undo a configuration built on a false assumption. By iterating on the requirements with stakeholder feedback, you ensure the foundation of your design is solid and agreed upon.

ChatGPT Prompt (File Upload): *Imagine you've compiled a spreadsheet listing dozens of user stories and pain points gathered from various departments.* You can upload this document directly into ChatGPT and ask: "**Please analyze the attached requirements list. Identify any gaps or ambiguous requirements and suggest clarifying questions I should ask stakeholders.**" ChatGPT will comb through the text, find items that might be unclear or incomplete (for example, a user story that lacks a clear "why"), and provide you with a list of pointed questions or assumptions to validate. This acts like a second set of eyes on your requirements, helping ensure you haven't missed anything critical before you move into design.

Pro Tip: Before uploading any internal documents to ChatGPT (or any AI tool), double-check your company's data privacy guidelines. Make sure you have **enabled settings that prevent data from being used for AI training**, or sanitize sensitive information. Salesforce architects often deal with confidential business data, so treat your requirement artifacts with the same care you would source code or production data.

Pro Tip: Maintain a simple requirements-to-solution **traceability matrix** as you design and build. This is a table or document that links each requirement to the Salesforce configuration that satisfies it. If you plan a new field, automation, or any component, check it against the matrix to ensure it traces back to a real business need. If it doesn't, reconsider if that component is necessary. This practice keeps your implementation lean and purpose-driven, and it makes future audits or reviews much easier – you can demonstrate exactly why each element exists.

At this point, you have a well-vetted set of business requirements with priorities and a clear definition of success for each. You're equipped with a business-aligned blueprint that sets the stage for design. The next step is to translate that blueprint into actual Salesforce solutions. How do you decide which declarative tool fits each requirement best? With the *what* and *why* in hand, you're ready to explore the *how* – choosing the right declarative tools to bring these requirements to life in Salesforce.

CHOOSING THE RIGHT DECLARATIVE TOOLS

With a clear set of requirements in hand, the next challenge is deciding *how* to implement them using Salesforce's declarative toolkit. Salesforce offers many "clicks not code" options – from **Flow Builder** and **Validation Rules** to **formula fields**, **custom objects**, and more. But each tool has strengths and limitations. A skilled architect doesn't just randomly pick a tool; they evaluate which option best fits the job at hand. In this section, we'll develop a practical approach to selecting the right declarative tools (and knowing when to use programmatic solutions). It's about making intentional choices so that your solution is efficient, maintainable, and aligned with the business needs you've gathered.

Start by reviewing each requirement and ask: *Can this be solved with configuration alone?* In Salesforce, the general guideline is to prefer declarative solutions for speed and simplicity **unless** there's a compelling reason to write code. For each requirement, consider the core Salesforce feature areas:

Data Model: Does meeting this need require new fields or objects? If a sales process needs to track something entirely new (say, customer onboarding steps), you might introduce a custom object or field. On the other hand, if Salesforce already has a standard object that fits (for example, using the standard **Case** object for an IT ticketing requirement), leverage what exists. Using standard objects can save effort and benefit from built-in functionality, but only if they truly align with what the business needs.

Automation: If the requirement involves automating actions or enforcing logic, Salesforce Flow is the powerhouse tool. Within Flow, decide the type: **Record-Triggered Flows** for behind-the-scenes automation when records change, **Scheduled Flows** for batch jobs (like nightly data clean-ups), or **Screen Flows** for guided user interactions. If a requirement is simple – e.g., "when field X is checked, update field Y" – consider if a **Formula Field** or **Validation Rule** could accomplish it with less complexity. A formula field could calculate Y from X in real time (no automation needed), or a validation rule could simply prevent a bad combination instead of a Flow that later has to correct it.

User Interface: Some needs are about how data is presented or collected. For instance, if a requirement is "Sales reps need a tailored view of key fields during lead conversion," you might use **Dynamic Forms** or the **Lightning App Builder** to adjust page layouts and components, rather than writing a custom Lightning Web Component. Declarative UI tools (App Builder, Path, Dynamic Related Lists) can often deliver a great user experience without a single line of code. Always ask: can the user interface be configured declaratively to meet this need?

Approval and Collaboration: If the business process involves approvals (like a deal discount needing manager sign-off), **Approval Processes** might be the right declarative tool. They are point-and-click to configure and handle routing and notifications automatically. However, if the approval logic is very dynamic or conditional beyond what the standard approval process can do, you might combine it with Flow (for pre-checks or post-approval actions). But the pattern is to use declarative approvals first for any straightforward approval chain.

Reporting & Analytics: Sometimes the requirement is not transactional at all but about insights, like "we need to identify top-performing products" or "show me all cases that breached SLA." Before building an automation or new fields, consider if a **Report** or **Dashboard** could fulfill this need by querying existing data. Often, a well-designed report can replace what might seem like a need for new process logic. This saves effort and stays within the declarative ethos (reports are configuration too!).

As you match requirements to potential tools, also weigh any **technical constraints** and long-term factors. Key questions include:

Volume and Scale: Will this solution need to handle large volumes of data or high transaction counts? For example, a Flow that loops through hundreds of related records might hit governor limits or slow down under load. If you expect scale issues, perhaps an **Apex** solution (like a bulkified trigger or a scheduled batch) is more appropriate, or you break the work into smaller chunks (multiple flows or a combination of Flow and Apex). Conversely, if volume is low to moderate, a declarative approach is usually preferable for its simplicity and lower maintenance burden.

Complexity of Logic: How complex is the business logic? Declarative tools can handle quite complex scenarios, especially with the power of modern Flows (decisions, loops, formulas inside flows). But if you find yourself contorting a Flow into a maze of branches and elements, that's a sign the logic might be better implemented in Apex, where you can structure it in a more readable way (using code constructs). A good rule of thumb: if a flowchart of your process wouldn't fit clearly on a page, or you're hitting limits like "Too many SOQL queries" in a Flow, consider alternative implementations. You might simplify by splitting the flow, or if that's not possible, writing a portion in Apex could be cleaner.

Declarative Feature Gaps: Does the requirement involve something that declarative tools simply can't do (yet)? For example, calling an external web service from a Flow was historically impossible without Apex (though with the advent of External Services and HTTP callouts in Flow in recent releases, this gap is closing). If you hit such a gap, you might design a hybrid solution: the majority declarative, with a small Apex **Invocable Method** or **Lightning Web Component** to handle the part declarative can't. This way, you still maximize clicks while plugging the gap with code only where needed.

Choosing between declarative and programmatic isn't an all-or-nothing decision; it's often about finding the right mix. A classic scenario is **updating related records**. Say the requirement is: "Whenever an Account's rating is changed, update all the related Contacts to match that rating." You have options:

Use a **Record-Triggered Flow** on Account that loops through related Contacts and updates them – purely declarative. This works well if the number of Contacts per Account is reasonably small. It's simple and maintainable by an admin.

If some Accounts have hundreds of Contacts, that loop might be slow or hit limits. You could instead use an **Apex trigger** for efficiency (since Apex can handle larger volumes in a single transaction more gracefully than Flow, and you can control batching in code).

A hybrid approach: Use a Flow to detect the change on Account and put a message on a **Platform Event** or **Queueable Apex** job, which an Apex subscriber processes to update contacts. The Flow remains simple (just "fire event on change") and the heavy lifting is done in code that's optimized for bulk.

In making these choices, always consider the **team's ability to maintain** the solution. If you have a team of pro-code developers and the org already has a lot of Apex, writing a trigger might integrate well with your processes. But if the org is primarily maintained by an admin or declarative-centric team, introducing a lot of new Apex could create a dependency on developers for future tweaks. In that case, pushing declarative tools a bit further (within safe limits) to keep the solution in the admins' wheelhouse can be worth it. Maintainability and clarity are as important as initial build effort.

ChatGPT Prompt (Typing): You can leverage ChatGPT as a sounding board when you're unsure about tool selection. For example, ask: "*We need to enforce that when an Opportunity's Stage moves to 'Closed Won', a related Contract record is automatically created. Should I use a Salesforce Flow for this, or would it be better as an Apex trigger?*" ChatGPT can outline the pros and cons. It might respond: "A record-triggered Flow can handle this declaratively (create a Contract when Opportunity is Closed Won) and is easier to maintain without code. However, if you anticipate extremely complex logic around Contract creation or very high volumes, an Apex trigger could offer more control. In most cases, start with Flow for simplicity and maintainability." Essentially, ChatGPT's answer helps mirror the architectural thought process, ensuring you've considered the standard best practice (in this case, using Flow, since creating related records is well within its capabilities) before resorting to code.

Pro Tip: Always check if Salesforce has an **out-of-the-box feature** before building something custom. Many times, what looks like a need for a complex solution can be met by simply turning on a setting or using a standard feature. For example, if you're considering building a complex Flow to prevent duplicate records, first explore Salesforce's **Duplicate Rules** and **Matching Rules** – declarative features specifically designed for that. Or if you're about to build a custom approval mechanism, consider if the native Approval Process can be configured to fit. Using built-in capabilities not only saves you time, but those features are generally robust, well-tested, and familiar to other Salesforce professionals. The best solution might be no custom solution at all, if Salesforce already provides it.

As you decide on tools, document your reasoning. If you choose Flow over Apex, note why (e.g., "Requirement can be met with Flow alone; simpler to maintain by admins. No large data volume concerns."). If you choose to write Apex, be clear on the justification ("Complex calculation needed that Flow cannot do," or "High-volume scenario that exceeds Flow limits"). These notes are gold during design reviews and for future maintainers – they explain the intent behind your choices.

Remember, "*Clicks, but with Intent*" means every click (or line of code) has a rationale. By thoughtfully selecting the right approach for each requirement, you set yourself up for a solution that not only works but stands the test of time and change. Selecting a tool is just one side of the coin; using it in an architecturally sound way is the next. In that spirit, let's see how the **Well-Architected principles** can guide the way we implement our chosen tools.

APPLYING THE SALESFORCE WELL-ARCHITECTED FRAMEWORK

Great architects don't rely on ad-hoc intuition alone – they use guiding principles to shape their designs. Salesforce's **Well-Architected Framework** provides three core attributes to strive for in any solution: **Trusted**, **Easy**, and **Adaptable**. These pillars are like north stars for declarative configuration, ensuring your org's design remains healthy and balanced as it grows. Let's break down each pillar and see how to apply it in practice when configuring Salesforce declaratively.

Trusted: Designing for Security and Reliability

A *Trusted* solution protects the business and its stakeholders. In Salesforce terms, this means your configuration is secure, compliant, and reliable. How do we bake trust into declarative design? Start with **security by design**. For every object, field, and automation you create, think about data access. Are you using the proper **profiles, roles, and permission sets** so that users only see and do what they should? For example, if you create a custom Salary field on a custom Employee object for HR, ensure its properly marked as confidential and hidden from non-HR profiles via field-level security. A declarative solution is not truly "no-code" if an admin has to constantly put out fires from data leaks or permission issues – so set it up right from the beginning.

Beyond access control, trust involves **data integrity** and **system reliability**. Declarative tools like Validation Rules help keep bad data out – they enforce rules ("Close Date cannot be in the past for new Opportunities") that maintain data quality. Using a validation rule in this way is more reliable and transparent than, say, relying on a complex web of conditional logic in a Flow that might be bypassed or overlooked. Simplicity often supports trust: the fewer moving parts, the less chance of something breaking. If a requirement can be met with one well-written validation rule instead of a sprawling Flow, the rule might be the more reliable choice in the long run because it's straightforward.

Reliability also means considering failure modes. What if a Flow fails to update a record due to some unforeseen condition? A Trusted design would incorporate fault paths – for instance, configuring a **fault connector** in Flow that logs an error or sends an alert to an admin if something goes wrong. This way, errors aren't swallowed silently; you find out and can address them. Even though we're not writing try-catch blocks as in code, declarative design can still include checks and fall-backs (like using an **Email Alert** action when a Flow's fault path is hit to notify the team).

Compliance is another aspect of trust. If you're in a regulated industry, use declarative tools to enforce compliance requirements. For example, you might use **Approval Processes** to ensure no contract gets changed to "Active" without a manager sign-off, or use **Salesforce Shield** features (click-configurable encryption, field audit trail, etc.) to meet data privacy mandates. A practical Trusted principle practice is regularly running Salesforce's **Health Check** (in Setup) which scores your org's security settings against Salesforce recommendations – it's a mostly declarative way to spot security gaps and remedy them (like session settings or password policies that are weak).

In short, a Trusted declarative design is one where: a) **security** is front-and-center (right data, right people), b) **data integrity** is ensured (garbage in, garbage out is prevented), and c) the system is **reliable** (anticipating problems and handling them gracefully). If you build with those in mind, users and stakeholders will develop confidence in the system – they trust that Salesforce is doing the right thing.

Easy: Designing for Simplicity and Usability

An *Easy* solution delivers value fast and is, well, easy to use and easy to maintain. In declarative design, "easy" has a double meaning: easy for end-users and easy for admins/developers who work on the system.

From the end-user perspective, think about the **user experience** of what you configure. Suppose you built a complex Screen Flow with multiple steps for a service agent to log a case. Is it actually making their job easier, or did you inadvertently create a confusing wizard? Applying the Easy principle, you might simplify the flow: remove unnecessary steps, use clear language on screens, and only present fields that truly need user input. Utilize **dynamic form visibility** so that only relevant fields show based on earlier inputs. The interface should feel intuitive. A good test: if you have to train experienced users extensively on how to use your new form or process, it might not be as "Easy" as it could be. The best declarative solutions often feel obvious to the user.

Ease also means **delivering value quickly** to the business. Declarative configuration shines here because you can often build a prototype or initial solution in days, not weeks. We encourage an agile, iterative approach: get something out there that solves the core problem, then refine. For example, if users need a way to track customer feedback, you might first give them a custom object with a simple form and a report – that covers the basics. Later, you could enhance it with automation (like alerts for negative feedback) once the base process is validated. By not over-complicating the first iteration, you delivered value fast and kept the solution easy to understand. Then you layer on improvements as needed, which ties into our iterative philosophy.

From the maintenance perspective, an Easy design is one that new team members (or your future self) can understand without a detective work. This relates to things like **naming conventions and in-app descriptions** as we mentioned. When every Flow is named clearly by its purpose (e.g., "Opportunity – Post Closed Won Actions") and every custom field has a helpful description, admins find it much easier to navigate and modify the org. It's also about consistency: if you solve a problem one way in one part of the org, try to solve a similar problem in the same way in another. For instance, if you decide to enforce validation rules for data quality on one object, do it on comparable objects too, rather than sometimes using a validation rule and sometimes a flow for similar checks. Consistency makes the system easier to reason about.

Ease for maintainers also means not over-engineering. Favor declarative simplicity over clever complexity. A common mistake is building a very intricate solution for a simple requirement because the builder wanted to experiment or show off a new feature. That often backfires when someone else has to maintain it. The simplest solution that meets the requirements is usually the best. Salesforce provides a lot of ways to do the same thing – pick the one that other professionals would find most straightforward. That typically means using standard declarative features in standard ways.

In summary, applying the Easy pillar means always asking, *Is there a simpler way to achieve this?* Simpler for the user (fewer clicks, clear instructions, only necessary info) and simpler for the admin (well-organized configuration, not needlessly convoluted). When in doubt, err on the side of simplicity and clarity. A solution that is easy to use and easy to understand will have higher adoption and longevity.

Adaptable: Designing for Flexibility and Scalability

The only constant in business is change. An *Adaptable* Salesforce solution is ready to evolve with changing business needs and to scale as those needs grow. In declarative design, adaptability comes from building with flexibility in mind, avoiding hard-coded assumptions, and leveraging Salesforce features that allow for growth.

One strategy for adaptability is to use **metadata-driven design.** We touched on this in design patterns: using **Custom Metadata Types** or **Custom Settings** to store configuration data that your declarative logic (Flows, formulas, etc.) can read. This pattern turns what would have been a static rule into something admins can update. For example, if you have a Flow that routes leads based on region, instead of coding the regions and target queues into the Flow's decision paths, you could have a custom metadata type "Lead Routing" with records like Region = North, Queue = North_Sales_Queue. The Flow then simply looks up the matching record and routes accordingly. When regions change or a new region comes on board, you add or update metadata records – no need to alter the Flow itself. This makes your design far more adaptable to organizational changes.

Designing for scalability means anticipating **large data volumes and high usage.** Even if today only 1000 records flow through a process, if the business hopes to grow 10x, your design should handle 10,000 (or have a clear path to handle it). Adaptable declarative designs often use

best practices like indexing, selective queries, and efficient automation triggers. For instance, rather than having a Flow fire on *any* change to an object and then check a bunch of conditions (which means it runs every time, even when irrelevant), set the Flow to only fire when specific fields change or conditions are met (using entry conditions). That way, as record counts soar, your automation isn't doing needless work.

Also consider future functionality. When designing, ask "what if...?" questions. What if this process needs to handle multiple currencies or languages later (should I use **Translation Workbench** and **Custom Labels** now to prepare)? What if another department might use this app (maybe make the record types generic enough to extend)? It's a balance – you don't want to over-engineer for scenarios that might never happen, but a bit of forward thinking can save major refactoring later. A good example is naming things in a scalable way: don't hard-name a process "NA Order Process" if there's a chance Europe will need it – name it generically ("Order Process") so you can reuse or clone with minimal confusion.

The Adaptable pillar also reminds us to stay flexible in our mindset. Salesforce releases new features every release; an adaptable org and team will adopt these improvements to simplify or enhance existing processes. For instance, if you built something complex before Multi-Factor Authentication was a standard option, maybe now you can retire that complexity and use the out-of-box MFA. Keep an eye on release notes with an eye toward refactoring your declarative solutions to use newer, better approaches when they become available.

ChatGPT Prompt (Voice): *(Speaking aloud to ChatGPT)* "I'm designing a solution for our support team to handle escalations. How can I ensure my design is Trusted, Easy, and Adaptable?" After you ask this via voice, ChatGPT might respond with something like: "To make it **Trusted**, ensure only the right people can trigger or see escalations (perhaps only managers), and include error handling if an escalation fails. To keep it **Easy**, maybe provide a simple screen flow with minimal steps for agents, and use clear field labels or guidance so it's intuitive. To stay **Adaptable**, consider storing escalation rules (like which team it goes to) in custom metadata so you can change them without editing the flow, and design the flow in a modular way in case the process grows or changes." This kind of answer reinforces your planning – acting as a quick checklist that you've considered each pillar for your design. It's like having an architecture assistant to double-check your work against Salesforce's core principles.

Pro Tip: Make the Well-Architected pillars a part of every design review. Create a short checklist with **Trusted – Easy – Adaptable** and have the team explicitly discuss each for a given design. Ask questions like, "How have we made this feature Trusted? Are there any security or reliability gaps?"; "Is this solution Easy for users to understand and for us to maintain?"; "What happens if the business doubles in size or changes policy – will this design adapt or will we be scrambling?" By routinely evaluating your designs through this lens, it becomes second nature to catch issues early. Over time, you'll find that thinking in terms of Trusted, Easy, Adaptable leads to inherently better declarative solutions that require less rework and instill more confidence.

Armed with these guiding principles, you can evaluate design choices not just on immediate fit, but on long-term quality and resilience. In practice, this means you'll start to recognize certain patterns in your Salesforce configurations as clearly good or bad. Some approaches will align with the pillars – those are patterns to emulate. Others will violate them – those are patterns to avoid. Before we move on to the positive patterns to follow, let's shine a light on those common **anti-patterns** in declarative design that threaten Trust, Ease, and Adaptability, so you know how to spot and eliminate them.

RECOGNIZING AND AVOIDING ANTI-PATTERNS

Even with good principles in mind, it's easy to slip into certain **anti-patterns** – common configuration habits that seem to solve a problem but create bigger issues down the road. Let's expose some of the usual suspects in Salesforce declarative design and discuss how to avoid or fix them:

The "Monster" Flow (Overly Complex Flows): This is a Flow that tries to do *everything* – it updates records, sends emails, loops through related lists, branches into dozens of paths – all in one automation. While it might have started simple, over time additional requirements turned it into a behemoth. The result? A Flow that's hard to understand, hard to test, and prone to hitting limits or unexpected errors. **Better approach:** Break it up. If you find a Flow getting unwieldy, see if it represents logical sub-processes that can be modularized. Use **subflows** to compartmentalize parts of the logic (e.g., one subflow purely for sending notifications, another for calculations). Also consider splitting triggers: multiple smaller record-triggered flows might be clearer than one huge one with many branches – especially now that Salesforce provides the Flow Trigger Explorer to help manage order. Each flow should ideally have a focused purpose. This makes maintenance far easier (you know exactly which Flow to tweak for a given change) and reduces the risk that a change in one part will accidentally break something in another.

Formula Field Overload: Formula fields are fantastic for real-time calculations, but using too many or making them overly complex can be an anti-pattern. Signs of trouble include formula fields referencing other formula fields in long chains, or giant formulas with nested IF statements that practically encode business logic. While each individual formula might work, collectively they can slow down record save and edit performance (since formulas calculate on the fly) and can be a nightmare to debug if something goes wrong. **Better approach:** Use formula fields for *simple or essential real-time calculations* that you can't easily do otherwise. If a formula grows too complex (nearing Salesforce's 5,000 character limit or going 10 levels deep in logic), consider alternatives like a before-save Flow that computes the value once and stores it in a field. That way the heavy lifting happens on save rather than every time a record is viewed. Also, avoid duplicating logic in multiple formulas; see if you can combine into one formula or use helper fields to simplify. Keep an eye on Salesforce Optimizer reports, which can flag overly complex formulas.

Object Cramming (Misusing Object Relationships): This anti-pattern happens when the data model is bent out of shape for convenience. For example, using one custom object to store multiple unrelated types of records distinguished only by a "Type" field – resulting in one object with dozens of fields, many of which are irrelevant for a given record type. This can make security and page layouts messy and lead to inefficiencies. Another example is not using a **junction object** for a many-to-many relationship and instead hacking it with a multi-select picklist or multiple lookup fields (like having 5 lookup fields for "Related Product 1...5" on an Opportunity). These approaches might work initially but don't scale and complicate reporting and logic. **Better approach:** Adhere to solid data modeling practices. Use separate objects or at least record types for distinctly different data sets. If you have a many-to-many scenario, introduce that junction object (e.g., a "Enrollment" object between Contact and Class for a training app, instead of a multi-select of Contacts on Class). Yes, it's a bit more configuration upfront, but it pays off in query ability and avoiding limits. Remember, Salesforce's relational database is powerful – trust it to manage relationships rather than trying to pack everything into one object or field.

Duplicate or Siloed Automation Logic: In a growing org, different admins or teams might unknowingly create automations that overlap or conflict. For instance, one admin sets up a Process Builder (legacy) to auto-close dormant cases, and another later builds a Flow to do a similar thing. Now you have two automations firing on the same condition, potentially stepping on each other or doubling actions. Or you have validations, flows, and Apex all enforcing related rules, which can confuse debugging ("Which one prevented this record save?"). **Better approach:** Establish clear ownership and documentation of automations on each object (governance helps here). It's often wise to consolidate automations: for example, move all logic into Flows if possible, and turn off the old Workflow Rules/Process Builders once replaced. If duplicate logic is found, decide on one place to manage it and remove the redundancy. Additionally, commenting your Flow (with description text in the Flow) to explain what it covers can help avoid someone creating a parallel process out of ignorance.

Hardcoding and Magic Values: This is when specific IDs or values are embedded in your declarative logic. Maybe a validation rule references the Id of a specific record (like an HR record Id that might differ between orgs), or a Flow assigns a record to a user "Jane Doe" by name. These assumptions break easily – if Jane leaves or the org is deployed to another environment, things fail. **Better approach:** Use dynamic references or custom metadata for such things. For example, instead of hardcoding record Ids, use a query by some unique field (e.g., lookup "HR

Dept" record by Name rather than Id), or better yet, a custom metadata that stores that Id per environment (which you can reference in Flow via a Get Records). In validation rules, avoid literal values if possible; if you must, clearly comment why and consider using a custom label or custom setting that can be changed without editing the rule. The goal is to eliminate "magic" constants hidden in your config that might change one day unknown to you.

Outdated Tools or Approaches: Salesforce evolves, and sticking with old practices can become an anti-pattern. A prime example is continuing to build new automation with **Workflow Rules** or **Process Builder** instead of Flow (Salesforce has encouraged moving to Flow as the single automation tool). Using older tools means future maintainers have to check multiple places for logic and might miss things. Another example: not using newer capabilities like record-triggered flows in "before-save" mode for simple field updates, and instead using slower after-save patterns out of habit. **Better approach:** Stay current with Salesforce best practices. Standardize on **Flow** for new automations (and consider migrating critical old ones to Flow when possible). Learn and use the latest declarative features – for instance, if a new release lets you do something in a formula that previously required a Flow, update your approach. This keeps your org efficient and aligned with Salesforce's roadmap, and reduces the clutter of legacy components.

These are just a few common anti-patterns; there are certainly more (like overusing multi-select picklists, creating too many fields "just in case," etc.). The key to avoiding anti-patterns is awareness and a willingness to refactor. If something in your org is causing pain – slow performance, confusion among the team, frequent bugs – step back and ask if it might be an anti-pattern that needs redesign. The earlier you catch and correct an anti-pattern, the easier it is to fix.

ChatGPT Prompt (Browsing): *Let's say you're not sure if a particular approach is considered an anti-pattern.* You might ask ChatGPT (with browsing enabled) to **search for common Salesforce configuration mistakes.** For example: "**Search the web for common Salesforce Flow anti-patterns.**" ChatGPT could then find a relevant blog post or Salesforce discussion, and summarize guidance from it. It might come back with insights like, "Experts advise against using overly generic 'one size fits all' flows that handle unrelated processes, as this is hard to maintain. It's better to have focused flows per object or task." By leveraging ChatGPT's browsing, you get up-to-date insights from the community on what practices to be wary of. This can validate your hunch that, say, using 15 separate record-triggered flows on the same object is problematic, and it will often provide recommendations on a better approach.

Pro Tip: Make use of Salesforce's own tools to detect potential anti-patterns. **Salesforce Optimizer** (available in Setup) can scan your org and highlight issues like too many fields on an object, automation rules that might be redundant, or profiles with excessive permissions. Run it periodically and review the findings – many of them point directly to things we'd call anti-patterns. Likewise, consider an occasional **design audit**: pick a major process or object in your org and review all related configuration for simplicity and clarity. Bring in a colleague who wasn't involved in building it – can they understand it easily? If not, that's feedback to streamline. When you do catch an anti-pattern in your org, plan to refactor it sooner rather than later. Technical debt in declarative configuration can sneak up on you (just like in code). A little time spent to simplify now can save a lot of time troubleshooting later.

Having identified what *not* to do, you can more easily focus on what *to do*. In the next section, we'll flip our perspective and look at **proven declarative design patterns** – the smart, tried-and-true approaches that will help you build Salesforce solutions that are robust and elegant from the get-go.

LEVERAGING PROVEN DECLARATIVE DESIGN PATTERNS

With pitfalls identified, let's turn to the positive side: **design patterns** that have proven to work well in Salesforce declarative development. These are approaches that experienced architects and admins use repeatedly because they lead to cleaner, more scalable solutions. By adopting these patterns, you don't have to reinvent the wheel for every requirement – you can apply known good strategies to common scenarios. Here are several key patterns to add to your toolbox:

Modular Flow Design: Rather than one gigantic flow trying to do everything (the anti-pattern we discussed earlier), structure your automations into smaller, modular pieces. A common pattern is the **"One Flow per Object per Trigger"** approach: for each object and context (e.g., before-save, after-save), have a single flow orchestrator that may call subflows for specific tasks. For example, you might have an "Opportunity – After Save" flow that, when an Opportunity is closed, invokes one subflow to handle post-close tasks (like creating a renewal opportunity or updating a customer status field) and another subflow to send related notifications. Each subflow is focused and potentially reusable. If another process also needs to send that notification, it can call the same subflow rather than duplicating logic. Modular design prevents the "monster flow" scenario and makes testing easier – you can test each piece independently. It also means when a requirement changes, you often only adjust one subflow rather than a huge flow. Embracing subflows and the concept of single-responsibility flows is a design pattern that leads to clarity and reuse in declarative automation.

Reusable Components (Declare Once, Use Often): Reusability isn't just for code – you can apply it declaratively too. For instance, if multiple parts of your business need a similar **Quick Action** or screen flow, create one and deploy it in multiple contexts. A practical example: if you have a "Submit for Approval" screen flow that collects some inputs, use that same flow for different objects or scenarios by making it generic and passing in parameters (or using it as a template). Another form of reusability is creating a **generic subflow** that performs a common operation (like "Send Email to Account Owner") and call it from various parent flows. In the UI arena, think of building a **Lightning App Page** or home page template that serves multiple user profiles with slight variations through dynamic components. Invest time in a well-crafted component or flow that can serve many masters; it reduces maintenance (fix or improve one component and all uses benefit) and ensures consistency (every department gets the same quality of solution for the same need).

Custom Metadata for Dynamic Logic: One of the most powerful declarative patterns is to externalize decision logic using **Custom Metadata Types** (or Custom Settings, for simpler needs). Instead of embedding business rules and thresholds directly into Flows or validation rules, store them as data that your configuration can reference. For example, say you have a discount approval process where the approval required depends on the discount percentage. Rather than hardcoding "if discount > 20% then require VP approval" in a Flow, create a custom metadata type "Discount Approval Rule" with fields for Min_Discount, Max_Discount, Required_Approver_Level. Now your Flow simply queries this metadata to find which rule the discount falls into and acts accordingly. When the business inevitably changes the numbers or levels, you update the metadata record – no Flow edits needed. This pattern greatly improves maintainability and adaptability (aligning with our Adaptable pillar): admins can adjust business rules on the fly, and you reduce the risk of errors from having to modify logic in multiple places. We've effectively turned what would have been "hard-coded" logic into "data-driven" logic.

Event-Driven and Decoupled Processes: In some situations, especially in larger implementations, you want to decouple processes to make them more scalable and to avoid one change having ripple effects. A pattern here is to use **Platform Events** or custom notifications as a means for different components of your system to communicate without being tightly linked. For example, when a case is closed, you might publish a "Case Closed" platform event. You then have one Flow that listens for this event and handles customer survey outreach, another that maybe updates a related account health score, and yet another that, say, notifies the account team via chatter. Each of these is a separate flow triggered by the same event. The beauty of this pattern is that each flow (each subscriber) is independent – adding or changing one doesn't directly affect the others, and the publisher (the case closure logic) doesn't need to know about all the subscribers. This is a more advanced pattern and not needed for every org, but it's a scalable way to design complex, multi-step processes declaratively in an enterprise environment.

Consistent Naming and Structure: This might not sound like a "design pattern" in the traditional sense, but in practice, having a consistent approach to naming and organizing your declarative artifacts is a huge best practice. For example, adopt a naming convention for Flows that includes the object and trigger ("Account – After Save – Set Territory") or for formulas that clearly state their purpose ("IsVIPCustomer__c"). Consistency in naming is like creating a pattern language for your org – anyone who works in it can quickly understand what things are by name. Similarly, maintain a logical structure in how you compose things: if you always use subflows for certain tasks (like sending emails or posting to Chatter), that becomes a pattern others can follow, and your org's configuration will have a familiar "feel" everywhere. This reduces the learning curve for new team members and reduces mistakes (because people aren't inventing new ways to do the same thing each time).

Scalable Data Model Patterns: Certain data model designs are known patterns that work well in Salesforce. One common one is the **master-detail with rollup summary** for roll-up calculations instead of building a custom rollup logic. Another is the use of **Person Accounts** vs separate Contact/Account for B2C scenarios (a pattern Salesforce provides out-of-the-box when suitable). There's also the pattern of using a **Hierarchy Custom Setting** or **Role hierarchy** for territory/role-specific defaults – which can be considered data model or logic, but it's a pattern: storing per-profile or per-role configuration in a structured way. Recognizing when a requirement fits a known data model pattern (like needing a junction object, or a self-lookup hierarchy) allows you to implement a robust solution faster because the groundwork is well-understood. Always ask: have we solved something similar before, or has Salesforce provided a standard pattern (in documentation or Trailhead) for this kind of requirement?

Each of these patterns enhances maintainability and scalability. **Modular flows** make it easier to update automation without side effects. **Reusable components** mean less duplication and more consistency. **Custom metadata-driven logic** makes adapting to change as simple as data entry, not re-engineering. **Event-driven processes** allow different parts of the system to evolve independently yet cohesively. **Naming conventions and structured design** impose a helpful uniformity that makes the org self-documenting in many ways. And solid **data model patterns** ensure that as data volumes grow or relationships get more complex, your foundation can handle it.

ChatGPT Prompt (Image/Wireframe Generation): Design discussions often benefit from visuals. Suppose you want to illustrate the **modular flow design pattern** to your team. You might ask ChatGPT to *generate a simple diagram*: "**Draw a flowchart showing a main Opportunity flow that calls two subflows – one for sending an email and one for creating a follow-up task.**" ChatGPT can use an integrated image generation tool to produce a diagram of this pattern. The image might show a box labeled "Opportunity (After Save) Flow," leading to two separate boxes: "Send Email Subflow" and "Create Task Subflow," which then complete. By visualizing the pattern, stakeholders and team members can quickly grasp the concept of breaking a flow into parts. Similarly, you could request a wireframe for a "Reusable Feedback Form" to show how one screen flow could serve multiple departments with slight variations. Having AI generate these visuals saves you time and helps ensure everyone has a shared understanding of the design pattern before implementation.

Pro Tip: Keep a **library of patterns and best practices** for your org. Document (in a wiki or even a simple PDF) the key declarative patterns you use, with examples. For instance, write a short entry on "Approval Process vs. Flow for approvals – criteria for when we use each" or "Template for record-triggered flow design with subflows." This becomes a go-to reference for your team. New project starting? Check the library to see if similar use cases have a defined pattern. Over time, this library can include snippets, screenshots, or descriptions of things like "How we set up sales process branching with record types and path" or "Preferred way to do scheduled mass updates (using scheduled flows)." Maintaining such a reference guide accelerates design work, onboards new team members faster, and promotes consistency across your Salesforce project. Essentially, you're creating an internal knowledge base of Salesforce declarative architecture for your company – an invaluable asset as you scale.

By leveraging these design patterns, you build solutions that are robust by design. The work you do upfront to follow patterns pays off in reduced maintenance effort and increased agility. However, to truly benefit from them, you need to ensure others (and future you) can understand and follow what you've built. This is where **documentation**

and communication of your declarative design become vital – that's our next focus.

DOCUMENTING DECLARATIVE DESIGN DECISIONS

The only thing worse than a bad design is a good design that no one knows about or understands. Documentation might not be the most glamorous part of solution building, but it is absolutely critical. In the Salesforce declarative world, where a lot of the "code" lives in clicks and configurations, **documentation bridges the gap** – it tells the story of your system: the why, how, and what of your design choices.

Start with capturing the **rationale behind key decisions**. Whenever you choose one path over another (say, using a custom object vs. extending a standard object, or deciding between Flow and Apex for a particular automation), take a moment to write down why. Earlier, we saw how ChatGPT can help weigh options like Flow vs. Apex in context. Prompts like that can double as lightweight decision tools—useful not just at the start but also when documenting rationale later in the design process. You can use a lightweight format for this, often called an *Architecture Decision Record (ADR)*. For example, an ADR entry might be: *Decision:* Use Flow for order fulfillment logic instead of Apex. *Context:* Team has strong admin skillset; requirement can be met with out-of-the-box Flow features. *Options considered:* Apex trigger, Flow, or combination of both. *Decision:* Flow chosen to maximize maintainability and minimize code, as performance impact is negligible for current order volume (50 orders/day). *Consequences:* Monitoring will be needed; if order volume grows dramatically, may refactor to Apex for efficiency. Writing a few sentences like that whenever you make a major design decision helps anyone reading later to understand the trade-offs you weighed. It saves future architects from second-guessing "Why on earth did we do it this way?" – because the answer is recorded.

For more complex features, create a **Solution Design document**. This doesn't have to be overly formal, but it should outline the requirements being addressed, the high-level design (which objects, flows, fields are involved), and any diagrams that help explain the flow of information or the process. If you've built a multi-step process with several custom objects and flows interacting, diagram it! A simple flowchart or an Entity-Relationship Diagram (ERD) can speak volumes. Modern Salesforce teams often include these diagrams in design specs or technical documentation wikis. Even a screenshot of a Flow with annotations can be useful. The goal is to give readers a map of how the pieces fit together. If someone new joins the team, they should be able to read the design doc and quickly grasp the essence of how the solution works and why it was built that way.

Use the tools Salesforce provides to **embed documentation within the configuration** as well. We've mentioned using description fields on objects, fields, and Flow elements – it's a great practice to fill those in. For example, if you create a custom field "Customer Tier" that's used in several automations, note in its description something like "Used in VIP Discount Flow and Renewal Reminder Process" so future admins know touching that field has wider implications. Flow elements have description fields too; use them to describe what a tricky formula or decision is doing. Another tip: maintain a "README" or home page dashboard in Salesforce that gives an overview of the system to any logged-in admin (some orgs create a Home Page component visible only to admins that has links to documentation or key info). This way, the org itself contains pointers to documentation.

Communication is the twin of documentation. It's not enough to write design docs and decision logs – you should also **discuss design decisions with your team and stakeholders.** For stakeholders (like department heads), you might provide a high-level summary of the design in a meeting or email, focusing on how it meets their needs and any trade-offs (e.g., "We chose to standardize on the existing Opportunity object for renewals instead of creating a new Renewal object – this avoids data duplication and leverages familiar reports"). For the internal team, hold design review sessions where you walk through your approach and get feedback. These discussions often surface great ideas or catch potential issues early. They also ensure everyone is on the same page and can defend the design if questions arise later.

Another important aspect is **keeping documentation up-to-date.** An out-of-date document can be as bad as none at all, because it can mislead. Make updating docs part of your deployment checklist. For instance, if a change is made to a flow due to a new requirement, update the design spec or ADR with the new rationale ("Updated to include new condition X per change request on 2025-05-01"). If you maintain an org diagram, refresh it every few months as things change. This is where having documentation in an easily editable format (like a shared Confluence or Google Doc, or a version-controlled markdown file) helps – it lowers the barrier to making quick updates.

ChatGPT Prompt (File Upload): Documentation can be made easier with AI assistance. Imagine you have a diagram of your new Lead Conversion process – a PNG image exported from Lucidchart or Visio. You can upload this diagram to ChatGPT and ask: "**Please summarize the design shown in this diagram and list the key steps in the lead conversion process it depicts.**" ChatGPT, with its vision capabilities, will analyze the diagram and come back with something like: "*This diagram illustrates the lead conversion flow: First, a lead is qualified by checking that required fields A, B, C are present. Next, an Opportunity is created and linked to the new Account and Contact. Then, if the lead had product interests, a related Quote is automatically generated via a subflow. Finally, a notification email is sent to the account owner.*" You can then take that AI-generated summary, tweak it if needed, and paste it into your design doc. This saves you time writing out descriptions of processes and ensures you didn't overlook any element when explaining the diagram. Similarly, you could have ChatGPT review a complex Validation Rule or formula text you provide and output a human-readable explanation for your documentation.

Pro Tip: Treat documentation as a first-class deliverable, not an afterthought. One practical way to enforce this is to include documentation tasks in your project plan or sprint stories. For example, if you're following agile, have a user story like "Document case escalation flow" with acceptance criteria such as "All new fields and flows are described in the knowledge base; ADR updated with decision on escalation criteria." When planning sprints, allocate time for these tasks so they don't get skipped. Also, leverage team strengths: if one teammate is great at creating diagrams or writing, have them take the lead on documentation efforts (while others can peer review for accuracy). Good documentation and clear communication pay off by reducing miscommunication, speeding up onboarding of new team members, and helping stakeholders trust the system (because they see the professionalism in how it's managed).

With your design decisions well-documented and shared, your Salesforce implementation gains a long-term memory. This foundation empowers your team to manage and evolve the system confidently. And evolve it will – needs change, and even the best initial design will require adjustments. In our final section, let's explore how to embrace **iterative design and continuous improvement** so your declarative solutions stay effective and relevant as time goes on.

ITERATIVE DESIGN AND CONTINUOUS IMPROVEMENT

No Salesforce org is ever truly "finished." Business processes evolve, users come up with new requests, and Salesforce itself introduces new capabilities three times a year. A hallmark of a successful Salesforce practice is embracing an **iterative design mindset** – continuously refining and improving your declarative solutions rather than treating them as one-and-done deliverables.

Think of your Salesforce configuration as a living system. After you deploy a declarative solution, **observe it in action.** Gather feedback from end-users: Are there fields on a screen they never use? Is there a step in a Flow that confuses them or slows them down? Perhaps a report you built isn't showing exactly what managers need, causing them to export to Excel for analysis (a sign something might be missing). By creating channels for feedback (like a dedicated Chatter group for Salesforce suggestions or periodic user group meetings), you create a loop where users feel heard and you get valuable insights. Often, users will suggest minor tweaks that dramatically improve their experience – for example, "It would save me a lot of time if this form remembered my last input for field X" or "Can we add the customer phone number to this screen so I don't have to navigate away?" These are usually small declarative adjustments that can be rolled into the next iteration.

Schedule **regular design reviews** of your major features or processes. This might be every quarter or at the end of a project phase. In a design review, the team revisits what was built with fresh eyes: Did our solution actually solve the problem well? Is it causing any unintended side effects or maintenance pain? What can we improve now that we have real usage data? For example, you might review your case assignment rules and find that as volume grew, the round-robin assignment Flow you built is struggling – maybe now is the time to adopt the native Omni-Channel features (a more scalable pattern). Or you might find that a validation rule put in place is firing too often and needlessly preventing legitimate data entry – perhaps you refine its conditions or offer a better user guidance.

Iterative improvement also means staying **informed about Salesforce updates.** Each Salesforce release could bring features that allow you to simplify or enhance something you built. Maybe in the past you had an Apex trigger to handle something that now a Flow can do, or a new component can replace a custom workaround. Build time into your cycle to experiment with new release features in a sandbox and see if any of your existing processes could benefit. Continuously modernizing your org in small increments prevents it from becoming a museum of legacy practices.

Another key element is **measuring and monitoring.** Identify some KPIs or at least qualitative indicators of success for each solution. If you deployed a new lead assignment process, track metrics like lead response time or conversion rates pre- and post-implementation. If you rolled out a new Lightning app for service agents, maybe measure case resolution time or even user satisfaction via a quick survey. If those metrics aren't moving in the desired direction, that's a clue that further iterations are needed. Maybe the design is sound but training was insufficient (so an improvement could be more training materials or in-app prompts). Or maybe users found a loophole or workaround – find out why and refine the design accordingly.

Embracing iteration also means fostering a culture where change is expected and accepted. Encourage your team and stakeholders to view the first version of anything as a baseline, not the final product. This helps in change management – users know improvements are coming, so they're more forgiving of initial quirks. It also encourages continuous feedback. One practical method is to have a rolling enhancement backlog for your Salesforce org. As users work with the system, they or the admins can log ideas and pain points. Regularly review and prioritize these just like any product backlog. Some ideas you'll implement, others you might not, but keeping that backlog ensures good ideas aren't lost and signals to users that you're listening.

Importantly, continuous improvement should be balanced with governance. Not every whim change should be immediately implemented – you still evaluate against your principles (Trusted, Easy, Adaptable) and overall strategy. But within that framework, you iterate. Perhaps set a cadence, like a minor Salesforce enhancement release for your org every month, aside from major projects. This way, small fixes and improvements are continuously rolled out, keeping the system fresh and aligned with user needs.

ChatGPT Prompt (SORA Video Generation): Continuous improvement often involves retraining or informing users about changes. Instead of lengthy emails or meetings, you can use AI to create quick, engaging update videos. For instance, after you refine the case escalation process based on support team feedback, you might prompt ChatGPT: **"Create a 1-minute video (in an upbeat tone) to inform support agents about the new changes to the case escalation process and how it will help them."** Using the SORA video generation feature, ChatGPT can produce a short video highlighting the changes (perhaps using simple graphics or an avatar narrator). It might say, "Hey team! We've made some improvements to how case escalations work. Here's what's new: when you escalate a case now, you'll see a pop-up asking for urgency level – that helps our managers prioritize better. You'll also automatically get a task reminding you to follow up after 24 hours. These tweaks should save you time and ensure no urgent issue falls through the cracks. Thanks for the feedback that led to these changes – keep it coming!" Such a video can be shared in your internal channels, making continuous improvement feel exciting and keeping users in the loop in a digestible way.

Pro Tip: Set aside time and resources for iterative improvements in your planning. It's common in agile teams to allocate, say, 10-20% of each sprint for "tech debt and enhancements." In a Salesforce context, that could mean one day a week an admin focuses on org cleanup, optimization, or minor enhancements that aren't tied to a specific project but improve the overall system. Additionally, conduct periodic retrospectives not just on the project process, but on the **solution itself**. Ask questions like, "What design decision worked well and should we do again? Which caused pain and should we approach differently next time?" This practice of *learning as you go* means each iteration not only improves the system but also your team's know-how. Over time, an iterative approach will evolve your org from good to great, and your team from reactive to proactive.

By adopting iterative design and continuous improvement, you ensure that your declarative solutions never grow stagnant. They remain aligned with the business, incorporate the latest best practices, and can gracefully adapt to whatever the future brings. This completes the virtuous cycle of declarative development: gather requirements, design intentionally, build with best practices, document and deploy, then learn and refine continuously.

SUMMARY

In this chapter, we journeyed through the core principles and practices that transform Salesforce declarative configuration into a robust solution architecture. We began by stressing the importance of a disciplined, **architecturally informed approach** to "clicks, not code" development. By gathering deep business insights and planning before building, we ensure every configuration change has intent and purpose.

You learned how to **choose the right tools** for the job – evaluating when to leverage Salesforce's declarative power (Flows, formulas, rules, etc.) and when requirements or constraints might call for programmatic solutions. Making these decisions deliberately helps balance agility with control, ensuring we use "clicks" where they shine and reserve code for cases that truly need it.

We then introduced the **Salesforce Well-Architected Framework** as our guiding compass, with its pillars of Trusted, Easy, and Adaptable. Through examples, we saw how to infuse security and reliability (Trusted), simplicity and usability (Easy), and flexibility for the future (Adaptable) into our designs. These principles aren't abstract ideals – they translate into concrete design choices that make our configurations stronger and more future-proof.

Recognizing what to avoid is as important as knowing what to do. We exposed common **anti-patterns** in declarative design – those pitfalls like over-complicating Flows, overusing formula fields, misusing data relationships, or hardcoding values – and discussed strategies to steer clear of them. By spotting these red flags early, you can maintain a clean and efficient org, avoiding technical debt and performance issues that might otherwise accumulate.

On the flip side, we explored **proven design patterns** used by experienced Salesforce professionals. From modular flow structures and metadata-driven logic to event-driven processes and consistent naming conventions, these patterns act as reusable blueprints for solving problems elegantly. Applying them not only accelerates development (why reinvent the wheel?) but also yields solutions that are easier to maintain and scale.

Crucially, we emphasized **documentation and communication**. We highlighted how documenting your declarative decisions – whether through design documents, decision logs, or embedded descriptions – preserves the "why" behind the "what." Clear documentation and open dialogue ensure that your team and stakeholders understand the solution, can support it, and can evolve it intelligently in the future. It turns your Salesforce implementation from a black box into a transparent, collaborative project.

Finally, we embraced an **iterative mindset**. A Salesforce org is a living system that must continuously adapt. By regularly gathering feedback, reviewing design effectiveness, and staying current with new features, you can iteratively refine your configurations. This continuous improvement cycle means your solutions stay aligned with business changes and technological advancements, rather than degrading over time.

In closing, Chapter 3 armed you with an enterprise-thinking approach to declarative Salesforce design: start with intent, adhere to sound principles, avoid known pitfalls, leverage best practices, document thoroughly, and iterate to excellence. With this strong design foundation, you are well-prepared to tackle broader aspects of Salesforce success.

Next, we'll see how these design best practices plug into the bigger picture of managing a Salesforce environment at scale. In **Chapter 4**, we'll review governance, org strategy, and DevOps – ensuring that our well-designed declarative solutions are deployed, controlled, and evolved in a structured and efficient manner across the enterprise. The journey continues as we connect design brilliance with operational excellence for Salesforce.

4

GOVERNANCE, ORG STRATEGY, AND DEVOPS

Imagine your Salesforce org as a growing city – without a solid plan and clear rules, it could sprawl into chaos. New custom objects pop up like unplanned side streets, and critical processes end up overlapping or conflicting. **Governance** is the practice that prevents this chaos. It sets the direction and guardrails so you can introduce changes quickly **without** endangering stability and best practices. In other words, good governance lets your teams innovate in Salesforce with confidence, balancing agility with risk control. This chapter explores how to establish and scale that governance in a practical, no-nonsense way.

We'll start by looking at how to build a **Salesforce governance framework** from the ground up. This means defining a clear operating model for change – who makes decisions, how ideas are evaluated, and what standards everyone follows. You'll learn how to set up a Center of Excellence (COE) or governance board that brings together admins, developers, architects, and business stakeholders under one umbrella. By creating a governance charter and defining roles and responsibilities, you'll align everyone on the same strategy and avoid the "seven guest rooms, one bathroom" scenario where an org's features are imbalanced and poorly planned.

Next, we dive into **change management and release processes.** A key to governance is having a structured way to handle every Salesforce change – from a small field addition to a major feature rollout. We'll discuss how to implement submission forms or backlogs for change requests, evaluate them with defined criteria, and use Change Advisory Boards (CABs) for high-stakes decisions. With effective release management, you can schedule deployments in a predictable cadence and ensure quality through reviews and testing. This reduces the firefighting and uncertainty that happen when changes are made ad hoc.

Of course, none of these changes happen in a vacuum – that's where **environment management** comes in. We will examine sandbox strategies that support your governance. You'll learn the purpose of each sandbox type (Developer, Full, Partial Copy, etc.) and how to use them wisely for development, testing, and training. Effective sandbox management includes planning refresh schedules and scrubbing or masking sensitive data so you stay compliant. By designing a sensible path from development sandboxes to production, you set the stage for smooth deployments and fewer surprises on go-live day.

With a governance framework and environment strategy in place, we'll then apply **DevOps principles** to declarative Salesforce development. Even if you're "clicking" instead of coding, you can take advantage of modern DevOps tools to manage Salesforce changes. We'll explore how Salesforce's DevOps Center and version control systems (like Git) bring rigor to your release process. Automated testing and continuous integration aren't just for software engineers – they help Salesforce teams catch issues early and deploy updates faster with confidence. You'll see how embracing DevOps culture improves collaboration between admins and developers, making deployments more reliable and repeatable.

In large organizations, governance also means **coordinating multiple teams** and preventing conflicts. We'll discuss strategies for managing complexity within a Salesforce org that has many projects and contributors. This includes establishing communication channels, documentation standards, and conflict resolution mechanisms when two teams want to change the same object or process. We'll talk about using shared repositories and project tracking tools to give everyone visibility into upcoming changes, avoiding the dreaded scenario of duplicate work or broken configurations.

Governance isn't set-and-forget – you need to **measure its effectiveness** and adjust. Later in this chapter, we'll identify key metrics to gauge how well your governance model is working. Are your deployments succeeding without last-minute issues? How fast are change requests turned around? Are business stakeholders satisfied with the process? By tracking things like deployment success rate, backlog cycle time, and user satisfaction, you can find improvement opportunities. A data-driven approach ensures your governance evolves and continuously adds value.

Finally, we'll look at **scaling governance as your org grows**. A governance framework that works for 50 users might strain under 5,000 users or when your company acquires another business. We'll cover how to evolve from a centralized governance model to a federated or hybrid model when needed, spreading decision-making while maintaining consistency. We'll also discuss future-proofing your governance to incorporate new Salesforce features and products (like Slack integration or AI tools) smoothly into your org strategy.

By the end of this chapter, you'll have a practical blueprint for Salesforce governance and DevOps. The goal is to keep your Salesforce implementation **healthy, agile, and audit-ready** as it expands. From setting up the right people and processes, to leveraging sandboxes and DevOps automation, to tracking results and adapting – you'll be equipped to manage Salesforce changes with confidence. Let's get started by building the foundation: an effective governance framework that brings order to the chaos and sets your org up for sustainable growth.

ESTABLISHING A SALESFORCE GOVERNANCE FRAMEWORK

Setting up a governance framework is like laying the foundation for a sturdy house – it defines the structure on which everything else will be built. In Salesforce terms, a governance framework is the **operating model** for how your org is managed. It spells out who is involved, how decisions are made, and what guidelines everyone must follow. Rather than wild-west development where each admin or team does their own thing, you create clear "**rules of the road**" to keep the org organized. This not only prevents conflicts and redundant work, but also ensures the Salesforce platform aligns with your business's goals and compliance needs.

Defining Roles and Responsibilities: A crucial first step is identifying the key players in your Salesforce governance. Most organizations form a **Steering Committee** or **Governance Council**, bringing together stakeholders from IT and the business (sales, service, marketing, etc.). Think of this group as the ultimate decision-making body – they set high-level priorities, approve major changes, and ensure that Salesforce usage supports the company's strategic objectives. In practice, this might include an executive sponsor (to champion Salesforce at the leadership level), product owners for each major business function, a lead Salesforce architect, and representatives from the admin/developer team.

Alongside the steering committee, many companies establish a **Center of Excellence (COE)**. This is a dedicated team of Salesforce experts – senior admins, solution architects, maybe a developer lead – who drive best practices day-to-day. They are like internal consultants, making sure projects follow design standards, answering questions from various teams, and mentoring newer team members. If the governance council decides *what* should be done, the COE often decides *how* it's done in line with best practices. Clearly defining these roles prevents confusion. For example, if a sales manager requests a complex change, the COE might evaluate the solution options and present a recommendation to the governance council for approval.

Don't forget to outline the responsibilities of individual roles too. Your **Salesforce Administrator(s)** keep the org running – handling user management, basic configuration changes, and ensuring data integrity. **Salesforce Developers** (if you have them) handle programmatic customization and work on more complex automations or integrations. Both admins and developers typically execute changes **within the guardrails** set by governance. That means they adhere to standards like naming conventions, documentation requirements, and deployment processes defined by the council or COE. By documenting who is accountable for what (approvals, build tasks, testing, training, etc.), you avoid the scenario where everyone assumes someone else is handling an area (like data cleanup or security reviews) and it falls through the cracks.

Creating a Governance Charter: Once you have the team identified, formalize it with a governance charter. This document is essentially the constitution for your Salesforce org. It should state the **purpose and scope** of governance – for instance, "to ensure all Salesforce changes align with business strategy, follow best practices, and minimize risk." It will list the **members** of the governance groups (names and roles), how often they meet, and the **processes** they oversee (such as change request evaluation, release planning, org standards). Also, define **decision-making criteria**. For example, the charter might say that any new AppExchange package or any change affecting more than 50 users must be reviewed by the governance council. By contrast, a minor field addition might be delegated to the COE or admin team to approve. Setting these thresholds keeps the council from becoming a bottleneck for trivial issues while still giving oversight on big-impact changes.

ChatGPT Prompt (File Upload):

User: *(Uploads a draft "Salesforce Governance Charter" document)* Please summarize the key roles and processes outlined in this charter, and suggest any improvements or missing elements.

ChatGPT: Sure. **Summary of Governance Charter** – The charter establishes a **Governance Council** with executive sponsors, department heads (Sales, Service, IT), and a lead Salesforce Architect. They meet **monthly** to review major change requests and set org-wide standards. It also defines a **Center of Excellence (COE)** team (admins and developers) responsible for day-to-day best practices and user support. Key processes include a **change request submission** procedure (via an online form), bi-weekly sprint planning for Salesforce updates, and quarterly permission audits by the Security Officer.

Suggestions – The charter could be strengthened by specifying success metrics (e.g., "Deployments will have a 95% success rate") to measure governance effectiveness. It might also define a regular **training plan** for new Salesforce features, and include a data governance policy (how data quality and privacy are managed), which I don't see mentioned. Adding these would give a more complete picture of how the org is managed.

Within the charter, it's helpful to articulate guiding **principles**. These are high-level rules of the road like "Clicks before code" (favor declarative solutions unless there's a compelling reason to code), "Least privilege" (give users the minimum access needed, which ties into security governance), or "Standardization over customization" (reuse existing functionality and keep processes consistent across departments unless a difference is justified). Such principles help resolve debates – if one team wants a heavily customized process but a principle is to prefer out-of-the-box features, the governance group has a pre-agreed stance to refer to.

Policy Setting and Standardization: Governance will produce various policies and standards that everyone should follow. Common examples include:

Naming Conventions: Define how things are named in Salesforce (field API names, picklist values, role names, etc.) so that they are understandable and consistent across the org.

Development Lifecycle: Set rules for how changes move from idea to production. For example, *all* changes must be built and tested in a sandbox (no editing in production directly), and must be accompanied by test scripts or UAT (User Acceptance Testing) sign-off before deployment.

Documentation: Require that every significant customization (like a Flow or validation rule) has a brief description or a wiki entry describing its purpose. This ensures future team members understand why something was built. Good documentation is the "memory" of your org and is often spearheaded by governance efforts.

Security and Access Policies: Reinforce Salesforce's security model by defining who can grant permissions, how often roles/profiles should be reviewed, and how to handle exceptions. For instance, a policy might be that any profile permission change needs approval from the org's security officer or that all integrations use Named Credentials for better security and manageability.

Data Management Policies: Determine how data should be managed across teams (e.g., no spreadsheets for data exports without approval, use data loader vs. manual input guidelines, retention policies for old records, etc.). These tie into the data strategy which governance often oversees.

When establishing these standards, involve the people who will have to follow them. This is where **stakeholder buy-in** is critical. Governance is not about a small group dictating rules in a vacuum – it's about consensus and collaboration. If you're forming a new governance board, it pays to run "roadshow" meetings or workshops with different departments. Let them voice concerns and needs: Sales might say they need faster turnaround on small changes during quarter-end, whereas IT might emphasize not bypassing testing. By incorporating feedback, your policies will be more practical and get broader acceptance. Remember, the best governance framework is one that your users see as enabling them, not restricting them.

Building Alignment Across Teams: A well-structured governance framework brings together diverse teams under a common vision for Salesforce. It's worth defining an overall **vision and strategy** for your Salesforce program as part of governance. For example: "We will use Salesforce as the single source of truth for customer data and as a platform to streamline all sales and service processes." Having this vision in writing helps all teams pull in the same direction. Additionally, governance should address how ideas from different teams get consolidated and evaluated. One effective approach is maintaining a **business backlog** – a centralized list of requested features or improvements, coming from any department. The governance council or COE can regularly review this backlog, score items on impact and effort, and decide what to tackle in upcoming releases. This ensures that limited development resources are spent on the most valuable changes and that everyone knows how priorities are set.

Another alignment tool is a **communication plan**. Your governance charter can include how decisions and changes will be communicated out to end-users and stakeholders (for instance, via a monthly release newsletter or a Chatter group announcement). This transparency builds trust: users see that there is a thoughtful process and they get updates on the org's evolution, so they're more likely to cooperate with governance decisions.

Pro Tip: Start with a lightweight governance framework and **mature it iteratively**. For example, begin by convening a monthly governance meeting with key stakeholders and set 2–3 basic policies. As your org and team grow, expand the charter, add subcommittees (e.g. a data governance group), or increase meeting frequency. This way you won't overwhelm a small team with big bureaucracy, but you'll have the structure ready to scale.

At the end of the day, establishing a Salesforce governance framework is about creating a shared playbook for everyone involved in your Salesforce ecosystem. It provides clarity on how decisions are made and who makes them, which prevents misalignment and accelerates execution. With a charter in hand and a committed group of people enforcing it, you set the tone that Salesforce is a strategic platform managed with discipline, not a free-for-all. With this framework established, the next step is to put it into action. We now turn to structured change management and release control – the processes that ensure every Salesforce modification follows these guidelines and gets delivered smoothly.

CHANGE MANAGEMENT AND RELEASE CONTROL PROCESSES

Once a governance framework is in place, the day-to-day process of managing change needs structure. Change management in Salesforce is about having a clear **pipeline** for modifications – from the moment someone has an idea or request, all the way to deploying it in production. Without a defined process, you might get random emails or hallway requests for changes, leading to confusion and potentially risky ad hoc updates. A well-designed change management process brings order by capturing each request, evaluating it, and only then moving it into build and release.

Capturing and Evaluating Change Requests: Start by defining how business users or team members will submit requests for Salesforce changes. Many organizations use a **single intake form or backlog** in a tool like Jira, Trello, or even a Salesforce custom object to track these requests. The key is to make sure all requests go into one **visible queue** – if different departments track changes in separate silos, things will slip through the cracks or conflict. Ideally, the intake form asks for enough detail to understand the request: what problem are we trying to solve, who is impacted, urgency, etc. Some companies even create a simple business requirements template for requesters to fill out (covering the "what" and "why" of the change). This upfront info helps the governance team triage effectively.

Once requests are logged, establish a routine for **reviewing and prioritizing** them. The COE or a subset of the governance council can meet (say, weekly or bi-weekly) to go through new requests in the backlog. They might assign an initial **impact score** or categorize changes by complexity and value. For example, adding help text to a field is a "low effort / low risk" change that could be done quickly, whereas automating a complex business process with a new Flow might be "high effort / high impact" and require more analysis. It helps to define some criteria for evaluation: alignment to business goals, number of users affected, regulatory impact, etc. Applying consistent criteria ensures fairness in how you decide what to do next.

Approval Workflows and Change Advisory Board (CAB): Not every change will need formal approval from top brass, but you should delineate which types do. Minor changes might be approved by a product owner or the Salesforce platform manager as part of backlog grooming. However, major changes – like altering sharing rules company-wide, rolling out a new AppExchange package, or anything that carries significant risk – should get elevated scrutiny. This is where a **Change Advisory Board (CAB)** comes in. A CAB is a group (often overlapping with your governance council or key stakeholders) that gives a final thumbs-up on high-impact changes. The CAB's role is to assess risks and timing: Are we sure this change won't disrupt critical processes? Is the organization ready to adopt it? Do we have a rollback plan?

A best practice is to schedule regular CAB meetings (for example, once a week or bi-weekly) to review pending big-ticket changes. That way, approvals are timely and don't become a bottleneck. Some teams have "CAB lite" processes for standard, low-risk changes – these might be pre-approved as long as they follow certain templates. The idea is to apply formality where needed, but not over-engineer the small stuff. Remember that an overly rigid CAB that meets infrequently can slow down innovation, so calibrate the process to your organization's risk tolerance.

Pro Tip: Define categories of changes by risk level. For example, designate "Standard Changes" (routine, low-risk updates) that can be pre-approved and deployed without full CAB review, versus "Major Changes" that always go through CAB. This way, you keep agility for everyday tweaks while still applying proper scrutiny to high-stakes modifications.

Transparent Decision-Making: It's important that everyone in the organization understands how and why decisions are made on change requests. Transparency builds trust in the governance process. One way to do this is to maintain a **living backlog dashboard** that all stakeholders can view. It might show all requested changes, their status (New, In Review, Approved, Scheduled, Completed), and maybe a short comment on decisions. If users see that their request is "Under Review" for example, they know it hasn't been lost. Additionally, when a request is not approved or is deferred, the governance team should communicate back *why*. Maybe the idea was great but needs to wait for next quarter's budget, or it conflicts with an upcoming Salesforce release feature – explaining this prevents frustration. Some organizations even publish their decision criteria or scoring model so it's clear how priorities are set.

Release Planning and Scheduling: Once changes are approved and built (we'll talk about the building and testing process in a moment), you need a plan for deploying them. Ad-hoc, on-the-fly deployments are risky, so it's better to operate on a release schedule. This could be something like: minor updates go out every two weeks, major releases go out monthly or quarterly, possibly aligning with Salesforce's own seasonal releases. Having a regular cadence means users know when to expect new features or fixes (for example, "Salesforce changes are deployed the first Thursday of each month"). It also allows you to bundle related changes together, reducing the number of disruptions. Each release should have a **deployment checklist** – ensure all changes have been tested, training materials prepared, and a back-out plan exists in case something goes wrong.

Governance plays a big part in release management by overseeing a **release calendar**. They'll consider factors like avoiding critical business periods (don't deploy a huge change on end-of-quarter day for Sales), and coordinating with other initiatives. If your company has multiple systems, you might have a "release blackout" period where no systems change during certain times. The Salesforce governance board can enforce those rules for Salesforce deployments too.

Another aspect of release control is using **sandboxes and staging environments** appropriately (which we'll explore in the next section). For now, what's important is that your change management process explicitly includes steps for proper testing and user acceptance before anything hits production. Governance can set a policy that no change goes to prod without a UAT sign-off from a business tester, for example.

ChatGPT Prompt (Image/Wireframe Generation):

User: Create a simple flowchart of our Salesforce change management process, from a new change request submission to deployment.

ChatGPT: **Sure. The flowchart follows these steps:** A user submits a **Change Request** → The request goes into the **Backlog** (awaiting review) → The **Governance/COE Team Reviews** the request (assess impact and priority) → If high-impact, it goes to **CAB Approval** (else skip this for minor changes) → Approved requests are **Planned in a Release** → Admin/Dev team **Builds & Tests** the change in a sandbox → **UAT** (business users validate in testing environment) → **Deployment to Production** during scheduled release window. *(Imagine these steps arranged as boxes with arrows in between, looping back if a change fails a review or test.)*

With a structured change process in place, you significantly reduce the chaos and risk of Salesforce development. Teams know how to propose changes and what criteria will be used to accept them. There are clear checkpoints for quality and risk management – from peer reviews to CAB oversight. Perhaps most importantly, having this process means the business side and technical side of the house stay in sync: priorities are agreed upon, and surprises are minimized. As we move forward, we'll examine how managing your **environments and sandboxes** goes hand-in-hand with change management, ensuring that the path from development to deployment is smooth and governed at every stage.

ENVIRONMENT MANAGEMENT AND SANDBOX STRATEGIES

A solid change management process works hand-in-hand with a smart environment strategy. Salesforce provides multiple sandbox types and environments, each suited for a different purpose. Governance should define how your team uses these environments so that development and testing are done safely and efficiently **before** anything touches production.

Know Your Sandbox Types: It's important for everyone to understand the roles of each sandbox:

Developer Sandbox / Developer Pro: These are for individual development and unit testing. They contain only metadata (no real data, or very minimal data if any). Typically, each admin or developer can have their own Developer sandbox to build and experiment in isolation without affecting others.

Partial Copy Sandbox: This sandbox includes a subset of production data (based on a sandbox template you define). It's great for testing with more realistic data than a Dev sandbox, especially for things like validating reports or complex automations. Partial sandboxes are often used as integration testing environments or for UAT (User Acceptance Testing) when a Full copy isn't available.

Full Sandbox: This is essentially a replica of production – all the data and configuration, refreshed on a schedule (such as once every 29 days). Full sandboxes are ideal for performance testing, full regression testing, or a "dress rehearsal" of a major release. Many teams maintain a Full sandbox as a staging environment where final pre-production testing occurs with real data volumes.

Mapping out a clear **environment hierarchy** is a best practice. For example, you might decide that all development work happens in Dev sandboxes, then is merged into an Integration sandbox (which could be a Partial Copy) for combined testing, and finally everything is validated in a Full sandbox before deploying to Production. By establishing this pipeline, you ensure that by the time changes reach production, they've been through multiple checkpoints with increasing levels of data and users.

Refresh Cadence and Planning: Sandboxes need periodic refreshes to stay useful. A refresh updates the sandbox's metadata (and data for Partial/Full) with the current state of production. Governance should set guidelines on when to refresh and how to prepare for it. For instance, a common strategy is to refresh Full sandboxes shortly after a major production deployment or a Salesforce seasonal release – that way the Full copy has up-to-date configurations and data. Partial Copy sandboxes might be refreshed more often (they have a shorter refresh interval, e.g. 5 days minimum) to bring in recent data samples for testing new features.

Pro Tip: Choose a sandbox refresh window that minimizes disruption, and **communicate it early** to the team. For example, plan the Full sandbox refresh over a weekend or off-peak hours. Giving everyone notice helps prevent lost work and ensures admins/developers can prepare (by migrating their changes or pausing work) before the refresh.

After a refresh, there's usually some post-refresh setup to do: reassign user roles, reconnect integrations (since sandbox org IDs change), reload any test data that wasn't in the template, etc. Having a checklist for post-refresh tasks ensures nothing is forgotten (for example, if you use a sandbox for training, you might need to reload certain training data or reset user passwords).

Data Management and Masking: When using sandboxes that contain production data (Partial or Full), data governance and privacy considerations come into play. Customer and sensitive data in a sandbox can be a risk if not handled properly – sandboxes are accessible to developers and test users who might not normally see all production data, and they're not meant to be long-term data stores. A best practice is to **mask or anonymize** sensitive data in sandboxes. Salesforce offers a Data Mask tool (an add-on) that automatically scrambles fields like names, emails, and phone numbers upon sandbox refresh, so the data looks real but isn't actually real. Even without that tool, scripts or manual processes can replace real emails with dummy values (to prevent accidentally emailing customers from a test sandbox) and obfuscate personal info. The governance policy should clearly state that any personally identifiable information (PII) in non-production environments must be protected.

Furthermore, consider **sandbox templates** (for Partial Copy or Full sandboxes) to control which data gets copied. For example, you might exclude ultra-sensitive objects entirely, or limit a Partial copy to just the past 6 months of records for key objects to reduce volume. Laying out a data strategy for sandboxes means testers get enough data to work with, but not more than necessary.

Managing Multiple Environments: In a multi-team or multi-project scenario, it's common to have several sandboxes in play. Governance should define who "owns" each sandbox and its purpose. For example, Sandbox A is dedicated to Project Alpha testing, Sandbox B is for integration testing among all ongoing projects, etc. Label your sandboxes in Salesforce Setup with descriptive names (e.g., "UAT_InsuranceApp" or "DevOpsIntegrationTest") so everyone knows their intended use. Also, encourage teams to keep configurations in sync – if Team A updates a shared setting in an integration sandbox, Team B should refresh or retrieve those changes before they deploy, to avoid surprises. This kind of coordination falls under governance to prevent environment drift (where two sandboxes diverge so much that deployments fail due to missing pieces).

Another environment aspect the governance team can oversee is **settings and features management.** Salesforce orgs have various settings (like multi-currency, person accounts, etc.) that can be enabled. Typically, you want to test enabling any new significant feature in a sandbox first. The governance framework might require an approval before enabling a major feature in production, with verification that it was tried in a Full sandbox successfully.

Smooth Deployments Between Environments: Finally, an effective environment strategy ensures that moving changes from one sandbox to another – and then to production – is as smooth as possible. This often means using **consistent tools or processes for deployment**, such as change sets, metadata API tools, or DevOps pipelines. Governance can standardize on a deployment approach (for example, "we use Salesforce

DevOps Center to promote changes from Dev to UAT to Prod"). Having a consistent method reduces errors. It's wise to practice deployments in a sandbox that mimics production (like Full) – do a "dress rehearsal" of a deployment, see if all components deploy cleanly, and run all tests. That practice run can catch missing components (like "oops, we forgot to include that custom field in the change set") before the real day.

Also, ensure you have **backups or version control** of your metadata before deploying. If something goes wrong, you can quickly roll back or restore. While Salesforce doesn't provide a simple "undo" button for deployments, a source-controlled repository or an automated backup of the previous state can be a lifesaver if a deployment goes awry.

ChatGPT Prompt (Operator Browsing):

User: Find the official recommendations on Salesforce sandbox refresh best practices – especially how to handle data and timing.

ChatGPT (with browsing): I found a Salesforce knowledge article and a best-practice blog. **Key points:** Full sandboxes should be refreshed during a window that minimizes business impact, and teams must be notified well in advance (to avoid work loss during refresh). It's recommended to establish a refresh schedule (for example, after every major release). **Data Handling:** Salesforce's guidance emphasizes masking sensitive data in sandboxes for compliance. One article notes that inadvertently sending emails from a sandbox is a common risk – so all outbound email should be disabled or redirected in sandbox settings, and real customer emails should be anonymized. Additionally, using sandbox templates can limit data exposure by only copying necessary objects and records. In summary, plan your refreshes carefully, communicate them, and sanitize sandbox data as part of your strategy.

By managing your environments proactively, you create a safer space to innovate. Developers and testers can work concurrently without stepping on each other's toes, and stakeholders can confidently try out new features in UAT knowing it mirrors production closely. With your environments under control, you can now introduce DevOps practices to further accelerate and safeguard the release pipeline – bringing automation and advanced tooling into the picture while keeping governance firmly in place.

SALESFORCE DEVOPS — DECLARATIVE APPROACHES

The term **DevOps** often brings to mind code-heavy engineering, but its principles are just as valuable in a declarative Salesforce world. In fact, as your org grows more complex, having DevOps practices in place becomes a lifesaver for managing change at scale. DevOps is all about tighter collaboration between development and operations, continuous delivery, and automation – all aimed at making deployments faster and more reliable. In the Salesforce context, this translates to managing your metadata (the "clicks, not code" configurations) with the same rigor that traditional software teams manage code.

Source Control for Salesforce Metadata: One of the foundational elements of DevOps is using **version control** (like Git) to track changes. Historically, an admin might click around in a sandbox and make changes without a detailed record of every field added or Flow modified. With source control, you extract those changes (using Salesforce metadata API or DevOps Center) and commit them to a repository. This provides a clear history of who changed what, when, and why. It also means you can **branch** your work – for example, if two admins are working on different features, each can have a separate branch in Git. This helps avoid conflicts, since changes get merged explicitly. If something breaks, you can compare versions in the repo to pinpoint what changed. Modern tools have made source control much more accessible to Salesforce teams; Salesforce's own **DevOps Center** (released GA in 2022) integrates a friendly UI with Git behind the scenes. Admins can click to deploy changes through DevOps Center, and under the hood it's creating commits and coordinating environments, so you get the benefits of Git without needing to use command line.

Pro Tip: Even if your team is mostly "clicks not code," treat your metadata like code. Establish a practice of pulling configuration changes (profiles, flows, etc.) into a Git repository regularly. This creates an audit trail and enables rollbacks if needed. It might feel optional on a small team, but it pays off as soon as multiple people are building – you'll catch conflicting edits early and preserve org knowledge over time.

Continuous Integration & Automated Testing: In a traditional software project, continuous integration (CI) means developers frequently merge their code changes and run automated tests to catch issues early. For Salesforce, we can achieve a similar rhythm. When multiple people are configuring Salesforce, it's useful to have a **CI job** that periodically deploys all recent changes to an integration sandbox and runs the test suite. Salesforce, of course, has Apex tests for code, but even in a declarative org, you can have automation tests: for example, you might use Selenium or a testing tool to click through key user flows after each deployment to ensure nothing got inadvertently broken. At minimum, every deployment to production should trigger all Apex tests to run – this is out-of-the-box behavior when you deploy or use scratch orgs in CI. If a test fails, that's a red flag to investigate before users are affected.

Continuous integration for Salesforce might involve using services like GitHub Actions, Jenkins, or CircleCI integrated with Salesforce CLI. These tools can spin up a fresh scratch org or deploy to a sandbox, then run tests and report back status. The outcome is that you catch deployment problems or bugs in a safe environment. For example, if someone accidentally removed a needed field from a page layout, your automated test script might catch that the field is missing on the form, before it ever hits production.

Continuous Delivery (CD) and Automated Deployments: Alongside CI is continuous delivery – the idea that you can deploy to downstream environments (and ultimately to production) at the push of a button (or on a schedule) rather than through painful manual steps. If you've ever spent a weekend doing change sets and manually ticking checkboxes, the contrast is huge. Using tools like DevOps Center, Gearset, Copado, or Jenkins pipelines, you can automate the deployment of metadata from one org to another. For instance, upon approving a change in source control, an automated job could deploy those changes to the UAT sandbox for testing, and even schedule the production deployment for a pre-defined release window. This reduces human error (no more forgetting a component in a change set) and speeds up releases. It also encourages smaller, more frequent deployments – a core DevOps tenet – because automation makes it easy to deploy incremental changes.

One common fear is, "Will automated deployments deploy something I didn't intend?" But that's where controls like code reviews and validation environments come in. You can set your pipeline so that any deployment to production requires, say, two approvals (perhaps the admin who made the change and a tech lead), and that it does a **check-only deployment** first (in Salesforce, that means simulate the deployment and run tests, but don't save changes) to ensure everything will succeed.

Quality and Performance Benefits: Embracing DevOps for declarative configuration yields very tangible improvements:

Fewer Deployment Errors: When changes are tracked in a repository and deployed through scripts or tools, it's less likely to miss dependent components. The tool can warn you "hey, you modified a Flow but not its referenced custom metadata – include that before deploying." Many deployment tools also offer **backout options** or quick rollback if something fails.

Faster Cycle Times: What used to take days of manually prepping change sets can become a smooth flow of merges and automated deploys. Teams can deliver updates to users more frequently, even daily in some cases, because the overhead is reduced.

Increased Confidence: Knowing that every change goes through tests (and possibly a code review of the config in Git) means you catch mistakes early. Maybe an admin inadvertently gave too much permission in a profile – a peer review might spot it in the Git diff. Or your automated test might flag that a new validation rule blocks a critical process. Catching these in a sandbox rather than after release is invaluable.

Moreover, DevOps practices foster a culture of **collaboration and accountability**. When everyone can see the change logs and deployment history, there's less chance of the classic "who changed this?!" mystery. It's all documented. And because multiple team members contribute to the pipeline, knowledge about the org is shared rather than siloed.

ChatGPT Prompt (Voice):

User (speaking): "We want to improve our Salesforce release process. What are some DevOps best practices we can adopt for our declarative development?"

ChatGPT (responding verbally): Certainly! Here are a few key practices:

Use Version Control: Start tracking your Salesforce metadata in a source control system like Git. This creates a single source of truth for all changes.

Implement Automated Testing: Whenever you deploy, run all Apex tests, and consider adding end-to-end tests for critical user flows. This ensures new changes don't break existing functionality.

Continuous Integration: Merge changes frequently and have a CI job deploy them to a test sandbox regularly. This catches integration issues early.

Automate Deployments: Utilize tools or Salesforce DevOps Center to move changes between orgs. Aim for a one-click or scripted deployment to reduce manual errors.

Review and Monitor: Have a peer review process for changes (even declarative ones) and monitor deployments. Learn from any failures – for example, if a deployment fails due to missing components, improve your process to include those next time.

By adopting these, you'll make your releases faster and more reliable, with much less fire-fighting.

Bringing DevOps into Salesforce might introduce new tools and processes for your team, but the payoff is a more predictable and efficient release cycle. Your admins and developers spend less time deploying and fixing deployment issues, and more time delivering value. With the technical groundwork of governance and DevOps in place, the next challenge is coordinating when you have **multiple teams and complex projects** working in Salesforce simultaneously – which is where strong collaboration practices come in.

MANAGING COMPLEXITY AND TEAM COLLABORATION

As Salesforce usage spreads across multiple teams and departments, complexity grows – not just in the org's configuration, but in how people work together on the platform. Without coordination, you might see scenarios where two teams unknowingly build duplicative solutions, or a change from one group breaks functionality for another. Effective governance in a multi-team environment means establishing practices that promote visibility, communication, and consistency across all the moving parts.

Avoiding Silos and Duplicative Work: One of the biggest risks with multiple teams is the creation of **silos** – each team making changes in isolation. The governance framework should encourage (or mandate) that teams share their plans and review each other's designs for overlap. A common example is two different departments each asking for a custom field to track similar data. If they don't communicate, you could end up with two nearly identical fields on the Account object used by different groups, which is confusing and inefficient. By having a centralized process for evaluating new fields or objects, governance can catch these overlaps. Some companies implement an "architecture review board" where representatives from each project meet regularly to vet new proposals. This doesn't have to be heavy-handed; even a bi-weekly 30-minute review call can surface potential conflicts ("Oh, you're planning a new sales commission field – we were going to build something similar, maybe we should collaborate on one solution").

Using a **shared backlog** or work tracking tool is also critical. If every team logs their user stories or change requests in a common system, it's easier to spot duplicates and dependencies. For instance, the Sales Ops team might log a request to add a validation rule, and the Support team might separately log a request to adjust a similar process. A central product owner or architect can see both in the backlog and connect the dots. Maintaining this transparency in planning prevents surprises down the line. It also helps with prioritization – leadership can view all proposed Salesforce work in one place and allocate resources to the highest-value items across teams.

Communication Rhythms: In addition to tooling, establish regular communication channels among Salesforce practitioners. Many organizations have a **weekly Salesforce sync** meeting or a dedicated Slack/Teams channel for Salesforce changes. In the weekly sync (sometimes called a "Scrum of Scrums" if you're using Agile), each team can briefly share what they're working on and flag any help they need or potential impacts. These forums build camaraderie among admins and devs in different departments, and often an issue one team faced has already been solved by another who can share lessons learned. The Slack or Teams channel serves as a quick help line – e.g., an admin in Marketing can ask "Has anyone integrated Salesforce with XYZ system before?" and get guidance from an IT Salesforce dev who has. It's far better than every team reinventing the wheel.

Pro Tip: Set up a dedicated channel (e.g., a Slack or Teams channel) for Salesforce collaboration. Encourage team members to post a quick note there about any **significant upcoming change**. A simple message like "FYI, Sales Cloud team adding a new Contact field for project X" gives others a chance to voice concerns or say "we need that too!" before work begins. This low-effort communication can catch overlaps early and foster transparency.

Documentation is another crucial part of collaboration. When one team builds something, ensure its documented in a place accessible to all teams (a Confluence page, shared doc, internal wiki, etc.). For example, if the CRM team establishes a naming convention for picklist values or a standard way to implement Lead conversion processes, writing that down and sharing it will save other teams from accidentally deviating. Some companies maintain an internal "Salesforce Playbook" – a living document or knowledge base that captures design standards, common patterns, and recent changes in the org. Governance can spearhead this, but all teams should contribute to keep it current.

Resolving Conflicts in Configurations: Even with best efforts, conflicts can happen – two changes might collide (for example, two people edit the same Lightning page differently). How to handle these? Technically, source control will flag metadata merges, and a good deployment process will catch most direct collisions (you can't deploy if you don't have the latest changes from someone else, forcing a reconcile). But there are more subtle conflicts, like business logic overlaps: maybe Sales set up an automation that assigns leads one way, and Marketing set up another automation that reassigns leads another way. In such cases, governance should facilitate a resolution by bringing the stakeholders together quickly. Treat it as a learning opportunity to improve process: perhaps it indicates a need for a more explicit domain ownership (e.g., Sales owns lead assignment logic, so Marketing should have checked with Sales before adding their flow). Going forward, you might update your governance checklist to ask "Could this change impact another department's process?" and require consultation if yes.

It's wise to define a clear **escalation path** for conflicts: for instance, if two teams disagree on a solution approach, the issue goes to the Salesforce Architect or Design Authority Board for a decision. This ensures that debates don't linger and someone has the mandate to decide with the org's best interest in mind. Everyone might not get exactly what they want, but they'll know there is a fair decision process.

On the flip side, celebrate collaborative wins. When teams come together to build a shared solution that benefits multiple groups, give it recognition in your internal communications. It reinforces the behavior you want to see.

Collaboration Tools and Practices: We've touched on Slack and shared backlogs; another tool worth mentioning is a **shared repository** for any code or complex config. If you have developers writing Apex or building intricate formula fields, storing those in a repository (or at least a shared library) accessible to all prevents duplicate efforts. For example, two devs might be writing similar validation rules in different sandboxes – a shared repo would make it evident there's reusable code or patterns.

For declarative work, consider using Salesforce packaging or change sets to **modularize** projects if you have truly separate initiatives that rarely interact. Some large orgs use unlocked packages to compartmentalize parts of the configuration, which can allow teams to work more independently (this is a more advanced strategy and comes with complexity, but it's an option for orgs at scale).

Finally, emphasize a **culture of openness.** Encourage team members to reach out when they're unsure if a change might impact others. It's much better to ask upfront ("Hey, Service Team, are you using this Case field? We're thinking to repurpose it") than to discover an issue later. By creating an atmosphere where asking questions and sharing plans is the norm, you leverage the collective knowledge of your Salesforce community.

ChatGPT Prompt (SORA Video Generation):

User: We need to onboard new team members to our Salesforce governance process. Can you generate a short video script that explains how our teams collaborate and follow governance?

ChatGPT: [Video Script: "Working Together in Salesforce"]

Scene 1: (Animated team meeting) Narrator: "Welcome to how we **work together** in Salesforce! Our company has multiple teams making changes in Salesforce, and we've set up a process so we stay in sync."

Scene 2: (Icons of request forms and calendars) Narrator: "It starts with planning. All teams log their change requests in one place – nothing gets done in secret. We meet every Tuesday to review upcoming changes **together**."

Scene 3: (Two characters merging a puzzle piece) Narrator: "Have an idea? Great – you'll collaborate with our Center of Excellence to design it. They'll check if another team has a similar project, so we can join forces."

Scene 4: (Shield or safety icon) Narrator: "We also protect each other's work. Before deploying, we test changes in shared sandboxes. And we follow standard practices, like code reviews and UAT, so there are **no surprises** in production."

Scene 5: (Thumbs up, team high-five) Narrator: "By communicating and following our guidelines, we deliver better results for everyone. Teamwork makes the dream work – in Salesforce too! Welcome aboard, we're excited to build together."

When teams collaborate effectively, the Salesforce org evolves in a more unified and coherent way. You reduce redundant effort, minimize "left hand vs right hand" situations, and create an environment where everyone feels ownership of the whole system, not just their piece. This collaborative governance is essential preparation for scaling even further – whether that's onboarding more users, taking on new projects, or integrating additional systems – all of which introduce new complexities that must be managed.

MEASURING GOVERNANCE EFFECTIVENESS

You can't improve what you don't measure – that adage applies to your Salesforce governance just as much as to sales performance or customer service. After putting governance processes in place, it's important to track how well they're working. By defining **key metrics** and reviewing them regularly, the governance team can celebrate successes, spot problem areas, and continuously refine their approach.

Key Governance Metrics to Track: What should you measure? Here are some metrics that give insight into different aspects of governance effectiveness:

Deployment Success Rate: How often do your deployments succeed without rollback or hotfixes? If you deploy 10 changes in a month and 9 of them go smoothly, that's a 90% success rate. A high success rate indicates good preparation, testing, and coordination. A low rate (frequent failures) might point to gaps in testing or communication.

Change Lead Time: This is the time from when a change request is submitted to when it's actually delivered to users. Shorter lead times mean your process is efficient and responsive. If your lead times are long (say the average minor enhancement takes 3 months to go live), you might investigate where the bottlenecks are – perhaps in approval steps or development resource constraints.

Deployment Frequency: How frequently are you delivering updates? This can be measured as releases per month or even per week. It ties to your agility – higher frequency (with small, incremental changes) often indicates a well-oiled DevOps process and a culture open to continuous improvement.

Change Failure Rate: Out of changes deployed, how many resulted in issues (bug fixes needed, or had to be backed out)? This complements the success rate metric. If many changes cause incidents or require quick fixes, it suggests quality control issues in your process.

Mean Time to Recover (MTTR): When something does go wrong, how quickly is it resolved? If an issue in production is detected, do you fix it in hours, days, or weeks? Short MTTR indicates a prepared team and solid rollback plans.

These four metrics – deployment frequency, lead time, change failure rate, and MTTR – are famously known as the **DORA metrics** in DevOps (from the State of DevOps research). They provide a holistic view of speed and stability. Governance can use them to set targets (e.g., "let's get our change failure rate below 5%") and identify trends over time.

Beyond DevOps-centric metrics, there are governance-specific ones:

Backlog Health: Track the number of outstanding change requests and their aging. If you see hundreds of requests piling up or many requests sitting idle for 6+ months, it may indicate your governance process isn't keeping up with demand. You could measure the percentage of requests completed within a target timeframe.

Stakeholder Satisfaction: This one is more qualitative but invaluable. Consider doing a brief survey of key business stakeholders (department heads, power users) every six months. Ask questions like "On a scale of 1-5, how satisfied are you with the Salesforce change management process?" and "Do you feel you have visibility into the roadmap for Salesforce improvements?". A high satisfaction score means governance is seen as adding value; a low score might mean the process feels too slow or opaque to the business.

Adherence to Process: Measure how well teams are following the governance protocols. For example, the number of "unapproved changes" detected in production (hopefully zero!). Or track if all production deployments had proper CAB approval and UAT sign-off. If exceptions were made (e.g., an emergency change without CAB review), record those and later discuss if the process needs adjusting to handle such cases.

Audit and Compliance Findings: If your company has internal or external audits of Salesforce, note how many issues are raised related to configuration management, security, or process. A drop in audit findings over time can be a great indicator that governance improvements are working (for instance, an audit might have previously found lack of documentation for changes – with governance in place, the next audit finds everything well-documented).

ChatGPT Prompt (Typing):

User: We want to know if our new Salesforce governance process is effective. What metrics should we be looking at?

ChatGPT: To evaluate your governance, you should track a mix of speed, quality, and satisfaction measures. For example, measure your **deployment success rate** (how many deployments go out without issues) and **lead time** for changes (how fast a request goes from idea to delivered). Also monitor **change failure rate** (the percentage of changes that cause incidents or need rollback). Don't forget stakeholder feedback – maybe survey your business users to gauge their **satisfaction** with the process. Other useful metrics include the size of your request **backlog** and how long items stay there, as well as compliance checks (e.g., ensuring every deployment had required approvals). Reviewing these metrics regularly will show you what's working and what could improve in your governance.

Establishing a Metrics Dashboard: To make these metrics actionable, compile them in a dashboard or report that the governance committee reviews periodically (e.g., monthly or quarterly). For example, you might use a Salesforce dashboard or a spreadsheet where you log:

- Number of deployments, success vs. failed, per month.
- Average lead time of changes closed in that month.
- Backlog size and average age of requests.
- Survey results or anecdotal feedback from stakeholders.
- Any notable incidents (with root cause identified, such as "Change X caused issue due to inadequate testing").

Seeing the data visualized can quickly highlight trends. Maybe you'll notice "Wow, our deployments spiked in Q4 but so did our failure rate – maybe we were rushing changes for end of year." Or "Lead time has improved since we added an extra admin, but backlog count is rising again after a new department started using Salesforce – perhaps we need to incorporate a rep from that team into our governance meetings."

Pro Tip: Don't go overboard by tracking too many metrics. Focus on a handful of meaningful indicators that drive discussion and improvement. And remember that numbers alone don't tell the whole story – use them as a starting point for conversations. If deployment frequency dropped one quarter, it might be due to a strategic decision to do a big consolidation project (not necessarily a bad thing). Always interpret metrics in context.

Continuous Improvement: Metrics should feed back into refining your governance. Treat your governance process itself as a product that can be iterated. For example, if stakeholder satisfaction is low because people feel kept in the dark, you might increase communication by publishing a quarterly roadmap or holding "office hours" to discuss upcoming changes. If change lead time is too high, perhaps streamline CAB meetings or empower a broader group to approve low-risk changes (as we discussed earlier). On the other hand, if you notice a high change failure rate, double down on testing protocols or require design reviews for complex changes.

It's also valuable to celebrate positive trends. When metrics show improvement – say, three months with 100% successful deployments or a significant reduction in backlog – share that with the teams. A quick shout-out in a meeting like "Hey everyone, our deployment success was 100% last quarter thanks to your hard work preparing and testing changes" reinforces the benefits of following governance practices and motivates the team to keep it up.

By quantifying the impact of governance, you turn what could be seen as bureaucratic overhead into something tangible and results-oriented. Leadership will appreciate seeing that, for instance, "Since establishing the COE and new release process, our deployment success rate improved from 85% to 99%, and incidents related to changes dropped to near-zero." Those kinds of facts build ongoing support for governance initiatives. As the focus shifts to Salesforce data modeling and architecture in the following chapter, the same governance principles will guide us in designing a data model that is scalable, consistent, and aligned with enterprise standards – ensuring that the technical structure of the org is just as sound as its process governance.

SCALING GOVERNANCE FOR GROWTH AND ADAPTABILITY

Salesforce is a dynamic platform, and over time your organization's use of it will likely expand – more users, more teams, perhaps additional Salesforce products (like Marketing Cloud, Tableau, or Slack integrations), and maybe even multiple orgs after mergers or acquisitions. A governance framework that worked for a smaller, simpler setup must evolve to handle this greater scale and complexity. The good news is that the core principles we've covered remain the same; what changes is how you **structure and distribute** the governance effort.

Evolving the Governance Model: When starting out, many companies keep governance **centralized** – a single COE or governance board making decisions for the entire org. This is efficient for enforcing standards and ensuring consistency early on. However, as Salesforce usage spreads, a strictly centralized model can become a bottleneck or too detached from individual business units. That's when organizations often shift to a **federated** or **hybrid** governance model. In a federated model, you maintain a core central team (the primary COE) that sets overarching standards and best practices, but you also establish subcommittees or local COEs embedded in business units. For example, your Retail division, your Commercial Sales division, and your Service division might each have a "mini-COE" focusing on their specific needs. These mini-COEs include key power users or admins from that unit who understand local processes. They take the central standards and **apply them within their unit**, and feed unit-level requirements up to the central governance council.

This approach balances control and agility. The central governance group still ensures the overall org isn't splintering off in different directions (everyone still follows core data model standards, security policies, etc.), but the federated teams can move quickly for local optimizations. It's important to define how these groups interact. Typically, local COEs handle smaller changes autonomously, and bring larger, cross-cutting changes to the central council for alignment. The central team might also rotate members into the local groups to act as liaisons and ensure consistency. One thing to avoid is letting federated governance turn into silos by another name – maintain regular syncs between central and local governance bodies so that knowledge flows freely.

Pro Tip: Revisit your governance charter and structure on a regular basis (for example, annually or after major company changes). As you grow, what worked last year may need tweaking. Maybe it's time to add a new role (like a dedicated **Release Manager**) or to split one committee into two. Treat the governance model itself as a living design that can be optimized.

Including New Products and Technologies: Growth isn't just about size – it can also be about scope. Say your company acquires another firm and suddenly you have a second Salesforce org to govern, or you adopt a new Salesforce product like Experience Cloud or an AI-driven feature. Your governance framework should expand to cover these. Perhaps you create a **sub-team for Marketing Cloud** governance, staffed by both the Marketing Ops folks and a rep from your core Salesforce COE, to ensure that marketing automation changes go through a similar rigor. Or if your Salesforce strategy begins to include heavy use of **Slack integrations** or custom mobile apps, consider pulling those teams or product owners into the governance fold. The goal is to avoid "shadow Salesforce" projects running with no oversight. By bringing all major Salesforce-related initiatives under a governance umbrella, you maintain a unified direction and prevent fragmentation.

Another example: as AI features (like Einstein GPT or predictive analytics) roll out, governance might need to establish guidelines for their use (ensuring compliance with AI ethics or data privacy when using those tools). The specifics will vary, but the principle is that governance needs to flex to include whatever new capabilities Salesforce introduces that your company decides to leverage.

Scaling Processes and Automation: As the volume of changes increases, consider which manual governance processes could benefit from automation or tooling. For instance, if you're now approving dozens of changes a week, maybe the old spreadsheet or email-based approval tracking should move to a dedicated app or an automated workflow (you could even use Salesforce itself to track Salesforce changes!). Some organizations implement a "Governance Portal" – a simple Salesforce app where anyone can submit change requests, view status, and see metrics. The more you can systematize routine governance tasks, the more your human governance meetings can focus on strategy and exceptions rather than paperwork.

Scaling also means paying attention to performance and technical debt. Governance should regularly ask: are the practices that made sense with 100 users still effective with 1,000 users? For example, maybe you allowed each department a lot of freedom in creating reports, but now there are thousands of reports and it's hard to find the right one – governance might initiate a clean-up project or introduce a report catalog for better manageability. Or perhaps sandbox usage needs to expand: with more developers, you might lobby for additional sandbox licenses or use scratch orgs to give everyone their own environment.

Future-Proofing the Governance Framework: Finally, foster an attitude that governance is about enabling the future, not just controlling the present. Keep an eye on Salesforce's roadmap (new features, upcoming changes like the enforcement of MFA or API deprecations) and use the governance team to proactively plan for them. A forward-looking governance group might, for instance, prepare the org for Salesforce's next big release by assigning team members to explore beta features and report back. This way, when those features become available, your organization is ready to benefit from them rather than being caught off guard.

It's also wise to periodically benchmark your governance approach against industry best practices. Salesforce user groups, webinars, and community events can provide insight into how other large organizations handle governance. You might discover, for example, that many enterprises of your size have a formal **Salesforce Center of Excellence Charter 2.0** that includes things you haven't considered – like a subcommittee specifically for data quality or a playbook for crisis management if a deployment goes awry. Don't hesitate to evolve your model based on proven practices from peers.

ChatGPT Prompt (Typing):

User: Our Salesforce usage has tripled in the last year, and we're adding two new departments to the platform. How should we adjust our governance model to keep up?

ChatGPT: You'll likely need to **expand and decentralize** your governance. Consider forming a **governance sub-team or COE** for each new department – include their key Salesforce users/admins – so they can handle day-to-day decisions. Have them follow the standards set by your main governance council. The main council can meet a bit less often for strategic oversight, while these sub-teams tackle operational decisions quickly. Also, update your **governance charter** to add the new departments' roles and clarify which decisions are made centrally vs. locally. In short, spread out the responsibility (with rules in place) now that more teams are involved. This will prevent bottlenecks and keep everyone engaged. And as always, keep communication tight between all the teams so the org stays unified.

Scaling governance is an ongoing journey. By structuring your framework to accommodate growth – through federated teams, updated processes, and continuous learning – you ensure that increased complexity doesn't mean losing control or consistency. Instead, your governance becomes like a strong backbone that can flex and grow with the organization, keeping the Salesforce environment agile, secure, and aligned with strategic objectives no matter how large or sophisticated it becomes.

SUMMARY

In this chapter, we explored how strong governance acts as the glue holding a growing Salesforce implementation together. We began by establishing what a Salesforce governance framework looks like: a well-defined **Center of Excellence** or governance board that sets the ground rules, defines roles and responsibilities, and aligns Salesforce efforts with business strategy. By creating a governance charter and clear operating model, organizations ensure everyone – admins, developers, and business stakeholders – is on the same page and following a shared vision.

We then reviewed **change management and release processes**. We saw that handling change requests in a structured way (with intake backlogs, review meetings, and approval workflows) prevents chaos. Techniques like maintaining a single backlog, utilizing a Change Advisory Board for high-impact decisions, and releasing on a predictable schedule all contribute to smoother, lower-risk changes. The result is fewer surprises in production and more transparency for stakeholders.

Next, we addressed **environment management** as a foundation for successful change delivery. Using sandboxes strategically – from Developer sandboxes for individual work to Full sandboxes for staging – allows teams to build and test safely. We emphasized the importance of refresh planning and data masking in these environments to protect sensitive information and keep testing reliable. A thoughtful sandbox strategy ensures that by the time changes reach production, they've been vetted in conditions that mirror the real world.

Layered on that, we brought in **DevOps practices** tailored for Salesforce. Even declarative configuration benefits hugely from version control, continuous integration, and automated deployment. By tracking metadata in Git and automating test runs and deployments, teams catch issues earlier and ship updates faster and more frequently. Embracing Salesforce DevOps tools (like DevOps Center or third-party platforms) and approaches turns deployments from a pain point into a competitive advantage, all while upholding governance standards.

We also explored how to facilitate **team collaboration** in a complex org with multiple concurrent projects. The key is breaking down silos: sharing plans via common backlogs and communication channels, instituting cross-team design reviews, and fostering a culture where Salesforce teams collaborate rather than collide. With regular syncs and shared documentation, organizations avoid duplicate efforts and conflicting solutions. Everyone works more efficiently because they're leveraging collective knowledge and adhering to unified standards.

To ensure governance is not just in place but *effective*, we discussed **metrics and continuous improvement**. By measuring things like deployment success rates, change lead time, backlog size, and stakeholder satisfaction, the governance team can get a pulse on how well the process is serving the organization. These metrics inform adjustments – maybe the CAB process needs streamlining, or communication needs boosting. Governance becomes a feedback loop of ongoing refinement, backed by data.

Finally, we looked at **scaling governance** as the organization grows. We learned that governance models may shift from a centralized approach to a federated one to accommodate new teams and products while still maintaining coherence. The framework must flex – adding subcommittees, integrating new Salesforce products into its scope, and automating where possible – to support a larger user base and expanding Salesforce footprint. The underlying lesson is that governance is not static; it evolves with the business and the technology landscape.

With effective governance in place, a Salesforce org gains the freedom to innovate rapidly without descending into disorder. By enforcing good practices around design, change control, and collaboration, governance creates a stable platform for growth. This strong foundation will be crucial as the focus shifts to Salesforce data modeling and architecture in the following chapter. There, the same governance principles will guide us in designing a data model that is scalable, consistent, and aligned with enterprise standards – ensuring that the technical structure of the org is just as sound as its process governance.

5

DATA MODELING AND ARCHITECTURE BEST PRACTICES

Salesforce data modeling is the foundation of every successful implementation. In this chapter, we explain why a clean, scalable, and intuitive data structure is critical for long-term success. A well-designed data model isn't just about creating custom objects and fields – it's about shaping data **intentionally** so that your applications run efficiently, users find information easily, and the org can grow without breaking. When the data architecture is solid, system performance improves, data integrity is easier to maintain, user adoption increases, and future requirements can be met with minimal rework. In short, a thoughtful data model today sets you up for success tomorrow.

We'll begin with fundamental Salesforce data modeling principles, covering the building blocks of objects, fields, and relationships. Core concepts such as when to use standard objects versus custom objects, and the trade-offs between normalization and denormalization, will be discussed. Getting these basics right is essential – for example, knowing when to link objects through relationships instead of duplicating data can drastically simplify your design. Real-world examples will illustrate how following best practices at this stage leads to cleaner, more efficient configurations down the road.

Next, we'll explore one of the most important design decisions: choosing the correct relationship type for your objects. Salesforce offers two primary relationship types – **Master-Detail** and **Lookup** – each with distinct behaviors and use cases. We'll provide clear guidelines on when to use a Master-Detail vs. a Lookup relationship based on factors like data ownership, sharing, roll-up summary needs, and what should happen to child records if a parent is deleted. Understanding these differences helps you avoid common mistakes (like using the wrong relationship and then struggling with limitations later). From simple one-to-many links to many-to-many relationships using junction objects, we'll cover how to model complex scenarios without writing a single line of code.

Data integrity and quality are recurring themes throughout this chapter. Even the best data model can fail if data becomes unreliable. That's why we dedicate a section to maintaining data integrity using declarative tools. You'll learn practical techniques for enforcing data quality *within* the model – for instance, using validation rules to prevent invalid data entry, leveraging duplicate management to stop duplicate records, and employing formula fields or roll-up summaries to keep information consistent. We'll discuss how to put guardrails in place so that your carefully crafted data architecture remains clean as users interact with it day to day.

As the chapter progresses, we'll zoom in on schema optimization tools that Salesforce provides, such as picklists and record types. You'll see how picklist fields (especially when backed by shared picklist value sets) can enforce consistency and simplify reporting, and how record types can tailor the user experience for different business scenarios without needing separate objects. We'll share best practices for using these features effectively – for example, how to avoid common pitfalls with multi-select picklists, or how to decide if you truly need multiple record types on an object. These choices can significantly impact both usability and maintainability of your org.

We will also cover advanced declarative configuration options like **Custom Metadata Types** and **Custom Settings** for storing configuration data. These tools let you move hardcoded logic out of formulas and flows and into flexible data records that are easy to update and deploy.

By externalizing business rules and thresholds into custom metadata or settings, you gain agility – you can adjust how your app behaves by editing data rather than modifying automation or code. This section will include practical examples of when to use custom metadata vs. custom settings and highlight how they contribute to an adaptable architecture.

Finally, the chapter concludes with strategies for designing data models that stand the test of time. Salesforce orgs often start small and then grow rapidly in scope and data volume. We'll discuss how to anticipate that growth with scalable design patterns: thinking about large data volumes, archiving old records, indexing key fields for performance, and leveraging standard objects where appropriate to benefit from Salesforce's built-in capabilities. The goal is to help you **future-proof** your data model so that adding new features or accommodating more data feels like an evolution of your design, not a complete overhaul. We'll identify common scalability pitfalls and how to avoid them, ensuring your architecture remains flexible and efficient as requirements evolve.

Throughout the chapter, we'll also demonstrate how you can partner with ChatGPT to refine your declarative design decisions. You'll find special callouts showing ChatGPT in action – from answering quick questions by voice, to analyzing an uploaded data model diagram, to even generating a visual of a complex relationship. These examples will illustrate how AI can serve as a helpful co-pilot for Salesforce architects and admins, offering suggestions and insights in an interactive way. By the end of this chapter, you will have not only a firm grasp of Salesforce data modeling best practices but also creative ideas for leveraging AI assistance in your configuration work.

Now, let's start by grounding ourselves in the fundamental principles of Salesforce data modeling, which form the basis for all the best practices to come.

FUNDAMENTAL SALESFORCE DATA MODELING PRINCIPLES

A great data model starts with understanding the basic building blocks of Salesforce: **objects**, **fields**, and **relationships**. In Salesforce, standard objects (like Account, Contact, Opportunity, Case, etc.) come pre-built with rich functionality, while custom objects allow you to tailor the data structure to your organization's unique needs. One guiding principle is to **leverage standard objects whenever possible** before resorting to custom objects. Standard objects bring built-in connectivity (for example, Contacts link to Accounts, Opportunities roll up into Forecasts) and are well-tested for scalability. Over-customizing by creating a custom object that duplicates a standard object's purpose can lead to more work and missed out-of-the-box features. For instance, if you need to track customer issues, using the standard Case object is typically better than creating a "Support Ticket" custom object from scratch, since Cases already integrate with Salesforce's support processes.

When defining fields, choosing the right field type is crucial. Always use a data type that closely matches the nature of the data. If something is a date, use a Date field (not a text). For currency amounts, use a Currency field so multi-currency and roll-up features can work. If data should be one of several predefined values, use a Picklist instead of free text to enforce consistency. Thoughtful field choices prevent bad data from ever being entered – for example, using a Checkbox for a yes/no choice avoids the ambiguity of users typing "Yes", "Y", or "True" in a text field. A practical mindset is: **constrain data at entry** to ensure quality. This means making fields required where appropriate and using validation rules to enforce complex conditions (we'll discuss those in the Data Integrity section). It's much easier to validate data on entry than to clean it up later.

Another key concept in data modeling is understanding **relationships** between objects. Salesforce supports one-to-many relationships through Lookup or Master-Detail fields (which we will explore in depth in the next section). At the modeling stage, think about how your objects relate: Should a given record exist on its own, or only as part of a parent record? For example, if you have an object "Expense" that should always be tied to a "Project", that suggests a relationship where Project is the parent and Expense is the child. If the parent is deleted, should the child remain or also be removed? Questions like these guide your choice of relationship type and ensure your data model mirrors the real-world constraints of your business scenario. It's often useful to sketch an entity-relationship diagram (ERD) or simply list out objects and their connections to visualize the overall structure before building it in Salesforce.

Normalization vs. denormalization is an important consideration often borrowed from traditional database design. **Normalization** means structuring your data model to avoid redundancy – in other words, store a piece of information in one place and reference it elsewhere rather than duplicating it. In Salesforce, an example of normalization would be having a separate "Region" object that multiple Accounts reference, rather than typing the region name on every Account record. The advantage is consistency (you update the region in one place) and avoiding conflicting data. **Denormalization**, on the other hand, means duplicating data or flattening relationships for the sake of simplicity or performance. Salesforce reporting and filtering sometimes benefit from denormalization. For instance, you might store an "Account Type" on a related custom object record so that you can report on it without needing to pull in the Account in a report. The trade-off is that you must keep the duplicate data in sync (perhaps using automation like a Flow to update child records when a parent changes). Best practice is to normalize by default for data integrity, but selectively denormalize when it significantly simplifies user access to data or improves performance. A common scenario is using formula fields to bring a parent field's value onto a child object – this is a form of safe denormalization that maintains a single source of truth while giving easy access to the data where it's needed.

Let's consider a quick example to illustrate these principles. Suppose you're designing a Salesforce app to manage **Training Courses** and **Enrollments.** A Course is delivered many times, and each Enrollment ties a Course to a specific student or employee. Here's how you might apply fundamental modeling decisions:

Object choices: You'd likely use a custom object "Course" (if no standard object fits) and another custom object "Enrollment" to represent someone enrolling in a course. Perhaps you realize that the people enrolling are internal employees which are already represented by the standard Contact or User objects – if so, you might not need a separate "Student" object at all, you could use Contact.

Field types: On the Enrollment object, you'd have a date field for Enrollment Date, a lookup field to the Course, and a lookup to the Contact (student). You might have a Picklist for Enrollment Status (Enrolled, Completed, Cancelled) to keep those status values controlled.

Relationships: The Enrollment object clearly sits between a Contact and a Course, linking them. This is a one-to-many from Course to Enrollment (one course can have many enrollments) and also one-to-many from Contact to Enrollment (one person can enroll in many courses). We'll later see that this scenario is effectively a many-to-many relationship between Contact and Course, resolved by the Enrollment object acting as a junction.

Normalization: The course details (like Course Name, Description, Duration) belong on the Course object, not repeated on each Enrollment. The student info (Name, Email) is already on Contact. The Enrollment mostly just references those. If reporting needs a student's email alongside enrollment records, you can either fetch it via report join or use a formula field on Enrollment to pull Contact's Email (denormalizing just for convenience). The key is you wouldn't store the email independently on Enrollment because it would then diverge if the Contact's email updates.

By walking through scenarios like this, you ensure each part of your data model has a purpose and follows best practices. To summarize the foundational principles:

- **Use standard objects when suitable** (to save effort and use built-in features).
- **Choose the right field types** and make fields required or validated as needed to enforce good data.
- **Model real-world relationships** between objects (don't treat related data as separate, use lookups or master-detail).
- **Normalize data** to avoid duplication, but **denormalize when it clearly benefits** user reporting or simplifies complex joins, and manage the sync carefully.

Pro Tip: Resist the urge to capture every piece of data in one object "just in case." A hallmark of good Salesforce design is *modularity*. If something can be its own object (with its own lifecycle or attributes), consider making it one. Conversely, avoid splitting data into too many objects if they will always be used together. Strive for a balance that makes the data model easy to understand and work with.

With these fundamentals in mind, you are equipped with a solid foundation. In the next section, we'll apply these principles to one of the most important aspects of Salesforce data architecture: selecting the right type of relationship between objects, specifically Master-Detail versus Lookup. This decision will influence everything from security to reporting in your org.

ChatGPT Prompt (Typing):

User: "I'm setting up a new app in Salesforce. What are some key principles I should follow when designing my data model?"

ChatGPT: "Great question! Here are a few fundamental principles for a solid Salesforce data model: First, prefer standard objects over custom if they serve the purpose – this leverages out-of-the-box functionality. Second, make sure each field uses the proper data type (picklists for predefined values, lookups for relationships, etc.) to enforce data quality. Third, design clear relationships between your objects – use lookup or master-detail fields instead of storing references in text fields. Also, avoid duplicate data: try to store information in one place and reference it elsewhere (normalization), unless you have a good reason to duplicate it for reporting (denormalization). Finally, keep the future in mind – design your object structure so it can scale (for example, don't cram too many concepts into one object if they might diverge as you grow). Following these principles will make your app easier to maintain and extend."

SELECTING RELATIONSHIPS: MASTER-DETAIL VS. LOOKUP

One of the most critical design choices in a Salesforce data model is how to relate objects to each other. Salesforce provides two main relationship types to connect objects: **Master-Detail (M-D)** and **Lookup**. While both serve to link one object to another, they have important differences in terms of ownership, sharing, required fields, and the cascade of actions like deletions. Choosing the right relationship type up front will save you from headaches later. Let's break down the distinctions and best practices for using each.

A **Master-Detail relationship** creates a tight bond between two objects, akin to a parent-child relationship. The "Master" (parent) object controls certain behaviors of the "Detail" (child) object:

Ownership and Sharing: The detail record doesn't have an independent owner – it inherits the owner (and sharing settings) from its master record. This means if a user can see the master, they can see the detail. You cannot share a detail record without giving access to its master. This is great for when the child record should be considered part of the parent for security purposes.

Required Parent: With master-detail, a child record **must** have a parent. You cannot save a detail record without a master lookup filled in. This ensures the child is never orphaned. If a parent record is deleted, by default all its detail records are also deleted (cascading delete), enforcing that tight relationship.

Roll-Up Summaries: One big advantage of master-detail is the ability to create roll-up summary fields on the master. These can automatically calculate values like count, sum, min, max of numeric fields on detail records (for example, an Account could roll up the total value of all related Opportunities if Opportunity were in a master-detail with Account). This is *not* available with a lookup relationship.

Limits: Each custom object can have up to two master-detail relationships. Also, master-detail cannot be used if the detail object is already a detail in another master-detail (because one detail can't have two different masters unless it's specifically a junction object scenario which we'll discuss). Master-detail relationships also can't be created if there are already a bunch of records with no parent (you must associate existing records to a parent before converting a lookup to master-detail).

Behavior on Delete and Reparenting: As mentioned, deleting a master can delete the details (you can optionally allow or prevent this cascade in the relationship settings). Details can sometimes be reparented to a different master record (if you enable "Allow reparenting" in the relationship settings), but they can never exist without a master.

A **Lookup relationship**, in contrast, is a looser coupling:

Ownership and Sharing: The child record in a lookup maintains its own owner and does not inherit sharing from the parent. It's more like a pointer or reference. You can have a private child record that points to a parent record the user can't see, and vice versa – though that might not be ideal design, it's technically possible with lookups. In short, security for lookups is independent unless you write custom rules or have Apex sharing.

Optional Relationship: By default, a lookup field can be left blank (though you can mark it required at the layout or use validation to simulate a required relationship). The child can exist without a parent, which is useful for scenarios where the relationship is truly optional or not always known.

Deletion Behavior: If you delete a record that others are looking up to, those lookup fields typically get cleared (set to null) – unless you set the relationship option to "Don't allow deletion of the lookup record if it's part of a lookup relationship". That setting can prevent a parent from being deleted if any child is pointing to it, adding a bit of safety. However, there's no automatic cascade delete; the child records remain (just without a reference) if the parent is removed.

No Roll-Up Summaries: Lookup relationships do not support roll-up summary fields on the parent. If you need roll-ups but want to use lookups, you'd have to use other means (such as Flow, Apex, or the declarative roll-up summary tool from AppExchange) to achieve similar summarization. Generally, needing roll-ups is a sign that master-detail might be more appropriate.

Flexibility: You can convert a lookup to a master-detail later (with some constraints) if you realize you needed that tighter link, but you must have all records populated with a parent first. Conversely, a master-detail can be converted to a lookup if necessary (this can be helpful if your requirements change and you need a more flexible relationship).

So how do you decide which to use? Ask yourself a few key questions about the relationship between the two objects:

Dependency: Should the child record always require a parent? If yes, Master-Detail enforces that inherently. If the child can stand alone or the link is optional, a Lookup is safer.

Ownership & Security: Do the parent and child share the same lifecycle and security context? For example, if you restrict access to the parent, must the child be restricted as well? And if you delete the parent, does it make sense to also delete all children? If yes to these, Master-Detail fits because it binds the child's access to the parent. If you need independent sharing (maybe the child is visible to different people than the parent), go with Lookup.

Aggregation: Do you need to perform calculations (count, sum, etc.) of child records on the parent record? If yes, Master-Detail provides roll-ups out-of-the-box. If not, or if you're okay managing without roll-ups, Lookup might suffice.

Cascade Actions: If the parent record is removed or changes hands, what should happen to the child? Master-Detail will cascade deletes (and can optionally cascade sharing recalculations etc.), while Lookup will leave child records orphaned (or prevent deletion if configured). Consider business rules: e.g., if a project is deleted, should its tasks also go? Often yes – that suggests Master-Detail. But if a Contact linked to an Account is deleted, you might not want to delete the Account – that's a lookup scenario.

If you answered "yes" to most of the first set of questions (need strict parent, shared security, roll-ups, cascade delete), then a Master-Detail relationship is likely the right choice. Otherwise, a Lookup relationship will give you the flexibility you need without the extra constraints. It's not uncommon to start with a Lookup during early development (for flexibility) and later decide to convert it to Master-Detail once you're confident of the required dependency.

Let's solidify this with a concrete scenario. Imagine you have an object **"Project"** and an object **"Expense"**:

If every Expense should be tied to a Project (and never exist on its own), and you want to roll up total expenses on each Project, use Master-Detail (Project = master, Expense = detail). Deleting a Project would wipe out its Expenses, which might be desired to keep data tidy. Also, if a user shouldn't see an Expense without seeing its Project, master-detail covers that by sharing rules.

On the other hand, consider "**Account**" and "**Contact**". Salesforce uses a Lookup for this relationship (each Contact may link to an Account but it's not required to have one, and Contacts have independent sharing in some cases). You wouldn't want deleting an Account to automatically delete all Contacts in most cases – you might just orphan them or relate them to another account. The flexibility of a lookup makes sense here. (Salesforce actually has special behavior for Account-Contact: if you delete an Account, you can choose to delete related Contacts or not, but it's not a true master-detail under the hood.)

Pro Tip: When in doubt, start with a Lookup relationship. It's easier to loosen constraints initially and tighten them later if needed than the opposite. If you find yourself writing a lot of validation rules to enforce parent existence or triggers to perform roll-ups, that's a sign you probably needed a Master-Detail. Conversely, if a Master-Detail is causing sharing complexities or you realize the child needs to sometimes exist independently, it might have been better as a Lookup. Always document *why* you chose one over the other – this helps future admins or architects understand the design.

To summarize the best practices:

Use **Master-Detail** for closely coupled objects where the child's life cycle is tied to the parent (examples: Line Items on an Invoice, where each line is meaningless without the main invoice; or a Campaign and Campaign Members, which Salesforce actually implements as a special kind of detail).

Use **Lookup** for relationships that are more flexible or optional (examples: linking a Contact to a Referral Partner Account – nice to have the link, but the Contact still exists if you remove the partner).

Keep an eye on special needs like roll-up summaries and sharing rules as key decision factors.

Remember that you cannot have a detail without a master in M-D; plan for data load accordingly (you might need to load parents first, then children, etc., or temporarily use lookup then convert to M-D after data migration by filling all parent IDs).

ChatGPT Prompt (Voice):

User (speaking): "We're designing a Salesforce app for equipment maintenance. There's an **Equipment** object and a related **Maintenance Record** object. Should those be master-detail or lookup?"

ChatGPT: "Let's think it through. If every Maintenance Record is always tied to an Equipment and shouldn't exist otherwise, that leans towards Master-Detail. Also, if you want to easily summarize maintenance costs or counts on the Equipment record, Master-Detail's roll-up summaries would help. With Master-Detail, if an Equipment is deleted, all its Maintenance Records would be deleted too – which might make sense if those records aren't useful without their Equipment. On the other hand, if maintenance might sometimes be logged without a specific Equipment or you want separate sharing (maybe maintenance records are visible to a different team), a Lookup would be better. For a typical maintenance scenario, I would use Master-Detail: it keeps the maintenance records strictly dependent on an Equipment and lets you aggregate data. Just ensure to set "Allow reparenting" if you ever might move a Maintenance Record to a different Equipment, and remember you'll need to have an Equipment in place before creating its Maintenance Records."

After understanding basic one-to-one and one-to-many relationships, you might encounter scenarios where you need to model **many-to-many** relationships or more complex hierarchies. In the next section, we'll discuss how to handle those cases using junction objects and other techniques, expanding our relationship modeling toolkit for more sophisticated data architectures.

MODELING COMPLEX DATA RELATIONSHIPS

Real-world business data is rarely limited to simple one-to-many relationships. Often, you'll find the need for **many-to-many relationships** or hierarchical relationships within a single object. Salesforce's declarative tools can handle these scenarios through a combination of relationship fields and what's known as **junction objects.** In this section, we'll discuss best practices for modeling these complex relationships while maintaining clarity and data integrity.

A **many-to-many relationship** occurs when a record from Object A can relate to multiple records of Object B *and* vice versa. For example, consider a scenario with **Projects** and **Employees:** an employee can work on multiple projects and each project has multiple employees. How do we represent this in Salesforce? The answer is to introduce a third object – commonly called a **junction object** – that sits in the middle and has a master-detail (or lookup) relationship to each of the two objects. In our example, we might have a custom object called **Project Assignment** that links one Employee to one Project for each record. The Project Assignment record would have a field linking to an Employee and another field linking to a Project. Each assignment is like the intersection of one Employee and one Project, and by creating records of that object, you effectively map many employees to many projects.

Figure: An example of a many-to-many relationship modeled via a junction object. Here, the "Opportunity Product" object (center) acts as a junction between Opportunities and Products, allowing each Opportunity to be associated with multiple Products and each Product to relate to multiple Opportunities. In this Salesforce schema diagram, the Opportunity Product has a lookup to Opportunity and a lookup to Product, making it a detail of both.

When creating a junction object, it's common to use **two master-detail relationships** on that object – one for each side of the many-to-many. This makes the junction object a detail of both parents. One of those master-detail fields will be the "primary" master (the first one you create). The primary master influences some properties, like whose ownership the junction records inherit and which parent's look and feel (e.g., color in Lightning record pages) the child records adopt. The primary master's owner is considered the owner of the junction records. In many cases, it doesn't matter which is primary, but it's something to be aware of. If the relationship between the two objects is truly peer-to-peer (like Projects and Employees), just choose one to be primary arbitrarily (perhaps the one that makes sense for ownership or roll-up summaries). If one side of the relationship is clearly the "dominant" parent, make that the master (for example, if an "Order" has many "Products" through an OrderLine item, you might consider Order as primary).

It's also possible to have a junction object where one relationship is master-detail and the other is lookup. This might be done if you want one side to enforce ownership and cascade deletes, but the other side to be more optional. For instance, say **Students** and **Courses** with **Enrollments** as the junction: you could make Course–Enrollment a master-detail (if an Enrollment must always be tied to a Course and should be deleted if the Course is deleted), but Student–Enrollment a lookup (maybe an Enrollment could exist without a Student in some edge case, or you want to allow orphaned records for some reason). However, using two master-detail relationships is the typical approach for a full many-to-many because it tightly links the junction to both parents.

When modeling many-to-many, naming and clarity are important. The junction object should be named in a way that explains the combination, like "Project Assignment" or "Course Enrollment" – something that implies it links the two. Additionally, on each of the parent objects, Salesforce will show a related list of the junction records. Make sure to configure those related lists on page layouts and set appropriate fields and labels. For example, on the Project page layout, the related list might be "Assignments" showing which employees are on the project. On the Employee layout, it might be "Project Assignments" or just "Projects" showing which projects that employee is assigned to. This makes the many-to-many relationship visible and navigable to users.

Another complex scenario is hierarchical relationships within a single object. Salesforce provides a special relationship type called **Hierarchical** (a variant of lookup) that is only available on the User object (to model management chains, for example). But for custom objects (and even some standard objects like Account), you can create a self-lookup. A self-lookup is simply a lookup field on an object that references the same object. This can model hierarchies or parent/child structures within one object. A classic use case is the **Account hierarchy:** each Account has a field "Parent Account" (lookup to Account) which lets you build an Account hierarchy of unlimited levels. Similarly, you could have a custom object "Department" with a lookup to itself called "Parent Department" to create an org chart of departments.

When implementing a hierarchy via self-lookups, consider if you need to enforce any rules (for instance, prevent loops where A is parent of B and B parent of A, or overly deep nesting). Salesforce will naturally prevent direct recursion (you can't set a record as its own parent), but it won't stop complex loops if you're not careful. Some orgs implement validation rules to ensure a record's parent isn't in a set of children – but that can be complex to enforce declaratively beyond a single level. Usually, simple hierarchies don't need heavy restriction; users typically won't create loops intentionally. If hierarchy depth is a concern (for performance or usability), you might set a guideline (like no more than 5 levels deep of parent-child). Out-of-the-box, Account hierarchy page views show up to 500 child accounts in a tree, for example, and it's rare to go beyond that in practical terms.

Whether modeling many-to-many or hierarchies, **data integrity** remains a priority. For junction objects, you should often set up some rules to prevent bad data combinations:

Consider adding a **validation rule or a unique index** (using a text formula or Flow) on the junction object to prevent duplicate pairings. For instance, you wouldn't want two "Project Assignment" records for the same Project and same Employee – that would be a duplicate assignment. While Salesforce can't natively enforce uniqueness across two fields out of the box, you can create a formula field that concatenates the two IDs (like ProjectId__c & "-" & EmployeeId__c) and mark that formula as Unique, or use a before-save flow to check for an existing record with those two values. This will stop users from accidentally creating the same link twice.

Ensure that the junction records get meaningful names (the Name field). Sometimes, you can auto-number junction records, but it might be more user-friendly to use a formula as the Name that combines info from both parents (like a formula name = Project Name + " - " + Employee Name). That way, in related lists or reports, the junction records are easily identifiable by both of their connections.

Pay attention to **security implications:** with two master-detail relationships, as noted earlier, the junction inherits security from *both* masters. In fact, for a user to have access to a junction record, they need access to both associated master records. Salesforce will automatically handle this by, for example, if you give read access to the junction object on a profile, it will require read access to the masters as well. This prevents someone from seeing a junction if they shouldn't see one of the linked records. It's generally good, but be mindful if you ever switch one side to lookup (then that side's access won't be automatically enforced).

Speaking of security and access in complex relationships, let's touch on an earlier question: *Can a junction object use two lookups instead of master-detail?* Yes, it can, and sometimes you might do this if you don't want the strictness of master-detail. Maybe you have a many-to-many that is more advisory (not absolutely required data). Using lookups would allow junction records to exist without one of the parents, or let you delete a parent without deleting the link (though that link then becomes a dangling reference you'd probably clean up). However, if you go lookup-lookup, you lose roll-up summaries and cascade deletes entirely, and you must manage security and orphan logic yourself. In most cases, we prefer master-detail for genuine many-to-many relationships because it keeps data tight and consistent. Use lookup/lookup junctions only if there's a specific requirement that a junction might stand alone or if you hit some limit that prevents master-detail (e.g., maybe one of the objects is already a detail of something else and can't be a master).

Beyond many-to-many, sometimes **polymorphic relationships** come up – where a child could relate to one of several different parent objects. Salesforce supports polymorphic lookups in a limited set of standard objects (like the WhoId/WhatId on Activities can point to multiple types). For custom use cases, there's no declarative polymorphic field (you'd need custom development or multiple lookup fields).

A typical pattern is to use multiple lookup fields and perhaps a picklist to indicate which one is in use. For example, a "Comment" object might relate either to a Case or to a Project – you might include both a Case lookup and a Project lookup on Comment, and use a picklist or logic to ensure only one is filled for each record. This gets complex, so we mention it mainly to acknowledge that declarative options are limited – if you find you need polymorphic relationships often, re-examine if your data model can be adjusted or if an alternate approach (like using tasks or a more generic relationship object) is available.

Pro Tip: To visualize and verify complex relationships, use **Schema Builder** in Salesforce. It's a drag-and-drop tool that shows objects and their fields and relationships. This is especially useful after adding junction objects or self-lookups – the Schema Builder will display lines connecting objects (solid line for master-detail, dotted for lookup). It can be a quick sanity check to ensure, for example, that your junction object has the expected two relationships linking to the right objects. It's also a great communication tool to show colleagues or stakeholders how data entities are connected in your org.

ChatGPT Prompt (File Upload):

User: *[Uploads an Excel file DataModel.xlsx with a list of objects, fields, and relationships]* "I've uploaded a spreadsheet of our current data model design. Can you review it and suggest any improvements, especially concerning our object relationships?"

ChatGPT: "I've analyzed the provided data model spreadsheet. Here are my observations and suggestions: First, I see you have an object Partner and an object Deal, with a many-to-many relationship represented by a junction object PartnerDeal__c. This is a good approach. One improvement: ensure that PartnerDeal__c has a unique constraint to prevent duplicate Partner+Deal records – this will protect data integrity. Also, consider using Master-Detail for both relationships on PartnerDeal__c if not already; it looks like currently they might be lookups. Master-Detail will enforce that every PartnerDeal record is linked to a Partner and a Deal, and it will allow roll-up summaries (e.g., count of deals per partner).

I also noticed an object Event with a self-lookup field ParentEvent__c. That's a valid way to model sub-events or an event hierarchy. One suggestion is to add a validation rule to prevent recursive relationships – for instance, an event should not point to another event that eventually points back to the first. This can be tricky to enforce, but at minimum document the expected usage of this hierarchy to users.

Lastly, there's an object Order and a child object OrderLine. You used a lookup for OrderLine to Order. If an OrderLine should not exist without an Order, it might be safer to use Master-Detail, so that if someone deletes an Order accidentally, the OrderLines don't remain orphaned. Plus, Master-Detail would let you roll up totals on the Order (like total quantity or amount from its lines). Overall, your model is solid – these tweaks would align it even more with Salesforce best practices."

By utilizing a junction object for many-to-many and self-lookups for hierarchies, you can represent complex data relationships declaratively. The next step is ensuring that all this data remains *clean and reliable* as users interact with it. In the following section, we'll shift focus to maintaining data integrity and quality – using Salesforce features to keep your data model's content as robust as its structure.

MAINTAINING DATA INTEGRITY AND QUALITY

Designing a good data model is only half the battle – the data populating that model needs to be accurate and consistent for the system to truly succeed. Salesforce provides a suite of declarative tools to help maintain **data integrity and quality**. In this section, we'll cover how to enforce business rules and quality standards using validation rules, duplicate management, and other mechanisms, all without writing code. A key takeaway is that a strong data model *actively prevents* bad data from entering the system, rather than trying to fix it after the fact.

Validation Rules are one of the most powerful declarative features for data quality. A validation rule is essentially a formula condition that, when evaluating to true, stops the record from saving and displays an error message. Use validation rules to enforce your business constraints and assumptions. For example:

- Ensure important fields are filled in given certain conditions (e.g., "If Stage = Closed Won, then Winning Reason must be provided").
- Enforce formatting or value ranges (e.g., phone number must match a specific pattern, or discount percent must be between 0 and 100).
- Cross-field consistency (e.g., if Field A = "Yes" then Field B must not be blank).

Salesforce validation formulas can be quite sophisticated, incorporating mathematical operations, text checks, date comparisons, and even referencing related records or $Profile and $User details. A best practice is to centralize and streamline validation rules: it's better to have one rule with a complex formula catching multiple related errors than many small rules that could conflict or be harder to maintain. Also, craft clear error messages for users – they should know exactly what to fix when a validation error pops up.

One clever advanced tip is using a Custom Setting or Custom Metadata in validation rules to make them more flexible. For instance, you might have a validation that normally triggers but you want an easy way to bypass it in specific scenarios (like data loads or integrations). You can include a check in the validation formula that references a custom setting value (e.g., a checkbox "Enable_Validation__c" that an admin can uncheck temporarily). This way, admins could disable certain validations without modifying the rule logic, simply by toggling a config value. This pattern should be used sparingly, but it's a nifty trick for those edge cases where strict validation needs an off switch (for example, to allow loading a bunch of legacy records that violate a rule, you could temporarily bypass and then re-enable the rule).

Duplicate Management is another key aspect of data quality. Salesforce offers out-of-the-box duplicate prevention and handling tools. Using **Matching Rules**, you define how to identify potential duplicates (for example, two Contacts with the same email address might be considered dupes). Matching rules can be exact or fuzzy; Salesforce provides some pre-built ones (like exact email match or fuzzy name match) and you can customize your own. Then, using **Duplicate Rules**, you decide what happens when a potential duplicate is detected: you can just allow it and alert the user, block the save entirely, or even allow and auto-report it. A common approach is to alert on possible dupes for most users (so they can choose to proceed if it's really not a dup) but maybe block in certain high-risk objects or for certain fields.

Salesforce duplicate rules can compare across objects as well (for instance, flagging if a Lead is a dup of an existing Contact). They work in real-time on create and edit, and also asynchronously via **Duplicate Jobs** (which let you scan the database for dupes in bulk). By implementing duplicate management, you ensure your model's integrity isn't compromised by, say, the same customer being in the system twice, which can lead to scattered information and reporting inaccuracies. As the Salesforce blog succinctly puts it: duplicate management is made up of matching rules (which fields to compare, exact or fuzzy) and duplicate rules (what action to take when a match is found). A best practice is to start with a **loose** alert rule (catch obvious dupes, warn users) and refine over time. You can review how often dupes are detected and adjust the strictness of matching. Engaging business users in defining what constitutes a duplicate is important – they might say two accounts with the same tax ID are definitely duplicates, while two accounts with similar names might not be.

Another tool in the toolbox is the **Required Field** setting and **Unique Field** setting. Marking a field as Required (at the schema level or even just on page layouts) ensures a value is always provided. Do this for any field that truly must have data for the record to be valid – but be cautious not to over-require fields, as it can frustrate users if they're forced to input something that isn't always applicable. The Unique attribute on a field (available for text, email, number, etc.) can automatically prevent duplicate values in that field across records. For example, if you have a "Government ID" field on a Contact, marking it Unique will make sure no two contacts can have the same ID. This is a simpler alternative to duplicate rules for single-field uniqueness (with the limitation that it's case sensitive by default and doesn't offer "fuzzy" logic – it's purely exact match). Use unique fields especially for any natural key (like an external ID from another system, or something like a Social Security Number, etc.) to guard against double entry of the same entity.

Picklists and dependent picklists also contribute to data quality by restricting inputs. We'll cover picklists in more detail in the next section, but one quick example: if you have a Country picklist and a State picklist, making State dependent on Country (so that only valid states for the chosen country appear) will dramatically improve data consistency compared to free-text fields. This ensures, for instance, that users don't select "Canada" as country and type a state that's not a Canadian province.

Maintaining data quality isn't just about prevention – it's also about ongoing **monitoring and cleanup**. Salesforce admins often set up **reports or dashboards** to monitor data completeness (e.g., a dashboard showing what percentage of accounts have key fields filled in). Salesforce even provides a free app called **Data Quality Analysis Dashboards** on the AppExchange that gives you a head start in measuring data quality across standard objects. By regularly reviewing these metrics, you can identify when data starts to go off-track (maybe a new team is creating accounts without filling in "Industry" consistently, etc.) and take corrective action, such as training users or adding a new validation rule.

Lastly, let's not forget about **business logic** enforcement declaratively. Apart from validation rules, you have tools like **Workflow (legacy)**, **Process Builder (legacy)**, and **Flow (modern)** that can update fields or alert users based on conditions. For example, a Flow could automatically set a checkbox "High Value Customer" to true if an account's annual revenue goes above a threshold. That keeps data in sync with business definitions. Another example: an **automation could handle data cleanup** like formatting a text field (say you want all codes stored uppercase – a before-save flow can do that). While automation goes beyond pure modeling, it works hand-in-hand with the model to maintain integrity (just be careful not to overdo automation for data cleanup if a simpler fix like a validation rule on input will suffice).

Pro Tip: Implement a **data stewardship process.** Even with all the automated rules and constraints, having human oversight on data quality can catch issues rules don't. For instance, schedule a weekly or monthly review of newly created accounts by a data steward or admin – they might spot that users are selecting "Other" for Industry too often and realize a new picklist value is needed. They could also merge duplicate records that slipped past rules or correct typos. This proactive governance ensures your data model remains healthy and trusted by users.

ChatGPT Prompt (Browsing):

User: "Can you find official Salesforce guidance on setting up duplicate management? I want to make sure I follow best practices."

ChatGPT (with browsing): "Certainly! According to Salesforce's documentation on managing duplicates, duplicate management is achieved by defining matching rules to identify potential duplicates, and duplicate rules to specify actions when duplicates are found. For example, you might use an exact match on Email to catch contact duplicates, and set the duplicate rule to alert the user but allow save. Salesforce recommends starting with simple rules and gradually refining them. The help article also suggests using duplicate jobs for periodically scanning the database, and emphasizes combining multiple fields in matching criteria for higher accuracy. In practice, a common best practice is to use a combination of a unique field (for strict prevention) and duplicate rules (for more flexible, user-managed merging) to cover all bases." (Sources: Salesforce Help & Salesforce Ben blog on duplicate management)

With a strong handle on data quality and integrity measures, your data model will not only be well-designed but also well-maintained. Next, we'll look at how to further refine the schema itself for efficiency and clarity, using features like picklists and record types to optimize both the data and the user experience.

OPTIMIZING SCHEMA WITH PICKLISTS AND RECORD TYPES

An optimized schema makes your Salesforce configuration more user-friendly, more report-friendly, and easier to administer. Two key tools for schema optimization are **Picklists** (including dependent picklists and global value sets) and **Record Types**. These features help manage complexity by structuring data input and behavior in a declarative way. In this section, we'll explore best practices for using picklists to enforce data consistency and using record types to tailor processes for different scenarios – all while avoiding common pitfalls.

Picklist Fields allow you to define a set of allowed values for a field, ensuring data in that field is consistent across records. Whenever you find that a field should only have a specific range of values (like Status = New, In Progress, Closed), a picklist is usually the way to go. This not only makes data entry easier for users (no guesswork, just pick from the list) but also makes reporting and filtering much more reliable. For example, it's much easier to create a report filtering where Status = "Closed" if you know only valid options exist, versus if users typed in status text and you have to account for "closed", "Closed", "Close" typos, etc.

Global Picklist Value Sets are a best practice when the same set of values is used in multiple picklist fields (even across objects). Instead of entering the same list (say a list of Regions or Product Categories) in multiple fields, you define it once as a global value set. Then your picklist fields can **inherit** those values. This centralizes management – add a new Region in one place and all fields using that set get the new value. It also keeps values consistent across objects (preventing a scenario where "Northwest" is spelled differently in two picklists). Use global value sets for things like industry lists, country lists, common status values that appear in multiple objects, etc.

Dependent Picklists are another powerful feature. They let you filter the available values of one picklist based on the value of another. A classic case is Country and State/Province: you have a master picklist of all states/provinces and a controlling picklist of Country. When a user selects Country = "Canada", the State picklist only shows Canadian provinces. This enforces validity (no "California" if country is Canada) and improves user experience by shortening the list to relevant options. Setting up dependencies is declarative in the field settings – you mark one picklist as controlling and map which values are valid for each controlling value. One thing to keep in mind: multi-select picklists can only be controlling (not dependent), and they bring some complexity – more on multi-select shortly.

A few best practices for picklists:

Avoid long picklists if possible. If you have more than, say, 100 values, consider if those values should be records in an object instead (lookup) rather than picklist values. Extremely long picklists are hard for users to navigate and can be a pain to manage (though Salesforce can handle up to 1000 or so). As a rule of thumb, if the list of options frequently changes or has many entries (like a list of customer companies, or a list of SKUs), that probably should be a separate object. Picklists are best for relatively static sets of categorical data.

Use descriptive API names for values, especially if you use the values in formulas or automation. (The API name is the behind-the-scenes name of the value, which can differ from the label shown to users.) For example, a picklist "Priority" might have label "High" but maybe you use API name "High_Priority" or simply "High" consistently. If you ever rename a value's label (say "High" to "Urgent"), you can keep the API name the same to avoid breaking any case logic in formulas that check for "High".

Restrict picklist to the value set: Salesforce picklists have an option "Restrict picklist to the values defined" (for custom picklists). Always enable this unless you have a specific reason not to. If it's not restricted, users (or integrations) could actually push in values that aren't in your list (leading to "dirty" data that doesn't conform). By restricting, you ensure only the defined values are allowed, which is usually what you want for data consistency.

Plan for translations if you have a multi-language org. Picklist values can be translated using the Translation Workbench. Using picklists makes multi-language support easier because the underlying value is the same and you can provide labels per language.

Now, **Multi-Select Picklists** deserve a special mention. They allow a user to select multiple values in a single picklist field (values appear in a semi-colon separated list in the record). Multi-select picklists might seem convenient (e.g., "Select all applicable features: A, B, C"), but they come with significant drawbacks:

They are harder to filter on in reports (you can filter "contains X" or "includes X", but you can't easily do "has exactly these two and no others" logic).

In automation (Flows, validation rules), checking if a multi-select contains a specific value is doable with functions like INCLUDES(), but doing complex logic like "one of these values but not another" gets cumbersome.

You cannot use multi-select picklist fields in certain places, like controlling a dependent picklist (they can be controlling but not dependent), and some older features (like some formula functions or process builder criteria) had limited support for them.

Data loading can be tricky because you have to format the values exactly with semicolons etc.

Because of this, consider alternatives: sometimes a **junction object** or a **set of checkboxes** might be better. For example, instead of a multi-select for "Skills" on a Contact, you might have a related object Skill with a record per skill (that's a more normalized design, easier to query each skill separately, though it's a bit heavier to the user to manage multiple records). If the set of options is relatively small and fixed, multi-select can still be okay – just be aware of its limitations. If you do use multi-select, treat each value carefully in logic and documentation so that everyone knows how to interpret that field.

Moving on to **Record Types:** Record types allow you to have different "flavors" of records on the same object, primarily to show different **page layouts** and **picklist values** to users based on context. They are commonly used when one object is used for more than one business process. For example, on the Case object you might have "Support Case" vs "Product Feedback" record types – they may collect different fields and have different status picklist values. Or on Opportunity, you might have different record types for different sales processes (New Business vs Renewal, each with its own stage values and sales path).

When to use record types:

When you need to **show different fields or sections** on the page layout to different users or scenarios. If two groups use an object differently (e.g., a "Partner Account" vs "Customer Account" could have separate layouts highlighting different info).

When you need **different picklist values** for the same field in different situations. For instance, an Opportunity Stage picklist might have one set of stages for new sales and a different set for renewals. Record types let you filter the picklist values depending on the record's type.

When you have significantly different **business processes** that you want to manage distinctly but they still share the same fundamental object. Salesforce ties certain standard processes to record types as well (Lead Process, Sales Process, Support Process behind the scenes are linked to record types for Lead, Opportunity, Case respectively).

However, don't use record types just for the sake of it. They introduce complexity: every profile needs to be set up with access to specific record types, you have to manage multiple page layouts, and users might have to choose a record type when creating records (an extra step). If the differences are minor, consider using dynamic forms or conditional components for Lightning pages first (for example, a lightning page that shows certain components based on field values could sometimes eliminate the need for separate record types). That said, record types are often the cleanest solution for divergent processes on one object.

Best practices for record types:

Keep the number of record types reasonable. There's no hard rule, but if you find an object with, say, 10 record types, question if that object is trying to do too much. Maybe it should be two objects, or some of those variations could be handled with fields instead. Many orgs successfully use a handful of record types on core objects (e.g., 2-3 on Opportunity for different sales processes, a couple on Account for customer vs partner vs vendor, etc.).

Name record types clearly and use descriptions. Users will see the record type name when creating records (unless you have a default and skip that step for them). A clear name like "Opportunity – New Business" vs "Opportunity – Renewal" helps them pick correctly.

Align record types with profiles and permission sets thoughtfully. For instance, inside sales reps might only use the "Quick Lead" record type, while field sales use "Detailed Lead" – you can restrict visibility accordingly, so users aren't confused by options that don't apply to them.

Use record types to filter picklists: In setup for each record type, you can customize which picklist values are available for each picklist field. This is hugely helpful. For example, with Cases, if you have a "Type" field, certain types might only apply to one case record type. By filtering, you ensure users of that record type only see relevant values.

Record types also interplay with other features:

They can trigger **different processes** in Process Builder or Flow if you check record type as a condition (like, "if record type = X, do these actions").

They can control **different Lightning pages** (you can assign a Lightning record page by app and record type, meaning you could have a completely different UI for different record types if needed).

When converting leads, mapping to record types etc., Salesforce will prompt if multiple types exist (like if a lead converts to an account, you can specify which account record type the new account should be). So, consider the user experience of moving data between objects – too many record types can complicate these transitions.

A quick example of using record types effectively: Imagine a "**Request**" object that serves both IT support requests and HR requests. These are quite different in nature – IT requests might track software/hardware info, while HR requests might involve personal or payroll info. Instead of having two objects, you can have one Request object with two record types: IT and HR. Each has its own page layout (IT layout showing computer details fields, HR layout showing employee ID, etc.) and its own picklist values for, say, Request Category. This way, you unify reporting on requests in one object (good for admins and for any overlapped processes), but still give tailored interfaces and options to users. Just ensure that any user who needs to create a Request is assigned the appropriate record type, and maybe default it based on their department to streamline data entry.

Pro Tip: Too many record types can sometimes be a sign of over-engineering. If you realize that every single record is getting a different record type or you're branching logic all over the place based on record type, consider if those should actually be separate objects or if you could collapse some record types. Record types should simplify your life (segmenting data and UI logically), not complicate it. Use them with purpose – usually tied to distinct business processes.

ChatGPT Prompt (Video Generation):

User: "I need to explain to my team the difference between using a picklist field and a custom object for a list of values. Can you create a brief script for a 2-minute training video?"

ChatGPT: "Certainly! Here's a concise script outline for your video:

Scene 1: Introduction – (Narrator voice) "In Salesforce, when you need to provide users a list of choices, you have two main options: a Picklist or a Lookup to another object. Choosing the right one impacts both user experience and data management."

Scene 2: Picklist Explanation – (On screen: image of a form with a dropdown) Narrator: "Picklists are dropdown menus with predefined values. They're great for short, relatively static lists of options. For example, a 'Priority' picklist with High, Medium, Low. Picklists ensure consistency – users can only pick from values you define."

Scene 3: Custom Object Lookup – (On screen: diagram of two related objects) Narrator: "But what if the list of values is long, frequently changing, or has additional data associated? That's when a custom object might be better. Instead of a picklist of product codes that changes often, you could have a 'Product' object. Users then select a product via a lookup field. This way, when a new product is added, you just create a new record – no need to modify picklist settings."

Scene 4: Comparison – Narrator: "Use a picklist for simple, unchanging lists – it's quick and keeps everything in one place. Use a lookup to another object for complex or growing lists – it's more scalable and you can store extra info on those lookup records." (On screen: side-by-side checklist comparing features).

Scene 5: Conclusion – Narrator: "By choosing the right method, you make data entry easier and ensure your data stays clean. Picklists for simplicity, custom objects for flexibility – and now you know when to use each!"

(Video ends with your company logo or Salesforce logo and a thank you.)

This script should help your team understand the trade-offs in an engaging way. Good luck with your training!"

By using picklists and record types wisely, you've optimized the structure of your data and the user interface around it. Users see only relevant options, data stays consistent, and as an admin or architect you retain control over schema evolution. Now, we'll turn to another powerful way to manage complexity without code: using **Custom Metadata Types and Custom Settings** to store configuration and logical rules in a declarative fashion.

LEVERAGING CUSTOM METADATA AND CUSTOM SETTINGS

As your Salesforce org grows in functionality, you may find yourself needing to store configurations or preferences that guide how the system behaves. Hard-coding these values (in validation rules, formulas, or flows) can make maintenance cumbersome – every change requires modifying the rule or automation. Salesforce provides declarative objects specifically for these scenarios: **Custom Metadata Types** and **Custom Settings**. These act as storage for metadata and configuration data that you can reference in declarative logic. In this section, we'll explore the differences between them and best practices for using each to build flexible, maintainable declarative architectures.

Custom Settings come in two flavors: **List Custom Settings** and **Hierarchy Custom Settings**. A custom setting is like a custom object, but it's specifically meant for configuration data. One big advantage is that custom setting data is cached, making it efficient to access without hitting governor limits like normal SOQL queries (particularly in Apex, but even flows can benefit). Custom settings can be accessed in formulas (hierarchy settings can) and in validation rules via the special $Setup global variable.

List Custom Settings: This is essentially a simple table of data – you can think of it as a custom object where all users usually see the same values. There's no per-user or per-profile variation; any record in a list setting is like a constant. For example, you might have a "Sales Targets" custom setting that holds a target revenue number for the whole org. List settings are often used in flows or Apex to avoid hardcoding something like an "approved email domain" or a threshold value.

Hierarchy Custom Settings: These are unique in that they can have data at multiple levels: organization default, profile, and user. When you reference a hierarchy setting in a formula or validation rule via $Setup.MySetting__c.Field__c, Salesforce will first look for a value at the current user level, if none then at their profile level, if none then at org default. This is extremely handy to create per-user or per-profile preferences without writing code. For instance, you could have a hierarchy custom setting "Approval Limits" with a field "Max_Approval_Amount__c". For each profile, you set a different value (Managers = $5,000, Directors = $20,000, etc.). Then a validation rule on an Expense object could say "if Amount > $Setup.ApprovalLimits__c.Max_Approval_Amount__c, throw error". This way, the rule dynamically adapts to who the user is by looking at their profile's configured limit. If those limits change, you just update the custom setting records, not the rule itself.

Custom Metadata Types are a newer and often preferred way to store configuration **as metadata.** You define a custom metadata type (similar to creating a custom object, but flagged as metadata). The records of a custom metadata type are considered metadata themselves. What does that mean practically?

You can deploy custom metadata **records** from sandbox to production using change sets or packages (whereas custom setting data generally doesn't deploy – you'd have to load it separately in each environment).

Custom metadata records can be packaged and distributed (good for ISVs sharing config).

They can be referenced in formulas using the $CustomMetadata global variable (with some limitations, like you have to reference a specific record by name).

They can be accessed in Apex without consuming SOQL limits (there's a specific API for it), and in flows via a Get Records element.

Importantly, because they are metadata, you can treat configuration more like part of your application version. For example, a set of business rules or a mapping table can be versioned and deployed along with your app changes.

A typical use case for custom metadata: Let's say you have multiple regions and each region has slightly different discount rules. Instead of writing a bunch of IF statements in a formula or building a complex flow, you could create a custom metadata type "Region Discount Rule" with fields like Region (text) and Max_Discount_Percent. Create records for each region (NA = 20%, EMEA = 15%, etc.). Then in a validation rule or flow, you can reference "the Max_Discount_Percent where Region = (the record's region)". If these numbers change, you update the custom metadata records. If you add a new region, just add a record. No need to touch the validation rule or flow logic – they're data-driven by the custom metadata.

Another example: storing a list of values or settings that control a flow's behavior. Maybe you have a flow that routes cases to queues. Instead of embedding the queue names in the flow, you could have a custom metadata type with records like "Case Type = Inquiry -> Queue = Inquiry Queue". The flow just looks up the matching record and finds the queue. When you want to change routing, you edit the data, not the flow.

One more powerful aspect – custom metadata can have **relationships to other metadata.** You can create a relationship field on a custom metadata type that looks up to another custom metadata type or even certain Salesforce metadata like an Object or Field (in limited ways). This is an advanced feature, but it can let you create quite complex metadata-driven logic (like mapping fields between objects by storing field API names in custom metadata records, etc.).

When deciding between custom settings and custom metadata, consider:

Do you need the data to be deployable? If yes, custom metadata is the way (since custom setting data isn't included in deployments by default).

Volume and access patterns: Custom settings can be altered directly in prod by users/admins (like editing a list of values on the fly). Custom metadata can also be edited in prod (if you have authoring permissions), but often you treat them more like setup records. Technically, you can allow end users to edit custom metadata via a custom UI or something, but out-of-the-box it's more admin-facing (just like editing a custom object record in Setup).

Hierarchy needs: Only custom settings (hierarchy type) give you that per-user/profile layer. If you need that, you might use a hierarchy custom setting; there's no exact analog in custom metadata.

Formula access: In formulas and validation rules, you can reference hierarchy custom settings easily (Salesforce provides a syntax $Setup.-CustomSettingName__c.Field__c). For custom metadata, referencing them in formulas was historically not possible, but Salesforce introduced the ability to reference specific custom metadata records via $CustomMetadata.TypeApiName.RecordName.Field__c. That means if you have a single record you want to use (like an org-wide constant), it's doable. But you can't dynamically choose a record in a formula (no lookups in formula logic). So if your use case is something like "different threshold per record type", you can't have the formula dynamically pick a custom metadata record for that record type (because formula can't query). In such cases, you'd use a custom setting (with hierarchy or maybe a list custom setting and do a workaround). Or more commonly, you'd handle it in a Flow or Apex which can do the query on custom metadata.

Governance and packaging: For managed packages (ISV apps), custom metadata is gold because you can include some default records that customer orgs can then modify (if you allow) or not. Custom settings data in a package is not easily handled the same way.

In practice, **Custom Metadata Types** have become the go-to for most new declarative configuration storage needs because of their deployability and versatility. Many Salesforce features themselves now use custom metadata behind the scenes. However, **Hierarchy Custom**

Settings remain very useful for things like user-specific or profile-specific toggles (for example, a hierarchy setting to turn off certain validation rules for a specific integration user – you can set that integration user's custom setting record to bypass and have your validation rule check it).

A best practice is to externalize any "magic numbers", "constants", or decision tables out of your automations and into custom settings/metadata. That makes your automations simpler and your logic transparent. For example, if a discount approval flow has different approval steps based on product family, instead of building a giant decision element with every product family hardcoded, you could maintain a custom metadata table of "Product Family -> Approval Team". The flow just looks it up. When the approval team changes, no need to modify the flow, just change the data.

Pro Tip: Document your custom settings and custom metadata! Since these are by nature somewhat behind-the-scenes, make sure to include descriptions for each field and record explaining what they do. For instance, if you have a custom metadata "Validation Rule Controls" with a boolean field "BypassRuleA", note in the description which validation rule it influences. This way, if someone sees a record with BypassRuleA = true, they know it correlates to, say, the "Account must have phone" rule being bypassed for certain cases. Good documentation prevents confusion, especially as team members come and go.

ChatGPT Prompt (Image Generation):

User: "Can you generate a simple diagram showing how a custom metadata type controls a flow's behavior (for example, a flow that looks up a discount percentage from custom metadata)?"

ChatGPT: "Sure. Imagine a diagram with three elements: on the left, an icon representing a **Custom Metadata Type** called "Discount Setting" with a field "MaxDiscount". In the middle, a **Flow** icon showing a decision element or an assignment that fetches the MaxDiscount value. On the right, an icon of a **Process** or Outcome. The flow pulls the MaxDiscount from the custom metadata and then uses it in a decision (e.g., "if discount > MaxDiscount then route for approval").

Figure: In this conceptual diagram, a Flow (center) retrieves a value from a Custom Metadata Type (left) and uses it to determine an outcome (right). The Custom Metadata Type record might say "Max Discount = 20%" for a given context, and the Flow checks the discount on a record against that value to decide if approval is needed. This setup allows an admin to change the threshold in the custom metadata without modifying the flow itself.

"In the figure, you can see how the custom metadata acts as a configurable source of truth for the flow. If the business later decides the max discount should be 25%, you'd simply update the custom metadata record, and the flow will automatically use the new value on the next run. This removes any hardcoding from the flow."

By leveraging custom settings and custom metadata types, you introduce a layer of abstraction in your data model – separating what might change from how the system operates. This makes your configurations truly declarative and easy to adjust on the fly or deploy in a governed manner. In the final section, we'll discuss designing your data model with an eye towards **scalability and flexibility**, bringing together many of the themes we've touched on and preparing you for growth and change in your Salesforce org.

DESIGNING FOR SCALABILITY AND FLEXIBILITY

One thing constant in business (and in Salesforce orgs) is change. Data volumes grow, new requirements emerge, and what worked for 100 records may choke on 1,000,000 records. Designing your data model for scalability and flexibility from the outset can save you from painful refactoring later. In this final section of the chapter, we'll cover strategies to ensure your data architecture can scale in terms of volume, complexity, and evolving business needs.

Think Big (Data) but Start Small: When designing objects and fields, consider not only the records you have today but what you might have in a few years if the system is successful. If an object could realistically grow to millions of records (for example, a "Log" or "Transaction" object in a high-volume scenario), plan for how to manage that:

Indexes: Identify which fields are likely to be used in filters and queries (like an external ID, record owner, status, etc.). Mark them as **External ID** or **Unique** where appropriate, as that often indexes the field. Even without that, you can request Salesforce add custom indexes on selective fields via support (within certain limits). Indexed fields make queries and list views scale better as data grows.

Skinny Tables & Big Objects: For extremely large volumes on core objects, Salesforce has features like Skinny Tables (an internal performance optimization Salesforce support can enable) – you don't manage these directly, but being aware to ask Salesforce when needed is good. **Big Objects** are a special type of object for immutable, archival data that can scale to billions of records with some query limitations; if you anticipate needing to store *huge* data for historical purposes (like decades of logs), you might consider a Big Object for that use case early on.

Data Archiving Strategy: Design your model to accommodate archiving old data. Perhaps any object that will accumulate records should have a date field that you can use to filter out or move out old records after X years. Some orgs create parallel archive objects or use an external system or the Salesforce **Data Archive** (which at this time, Salesforce offers something called "Automatic Archive" for certain standard objects older than 18 months, and also big objects for custom archives). The key is, don't keep every record in the active tables if they're not needed – plan for cleanup jobs or archiving flows that run periodically.

Avoid Over-Complex Relationships: Salesforce has limits on how deep or how proliferated relationships can be in certain contexts (for instance, you can only go 5 levels deep in cross-object formulas, or in reporting you typically can join a primary object and a couple of related objects in one report type). If your data model has a very deep hierarchy (like 10 levels of parent-child), you might face challenges in reporting or just performance issues when aggregating. Sometimes a redesign where data is flattened or summarized can help. For example, instead of a chain of objects A -> B -> C -> D -> E, maybe you can combine a couple or use a different approach to avoid such depth.

Use Standard Objects When They Provide Scalability Benefits: Standard objects are tuned by Salesforce for large scale and come with special features. For instance, if you're dealing with sales data, using Opportunities (standard) means you can leverage forecasting, pipeline reports, etc., which are optimized. Standard objects often have better support for certain volume features (like Account and Contact have standard archiving for old tasks, etc.). Additionally, things like the Account object have the account hierarchy feature and built-in roll-ups (like total open opportunities on account), which you'd have to manually build if you used a custom "Company" object, for example. That said, don't force-fit into a standard object if it doesn't make sense (e.g., using "Opportunity" to store completely unrelated data just to get forecasting would be confusing). But do evaluate standard vs custom not just on current feature fit, but on future use cases. A good question is, "If we use a custom object instead of the standard one, what might we lose out on if later we want to use X Salesforce feature?"

Field Count and Page Performance: Salesforce allows up to 800 custom fields on an object (depending on edition). Just because you can doesn't mean you should! Objects with extremely high field counts can become unwieldy in page layouts and can slow down save times (especially if many formulas roll up or reference each other). If you find an object creeping into several hundreds of fields, consider if some of those should be moved to a related child object (particularly if they're rarely used fields or used by a subset of records). For example, if on the Account object you have a ton of fields related to "Survey Data" from customers, and it's not crucial for day-to-day, that could perhaps be a separate "Survey Response" child object rather than 50 survey fields on Account. Salesforce pages load all that data even if not shown, so slimmer objects mean faster performance. Lightning pages let you use tabs and conditional components to hide some content unless needed, which helps the user experience, but backend-wise the data is still fetched. Hence, organizing fields thoughtfully can contribute to scalability.

Flexibility through Configuration: We touched on this with custom metadata and settings. The more you can handle via configuration, the more adaptable your org is to change. If marketing decides to change what qualifies as a "Hot Lead" – if you've stored that threshold in a custom metadata rather than coded in 5 different places, you can adjust quickly. This concept extends to data model as well: for instance, say you're introducing a new product line that has some unique attributes – instead of adding a bunch of new fields just for those products on a generic object, you could consider using a related detail object for those attributes, which only gets used when needed. That way your main object isn't cluttered and performance stays better. The design should allow adding new data points without huge refactoring – think about using approaches like **Extensible Data Models** (Salesforce has something called Entity-Attribute-Value model where you have a generic "attribute" object if you need ultra-flexibility, though that can complicate querying – use sparingly).

Keep an Eye on Limits: Salesforce has various limits (like 15 lookup relationships per object, 40 total relationships per object including lookups and MDs , roll-up summary limits (25 per object via MD), etc.). While designing, be mindful of these. For example, if you foresee needing more than 25 roll-ups on an object, you might plan to do some via automation or reconsider your data model to reduce the need. If you think you'll approach the lookup limit, maybe some of those relationships can be consolidated or done via junctions. Hitting a limit can stall an otherwise straightforward expansion of functionality, so it's better to design knowing they exist.

Plan for Integration and External Systems: Scalability isn't just internal. If your data model will integrate with other systems, make sure you have stable unique identifiers (perhaps use meaningful External IDs on objects for matching with external data). Also consider using Platform Events or Change Data Capture for large-scale integrations instead of custom polling, as those are built for high volume. A flexible model might involve staging objects for imports/exports so that you can process data in batches. These are more integration architecture concerns, but they tie into data modeling (for instance, having an "Integration Log" object to track data coming in/out – if you design that early, you have monitoring in place for scale).

Regularly Revisit and Refactor: Finally, accept that an optimal data model today might need tweaks in the future. Build in time to do health-checks on your model. Salesforce provides the **Optimizer** and **Health Check** tools that can flag things like fields not used, or too many fields on layouts, etc. As your org grows, don't be afraid to refactor – merge two objects if they ended up being redundant, or split one object into two if it became overloaded. It's easier to refactor sooner than later, so when you see a scalability issue emerging (like reports timing out because an object has 5 million records and people are doing unindexed queries), address it with a design change promptly (maybe archive older records to a big object, or add that needed index, etc.).

Pro Tip: Establish a **data governance practice** that includes reviewing the data model. This might involve an Architecture Review Board that looks at new object proposals ("Do we really need a new custom object or can we extend an existing one? If new, let's confirm it doesn't duplicate something we have."). Also, consider using Salesforce's architect resources like the **Well-Architected Framework** and **Salesforce Optimizer** reports regularly to gauge your org's health. These can catch early signs of trouble, like too many fields with low fill rates (which might indicate you can remove or rework fields), or profiles with too many record types, etc.

By designing with scalability and flexibility in mind, you ensure your data model will gracefully handle growth and change. Your Salesforce org can evolve with your business instead of becoming a bottleneck. You've now covered everything from the basics of data modeling to advanced techniques for keeping your model clean and adaptable.

In the next chapter, you'll build on this solid data foundation and explore how to secure it and govern access – because a great data model must also be paired with robust security and permission strategies to truly succeed. Through security and access control best practices, you'll learn to protect the data architecture you've crafted, ensuring the right people see the right data. But for now, let's summarize what we've learned in this chapter.

SUMMARY

In this chapter, we explored Salesforce data modeling and architecture best practices from the ground up. We began with fundamental principles – leveraging standard objects when possible, choosing appropriate field types, and structuring objects and relationships in a way that mirrors real business entities while avoiding redundancy. We emphasized how normalization keeps data clean and how deliberate denormalization can improve usability in the right cases.

We then tackled the crucial decision of Master-Detail vs. Lookup relationships, learning how master-detail brings stricter coupling (with shared ownership and roll-up capabilities) while lookups offer flexibility. Guidelines were provided to choose the right model based on whether child records require parents, shared security, roll-up summaries, or cascade deletes. This set the stage for modeling more complex relationships.

For many-to-many needs, we introduced **junction objects** and saw best practices on how to implement them (like using two master-detail fields for full integrity, or validation rules to prevent duplicate junction entries). A visual example illustrated how a junction object connects two objects, enabling each to have many related records through the junction. We also touched on hierarchical relationships using self-lookups, noting how they enable structures like account hierarchies or organizational charts. Throughout, maintaining clarity and data integrity in these models was a key theme – for instance, adding unique constraints on junctions or using Schema Builder to double-check our relationships.

Data integrity and quality got a dedicated focus with declarative tools. We covered how **validation rules** act as gatekeepers to enforce business rules at the point of data entry, preventing bad data from ever hitting the database. We also looked at Salesforce's duplicate management features that use matching and duplicate rules to either block or alert on potential duplicates, thereby protecting the uniqueness of key records. Simple measures like required fields and unique field constraints play a big role in keeping data clean, and we highlighted using reports and dashboards to continuously monitor data quality over time. The message was clear: a well-modeled system must also proactively preserve the quality of its data.

We then considered schema optimization via **picklists and record types.** Picklists ensure consistency and drive better reporting by restricting inputs to valid choices. We discussed the use of global picklist value sets to manage options centrally, dependent picklists to create smart cascades of options (like Country -> State), and cautioned against the overuse of multi-select picklists due to their limitations. Record types emerged as a powerful way to tailor one object to multiple processes – enabling different page layouts and picklist values for different contexts, all while using the same underlying object. With record types, we underscored the importance of not going overboard: use them when needed for genuinely distinct scenarios, but avoid proliferating record types for minor differences to keep things simple.

The chapter also introduced advanced configuration storage through **Custom Metadata Types and Custom Settings.** We learned that custom metadata types are deployable, packageable, and act as part of your application's metadata (making them ideal for storing things like business rules, mappings, or constants that might change and need to be promoted through environments). Custom settings, especially hierarchy type, shine for per-user or org-wide preferences accessible in formulas (like setting threshold values or feature toggles). By moving logic out of hardcoded expressions and into these data-based tables, we gained agility – our org can adapt by editing a config record rather than rewriting a validation rule or flow. This approach makes our declarative solutions more maintainable and scalable in the long run.

Finally, we wrapped up with strategies for designing for **scalability and flexibility.** Anticipating growth means thinking about indexing fields for large data volumes, planning archival strategies for old records, and keeping objects lean to maintain performance. We stressed using standard objects when they provide a robust framework (to leverage Salesforce's optimizations) and revisiting the data model periodically as the org evolves. By being mindful of Salesforce limits and patterns (like the number of relationships or the depth of object hierarchies), and by implementing governance around the data model (ensuring no two teams accidentally create duplicate objects for the same purpose, for example), we set the stage for an architecture that can expand smoothly. In essence, we aimed to future-proof the data model – making it both sturdy and adaptable.

Throughout the chapter, we also saw how ChatGPT can assist at every step: from answering design questions in a friendly conversational way, to reviewing an uploaded data model for potential issues, to generating explanatory content like diagrams or video scripts to help train your team. This highlights an ongoing theme: while Salesforce provides the tools to configure without code, leveraging AI tools like ChatGPT can further accelerate solution design and knowledge sharing in your organization.

By mastering the content of this chapter, you've gained a 360-degree view of declarative data modeling best practices on the Salesforce platform. With a well-modeled and well-managed data architecture in place, you're ready to tackle the next challenge: ensuring that data is appropriately secured and that access is controlled. In the next chapter, we will build upon this foundation and dive into **security and access control**, exploring how to protect your data model with profiles, permission sets, role hierarchies, and sharing rules. A solid data structure paired with robust security forms the bedrock of any enterprise-grade Salesforce implementation, and you're now equipped to pursue that goal.

6

SECURITY AND ACCESS CONTROL

Security in Salesforce is **paramount** – in fact, "Trust" is Salesforce's number one value. This chapter focuses on how to uphold that trust by using declarative configuration to enforce robust security and compliance. We'll explore practical, *clicks-not-code* strategies to ensure the right people have the right access (and no more). In an enterprise Salesforce implementation, data security is not just an IT concern but a business imperative: customers and stakeholders expect that sensitive information is safe. A friendly, strategic approach to security can empower users to do their jobs efficiently **while** safeguarding data. By the end of this chapter, you'll have a toolkit of best practices to configure Salesforce security in a way that is both user-friendly and rigorously compliant.

At the core of Salesforce security lies the **principle of least privilege**. This principle – a cornerstone of modern cybersecurity – means granting users only the access **absolutely necessary** to perform their role, and nothing beyond. Embracing least privilege keeps your org locked down by default, opening access only as needed for business processes. In practice, this translates to carefully setting what data a user can see (or edit), which apps they can use, and which actions they can perform. Salesforce's declarative security model makes it possible to enforce least privilege with clicks: you define rules and permissions, and the platform consistently applies them. A key theme you'll see is that good security is **granular** and layered – we can finely tune access at the org, object, field, and record levels to meet any scenario.

We'll start with an overview of Salesforce's security architecture – the fundamental building blocks like **Organization-Wide Defaults** (OWD), roles, profiles, and permission sets. These components work in concert to determine who can access what. Understanding their interplay is critical: for example, just because someone is a manager in the role hierarchy doesn't mean they can see every field on an object – field-level security might restrict certain sensitive fields. Through real-world scenarios, we'll illustrate how these pieces fit together to enforce business rules (and how misconfigurations can inadvertently open holes). This foundation sets the stage for mastering more advanced tools.

Next, we'll dive deeper into **Profiles vs. Permission Sets and Permission Set Groups**. Historically, profiles were the main way to assign permissions, but today's best practice is to use profiles for baseline access and then layer on permission sets for specifics. We'll discuss why many orgs are moving to a "one profile per user type" model with granular permission sets granting additional access. Permission set groups help bundle those permission sets logically (for example, grouping all "Sales Team Access" perms together). You'll see how this approach offers flexibility and scalability, especially in large enterprises where managing dozens of profiles becomes unwieldy. We'll provide tips on transitioning to permission sets and how to decide what belongs in a profile versus a permission set.

From there, we'll tackle **advanced record-level and field-level security.** Salesforce provides multiple ways to control record visibility: starting from OWD (which set the default access for each object's records), then **role hierarchies** (to open up access vertically for managers), and **sharing rules** (to open access laterally based on criteria or ownership). We'll cover when to use each mechanism and how to design a sharing model that meets complex requirements without violating least privilege. On field-level security, we'll show how to lock down particularly sensitive fields (like salary or SSN) so that even if a user can see a record, they can't see or edit those fields. This section will include *practical techniques*, like using criteria-based sharing rules to share only "Need-to-know" records and using field-level security instead of relying on page layout tricks. You'll learn how to answer questions like, "Why can't User X see Field Y on Record Z?" by methodically checking each layer of access.

Salesforce is continually evolving its security features, so we'll also explore **Restriction Rules and other modern access control features.** Restriction Rules are a newer tool that let you *remove* access to certain records for specified users, even if those records would ordinarily be visible. This is a powerful way to fine-tune visibility in scenarios where the standard sharing model is too permissive. We'll explain how restriction rules work (for example, hiding certain records from interns or external users in a Public Read object), and provide guidance on using them safely. Additionally, we'll highlight other recent enhancements – such as scoping rules (which affect search and lookup visibility), and muting permission sets (which let you turn off specific permissions within a permission set group). These "modern" features give admins a more nuanced control over access without writing code. We'll walk through use cases to illustrate their benefits.

No security discussion is complete without covering **compliance and health checks**. In the fifth section, we'll show how to use Salesforce's built-in tools to **assess and improve your org's security posture**. The **Security Health Check** is your dashboard for org security settings: it evaluates your password policies, session settings, network access configuration, and more against Salesforce's recommended baseline (or your own custom baseline). We'll demonstrate how to interpret the Health Check score and address high-risk settings (for instance, if your password expiration is too long or API security settings are lax). Beyond Health Check, we'll discuss strategies for meeting common regulatory requirements declaratively – such as GDPR's data access rules, HIPAA's data protection standards, or SOX's audit trail needs. This includes things like setting up field history tracking for critical fields, using **Data Classification** to label sensitive fields, and scheduling regular permission reviews. Our goal is to show that being secure isn't a one-time setup; it's an ongoing, proactive process that you can manage with the right Salesforce tools and governance practices.

The chapter then turns to **Identity Management and Authentication** best practices. Even the best sharing model won't protect your org if bad actors can log in as valid users. We'll cover how to harden login and identity aspects of Salesforce declaratively. This means implementing **Multi-Factor Authentication (MFA)** (now a required standard for Salesforce users), setting up login IP ranges and login hour restrictions via profiles, and tweaking session settings to strike the right balance between security and usability. You'll learn, for example, how to enforce that certain high-privilege users can only log in from the office network, or how to require re-verification if a session is high-risk. We'll also discuss single sign-on (SSO) integration at a high level – while SSO involves outside systems, the configuration in Salesforce is largely declarative and can enhance security by centralizing authentication. Through scenarios, we'll illustrate the importance of a robust identity strategy: for instance, how MFA could have prevented a phishing-induced breach, or how an expired login IP range might block a traveling employee (and how to handle that gracefully). By the end of this section, you should be confident in configuring Salesforce's identity settings so that **only the right people, with the right verification, can get into your org.**

Finally, we'll address **Salesforce Platform Encryption and data protection.** Data-level protection is the last line of defense – even if someone manages to bypass other controls or access a database backup, encryption ensures they can't read sensitive data. We will explain how **Platform Encryption** (part of Salesforce Shield) lets you encrypt fields at rest with minimal impact on users. You'll learn when to consider encryption (for example, storing government ID numbers, health records, or other highly sensitive PII), and what trade-offs encryption brings (such as certain fields being un-sortable or un-searchable when encrypted). We'll guide you through enabling encryption on standard and custom fields declaratively, managing encryption keys (including options like BYOK – Bring Your Own Key), and the importance of key management policies like regular key rotation. Importantly, we'll discuss how encryption fits into a broader data protection strategy – complementing access controls, not replacing them. With real examples, we'll show how a company can protect sensitive customer data by encrypting it, while still allowing business processes to run smoothly. We'll also touch on related data protection tools like **Data Mask (for sandbox data)** and **Event Monitoring** for detecting suspicious data access, to give you a holistic view of keeping data safe.

Throughout the chapter, we will include **ChatGPT prompt callouts** to demonstrate how AI can assist you in applying these best practices. You'll see different interaction methods – from traditional typing queries, to voice commands, to uploading files for analysis – showcasing how ChatGPT (and tools like the new Operator agent and Sora voice interface) can be a co-pilot for Salesforce security configuration. These examples will range from getting quick explanations ("Why can't this user see that record?") to generating policy templates, analyzing org data exports for potential risks, and even creating simple diagrams of a sharing model. Keep an eye out for **Pro Tips** as well – these are insights for experienced admins and developers to level-up their security configuration, or to avoid common pitfalls that might not be obvious.

In short, this chapter is a comprehensive guide to declarative security on Salesforce. It balances **friendly guidance** (so you can explain these concepts to stakeholders in plain language) with **strategic rigor** (so you can design a security model that meets enterprise standards). We'll proceed step by step: starting from fundamentals and building up to advanced topics. By the end, you should feel confident that you can configure a Salesforce org where users have *exactly* the access they need, data is properly shielded at every layer, and compliance checkboxes are confidently checked – all without writing a single line of code. Let's begin with the fundamentals of Salesforce security and see how these building blocks form the foundation of a secure org.

UNDERSTANDING SALESFORCE SECURITY FUNDAMENTALS

To design a secure Salesforce implementation, you must understand the **layers** and components of its security model. Salesforce security is often described as a layered onion or pyramid – each layer addresses a different aspect of access control. At a high level, there are *four key layers*: **organization**, **object**, **field**, and **record**. In this section, we'll break down the foundational elements that correspond to these layers: org-wide defaults, roles, profiles, and permission sets. These are the basic tools every Salesforce professional should know cold. We'll clarify what each one does, how they differ, and how they work together to enforce the principle of least privilege. Consider this the "security anatomy" of a Salesforce org.

Organization-Wide Defaults (OWD) set the baseline *record-level* access for all users in your org for each object. An org-wide default is the most restrictive level of access for records that are *not* owned by the user. For example, if the Account object's OWD is **Private**, it means by default users cannot see each other's Account records (unless they own them or are granted access via another mechanism). If it's **Public Read/Write**, everyone can see and edit all Account records (bad for least privilege, generally!). OWD is where you implement the default "lockdown." Best practice: start as restrictive as possible (usually Private or at most Public Read-Only) and then open up access only as needed. OWD applies across the board, but it can be overridden by more granular sharing settings (which we'll discuss soon). When thinking about OWD, ask: "If I did nothing else, what's the minimum access everyone should have to each object's records?"

Roles and the Role Hierarchy come next and deal with record-level access in a hierarchical way. A *role* in Salesforce typically represents a position in an org chart or data access hierarchy (e.g., Sales Rep, Sales Manager, VP of Sales). The Role Hierarchy is a tree that often mirrors management chains. By default, Salesforce uses roles to allow managers to see records owned by their subordinates. If role hierarchy access is enabled (you can control this per object with "Grant Access Using Hierarchies"), then users in higher roles automatically get access to records owned by users in lower roles, if the OWD is Private or otherwise restricted. This makes sure, for instance, a support manager can see the cases that her support agents own, without needing each agent to manually share them. **Important:** The role hierarchy is about **data access**, not necessarily organizational prestige. Sometimes, data access hierarchies differ from the company org chart. You might have a role called "CEO" at the top, but you might also have separate branches for different data segments (like separate branch for "International Sales" vs "Domestic Sales" that meet at the CEO). When designing roles, focus on grouping users by who needs visibility to whose records, rather than just job titles.

Setting up a clear role hierarchy enables a lot of "automatic" sharing upwards. However, too broad a role (like everyone under one branch) might overshare. A Pro Tip for experienced architects: **don't overuse the role hierarchy**. Use it to reflect natural supervisory lines for data, but not for every nuance. If two roles at the same level need to share with each other, that's what sharing rules are for, not contorting the hierarchy. Also, minimal roles can sometimes suffice – e.g., if your org is small, you might give everyone the same role if peer-to-peer sharing isn't a concern, then introduce more roles as you grow.

Profiles are the cornerstone of object-level (and field-level) access, as well as other permissions. Every user **must** be assigned exactly one profile. A profile defines what the user can do *in general* in Salesforce: which apps they see, which objects they can access (read/create/edit/delete permissions on each object), which fields on those objects they can see or edit, and a host of system permissions (like "Run Reports", "Mass Email", or "Modify All Data") and login settings. You can think of a profile as a user's job description in the org. A *Salesforce developer* analogy: if the org were an operating system, the profile is the user's access level (like Admin vs Standard User) plus their application permissions.

Historically, profiles were the only way to grant object and field permissions – leading to many profiles (Sales, Sales Manager, Support, Support Tier 2, etc.) each slightly different. Nowadays, we aim for a simpler use of profiles. A common best practice is to create profiles that grant a **minimum baseline** of access for a broad role or for the entire org, and then layer on additional access with permission sets. Salesforce even provides a standard "**Minimum Access – Salesforce**" profile which has only the very basic permissions to log in and view the org's structure, with no object permissions at all – this is a great starting point to clone for custom minimal profiles. The idea is to reduce the number of profiles to a few "base personas" and handle everything else through permission sets (which we'll get to in a moment). Profiles also control some things permission sets do not, like which page layouts or record types a user gets on an object, or login hours and IP range restrictions.

Thus, profiles still play an important role in security and shouldn't be eliminated entirely, but we try to keep their permissions limited to avoid profile sprawl.

For example, imagine a profile called **"Sales User"** that grants access to core objects like Accounts, Contacts, Opportunities (perhaps read/create/edit on these) and basic user permissions. All salespeople get this profile. Now, some salespeople might need extra permissions (maybe some can import leads, or some are also part of a beta program requiring additional object access). Instead of creating a new profile variant, you'd give those specific users a permission set with the extra access. That way "Sales User" profile remains the same for everyone, and exceptions are handled granularly. This approach greatly simplifies administration over time.

Permission Sets are the flexible, granular way to assign additional permissions to users **without changing their profile**. A permission set can contain many of the same settings as a profile (object perms, field perms, system perms, app and Apex class access, etc.), but users can have **multiple** permission sets. This is the crucial difference: one user = one profile, but one user can accumulate multiple permission sets. Permission sets are how you adhere to least privilege while still accommodating varying needs. The mantra is often *"profiles for base access, permission sets for everything else."* If a user's job changes or they take on extra responsibilities, you can just add or remove a permission set, rather than juggling profiles.

Permission sets shine in complex orgs. Let's say you have a group of users who need access to the **Inventory** object. Instead of creating a special "Profile with Inventory Access", you can create an "Inventory Access" permission set and assign it to those users (who otherwise might have different profiles). This decouples the special access from the base profile. It also encourages reuse: maybe some Customer Support reps and some Sales reps both need Inventory access – they might have different profiles, but you can give them the same permission set. This avoids duplicating a lot of settings in new profiles.

Another advantage: permission sets (and their big sibling, permission set groups) are deployable and packageable. Profiles can be trickier to manage across orgs because they're so monolithic. Permission sets allow modular deployment of permissions in changesets or packages, which is friendlier to ALM (Application Lifecycle Management).

It's worth noting that some older orgs have a lot of legacy profiles with overlapping permissions – cleaning that up with permission sets can be a project, but it pays dividends in manageability. As a **Pro Tip**: Consider using the **"Permission Analyzer"** (a free tool in setup) or simply a spreadsheet export of profile vs permission set matrices to plan a consolidation. Salesforce's User Access and Permissions Assistant (if installed) can help find redundant profiles and suggest permission sets.

Permission Set Groups are an advanced feature that allows you to bundle multiple permission sets into a single group for easier assignment. Think of it as creating a "role-based bundle" out of granular permission sets. For instance, you might have 5 permission sets related to different aspects of a Sales Manager's access (Lead Management, Opportunity Approvals, Dashboard Viewing, etc.). Assigning all 5 to each user is error-prone and tedious. Instead, you create a permission set group called "Sales Manager Full Access" and add those 5 permission sets into it. Now you assign just the group to the user. Under the hood, the user gains all the perms from those 5 sets. The benefit is not only convenience – you can also **mute** permissions within a group using a Muting Permission Set. Muting lets you turn **off** specific permissions that are included in one of the group's sets. For example, maybe one of the 5 permission sets grants "Delete" on Opportunities, but you want to disable deletes even for Sales Managers except a few. You could mute the delete permission in the group, so none of the users with the group have it, even though it's technically in one component permission set.

In summary, profiles, permission sets, and permission set groups collectively implement object and field security in a layered way: profile = base, permission sets = add-ons, groups = convenient collections of add-ons. These work alongside OWD and roles which, as we discussed, govern record access at large. We'll soon see how when a user tries to view a record, Salesforce checks *all* these rules: **Does the user's profile/perm sets allow access to the object and field?** If yes, **do they have access to that specific record (via ownership, role hierarchy, or**

sharing)? If yes, they see it; if any check fails, they don't. It's a multi-layer filter.

Let's bring this to life with a scenario. Suppose Maria is a new **Inside Sales rep** at a company. By policy, all Inside Sales reps should only see their own Opportunities (OWD for Opportunity = Private). They report to a Sales Manager, who should see all Opportunities of their team (role hierarchy will grant that). Maria's profile (perhaps a clone of "Sales User") gives her Create/Read/Edit on Opportunity object and relevant fields, but **no access to the "Discount Approver" field** on Opportunity (that field is hidden via field-level security because reps shouldn't set or even see the approval status). Maria also has a permission set called "Leads Management" because she's handling leads too, which grants Create/Read on Leads (her profile might not include Leads by default if not all reps handle them). Now, when Maria logs in, she can only create and see her own Opportunity records. If she tries to view another rep's Opportunity, she'll be blocked – OWD Private stops her, unless a sharing rule says otherwise. Her manager, however, in the role hierarchy above her, can see Maria's Opportunities automatically. Neither of them can see or edit the "Discount Approver" field because of field security (even the manager's profile doesn't grant that field, perhaps only Finance has it). This simple example shows multiple fundamentals working together: OWD (Private) + Role Hierarchy + Profile perms + Field security + one permission set. We'll expand on scenarios like this as we go.

Before moving on, let's pause to leverage ChatGPT as a quick helper for these fundamentals. ChatGPT can summarize and clarify concepts or even analyze our current org's use of these components if we provide some data.

ChatGPT Prompt (Text Prompt):

User: "In Salesforce, what's the difference between a Role and a Profile in terms of security, and how do they work together?"

ChatGPT: "Roles and Profiles serve different purposes in Salesforce security. A **Profile** is a user's base set of permissions – it defines what objects, fields, and general actions the user can access. Profiles control things like 'Can Jane create Leads? Can she view the Amount field on Opportunities?' Every user has one profile. On the other hand, a **Role** is part of the role hierarchy, which primarily affects **record access**. Roles don't grant permissions to do things with objects; instead, they open up visibility to records owned by or shared with users lower in the hierarchy. In essence, profiles say 'what' a user can do, while roles often determine 'whose records' they can see. The two work together: a user's profile might let them read Account records, but whether they can see a particular Account record could depend on their role (if that record is owned by someone below them, for example). Both must permit access for the user to actually view or edit a record – the profile gives the ability, and the role/hierarchy gives the opportunity by making the record visible."

In the scenario above, ChatGPT neatly explains that you need both the functional permission (from profile) and the data visibility (often from role/hierarchy or sharing) to truly access a record. This dual requirement is fundamental to Salesforce's security architecture.

Another way ChatGPT can help is by analyzing a file that describes your security setup. For instance, you might export a CSV of all profiles and their user counts, or a diagram of your role hierarchy. By uploading that to ChatGPT, you could ask for observations or simplifications.

ChatGPT Prompt (File Upload): The admin uploads a spreadsheet file listing profiles, number of users in each, and last login date per profile.

User: "I've uploaded a spreadsheet of our profiles and user counts. Can you identify if any profiles might be redundant or over-privileged?"

ChatGPT: "Looking at the data, I see you have 25 profiles, but 5 of them have only 1 user each (Profiles named 'Custom Sales Profile 2', 'Custom Sales Profile 3', etc.). Those single-user profiles could indicate over-specialization – perhaps those users could be consolidated into a common profile with permission sets handling differences. Also, the profile 'System Administrator Clone' has 10 users, which is quite high for an admin-level profile – best practice is to limit System Admin privileges. You might review if all 10 need full admin access or if some could be in a lower-privileged profile with elevated permission sets for just the tasks they need. In summary, consider cleaning up the multiple custom sales profiles by merging them using permission sets, and audit the users with the admin clone profile."

Pro Tip: Regularly review your Roles and Profiles structure. As your org evolves, you might find roles or profiles that were created for a specific purpose are no longer needed. Eliminating those (or merging them) reduces complexity. A leaner role/profile hierarchy is easier to secure. Use Salesforce's "Role Hierarchy" setup UI and Profile list view to identify anomalies (like a role with no users, or profiles with very few users that might be merged). Also, document the purpose of each profile and role – that way, when someone asks "why does this profile exist?", you have an answer, or you realize it can be removed.

Now that we've covered the fundamentals – OWD, roles, profiles, and permission sets – you should have a solid base. We've seen **what** each component controls and shared some guiding principles (like keep OWD restrictive, minimize profiles, use roles for hierarchy-based visibility, etc.). In the next section, we'll build on this by examining the modern best practices around Profiles and Permission Sets in more depth. Specifically, we'll contrast the old approach of profiles-for-everything with the new approach of fine-grained permission sets, and introduce permission set groups more formally. This will set us up to manage user permissions in a scalable way as your org grows.

Now that the basics are clear, let's go deeper into how **Profiles** compare with **Permission Sets** in practice, and why many organizations are shifting to a permission-set-centric model for access control.

PROFILES VS. PERMISSION SETS AND PERMISSION SET GROUPS

As Salesforce implementations mature, a common question arises: *How do we manage user permissions at scale without losing our minds?* In early stages, admins often create profiles for each distinct role ("Sales", "Support", "Manager", etc.). Over time, this can explode into dozens of profiles to meet slightly different needs, which becomes a headache to maintain. Enter **permission sets** and **permission set groups**, which have changed the game in how we assign and organize permissions. In this section, we'll do an in-depth comparison between using profiles versus using permission sets (and groups) as the primary way to grant permissions. Our goal is to articulate the modern best practice: use profiles only for a minimal baseline and any settings that can't be handled elsewhere, and use permission sets for everything else. We'll explore scenarios illustrating the flexibility gained, and we'll provide guidance on migrating to this approach if your org is currently profile-heavy. We'll also highlight how to effectively use permission set groups to simplify user permission management even further.

Let's start by looking at the **limitations of profiles** when used alone. Imagine a mid-sized company with Sales reps, Sales managers, Support agents, and Support supervisors. If we rely solely on profiles, we might end up with profiles like: *Sales Rep*, *Senior Sales Rep*, *Sales Manager*, *Support Agent Level 1*, *Support Agent Level 2*, *Support Supervisor*, etc. Each one may differ by a handful of permissions (e.g., Senior Sales Reps can delete Opportunities, Sales Reps cannot; Level 2 Support can export reports, Level 1 cannot). With each new slight variation, someone clones a profile and tweaks it. Before long, it's a maze – when a new hire comes in, which exact profile do they get? Are we sure those profile permissions are up to date with policy? The more profiles, the harder it is to audit and ensure consistency (not to mention deploying changes across many profiles is tedious).

Permission sets solve this by breaking down permissions into bite-sized pieces that can be mixed and matched. Using the same example, we could instead have a core profile for *All Sales Users* (covering common needs like read/write on Accounts, Contacts, basic rights) and a core profile for *All Support Users*. Then create permission sets for the variations: e.g., "Opportunity Delete Right" as a perm set that grants Delete on Opportunity; "Export Reports" perm set that grants the Export Reports permission; "Advanced Case Handling" perm set that maybe grants Edit on certain case fields or access to a QA app, etc. Now, a **Sales Manager** could be simply a person with the Sales User profile **plus** the "Opportunity Delete Right" permission set (if that's the key difference from reps), and maybe "Manage Team Reports" permission set if they have extra reporting rights. A **Support Level 2** could be a Support User profile **plus** "Export Reports" perm set. In effect, we modularize permissions.

The benefits of this modular approach are huge:

- **Flexibility:** If tomorrow you create a new role, you often can compose existing permission sets to fit them, rather than crafting a brand new profile from scratch.
- **Least Privilege Maintenance:** It's easier to review a small permission set for one specific access and decide if a user truly needs it, than to comb through a giant profile with hundreds of permissions to find the one thing to toggle. Permission sets encourage more intentional granting of additional rights.
- **Reduced Redundancy:** Many profiles often contain overlapping permissions. By extracting common permissions into a base profile or widely-used perm sets, you reduce duplication. For example, if "Export Reports" should be allowed for a bunch of different roles, giving them all one common perm set is simpler than ensuring each relevant profile has that box checked (and updating multiple profiles if you change your mind).
- **Ease of Deployment:** If you've ever tried to deploy profile changes via change sets, you know it's tricky – profiles tie to many metadata items and can cause issues if target org doesn't have exactly the same setup. Permission sets are self-contained and deploy much more cleanly between orgs (e.g., from Sandbox to Production).
- **User-Specific Fine Tuning:** Sometimes two users with the same job function still need a slight difference in permissions (maybe one person is temporarily filling in a higher role). Rather than clone a profile just for them, you can give them a unique combination of permission sets. This *personalization* is controlled and visible (you can always see which perm sets a user has in their User record), and can be removed when not needed.

One might ask, **"Why not just one profile for everyone?"** In theory, you could give everyone a super minimal profile (like that "Minimum Access" profile, which has almost nothing) and then build up **everything** with permission sets. In practice, profiles still handle a few things permission sets can't, and separating very distinct user categories via profiles can be convenient. For example, profiles are still used to assign

Page Layouts and **Record Types** by app, which permission sets don't do. If Sales and Support need different layouts on the Case object, you often use profile differences to achieve that. Also, profiles can restrict login by IP or hours – you might want a different policy for internal vs external users, achieved via profiles. And some orgs use profile to separate users by license type (e.g., a profile for Partner Community users vs internal users), which is necessary because a profile is tied to a user license. So, we're not eliminating profiles – we're right-sizing their role.

A modern best practice is to use as few profiles as possible, often aligning them to user *licenses* or broad categories. For instance, one approach is: all internal standard users get a base profile (with minimal access), all community/Experience Cloud users get another (since they often have a different license), all integration users (APIs) maybe another. Then manage the rest by permission sets. This way, if a new regulation comes that requires (say) all internal users to have a certain restriction, you have one or two profiles to update instead of 50.

Permission Set Groups further simplify the admin's life. Let's say, after going crazy with granular permission sets, you end up with 100 permission sets for all kinds of features (it happens over years of adding features). An individual user might need 10 of those to do their job, which is a lot of clicking to assign. Permission Set Groups let you bundle those 10 into one container. A great strategy is to create groups that map to logical roles in the business. For example, create a group "Call Center Agent Perms" that includes all permission sets a call center agent needs (Cases access, phone integration perms, etc.), and another "Call Center Supervisor Perms" group that maybe includes all of the agent perms plus some supervisor-only perms like quality dashboard access. Now assigning one group gives everything needed. This is very role-based access control (RBAC) oriented, and many compliance standards prefer RBAC because it's easier to audit – you're granting people job-based bundles instead of many individual entitlements. In Salesforce, we implement RBAC by using permission set groups as the representation of a job role's access. Internally, it's still lots of granular bits, but it's packaged neatly.

One extremely useful feature of permission set groups is **muting permission sets**. This is like a kill switch for specific perms within the group. Suppose in general, Salespeople have edit access on Opportunities, but an auditing rule says new Sales interns should have read-only access to Opps for their first month. You could have a group "Salesperson Perms" which includes a permission set that normally grants edit on Opportunities. For the interns, you might use the same group but apply a muting set in their group that removes the edit permission. In reality, you might also just create an "Intern" group separate, but muting can address finer adjustments without duplicating a whole new set just to subtract one permission.

It's worth highlighting that permission set groups and muting sets are declarative but advanced – they require careful governance. If overused, you could end up with as many groups as you had profiles. We advise using groups for truly standard bundles that correspond to well-defined roles or user clusters. Continue to use individual permission sets for one-off or very narrow grants that cross roles.

Let's illustrate with a **complex org scenario**: CloudCo has 1,000 users across Sales, Support, and Marketing. Initially, they had 15 profiles (Sales Rep, Sales Manager, Support Tier 1, Tier 2, etc., Marketing, etc.). Over time, exceptions led to 30 profiles. Managing this was painful – e.g., to give **Field X** on Account object to all sales and support, they had to edit 10 profiles. CloudCo decided to simplify. They introduced just 3 profiles: "Base Internal User", "Base Partner User", "System Administrator" (for actual admins). Every internal user moved to "Base Internal User" which has very limited access (no object permissions beyond maybe read on Accounts/Contacts). Then they created permission set groups: "Sales – Standard Access", "Sales – Manager Access", "Support – Tier 1", "Support – Tier 2", "Marketing – Standard", etc. Each group aggregates the necessary permission sets for that role. Now when a new Sales Manager joins, instead of picking one of 5 subtly different profiles, the admin: sets them on "Base Internal User" profile (which everyone has) and assigns two groups: "Sales – Standard" and "Sales – Manager". Those groups behind the scenes give read/edit on opportunities, some Apex classes, etc., and in the Manager case maybe extra dashboard viewing or approval authority. CloudCo also created a few stand-alone permission sets for things that cut across roles, like "Access Billing App" (some people in both Sales and Support need it). Those are assigned individually as needed. After this, CloudCo reduced profiles from 30 to 3, and the admin can easily see what access any user has by looking at their assigned permission sets and groups (which is conveniently listed in Setup > Users). Training the admin team on this model was important, but after that, it became much easier to audit permissions. If Security asks "who can delete Cases?", the admin can now check which permission set includes "Delete" on Case and see its assignments, rather than opening 30 profiles to find that permission.

One challenge in migrating to this model is **making sure no access gaps or overlaps appear**. Tools like the Salesforce Optimizer or third-party profile comparison tools are helpful. A measured approach is to start by identifying a profile that could be broken into profile + perm sets as a pilot. For example, find a profile that only differs from another by a couple permissions, then remove those extra perms from the profile and provide them via a permission set to only the relevant users. Over time, you can collapse multiple profiles into one.

Let's use ChatGPT to help craft a plan for such a migration:

ChatGPT Prompt (Operator Mode):

User: "We have 10 profiles for various sales roles. I want to consolidate them using permission sets and permission set groups. Can you outline a step-by-step plan?"

ChatGPT (as an AI Operator agent): "Sure! Here's a structured approach:

Inventory Current Profiles: List each sales-related profile and note their differences in object permissions, field permissions, and system permissions.

Identify Common Denominator: Determine the permissions *all* sales users need – that will form the basis of a new base profile (or confirm one of your profiles can serve as base after removing extras). For instance, all might need read/write Accounts and Contacts, create Leads, etc.

Create Base Sales Profile: Configure a profile with only the common permissions (perhaps clone a profile that was most restrictive among the group, or clone Minimum Access and add needed core perms). Assign a few test users to it in a sandbox to ensure it covers essential access.

Create Permission Sets for Variations: For each unique permission or set of permissions that only some roles had, create permission sets. For example, if only the "Sales Director" profile had Modify All on Opportunities for pipeline oversight, make a perm set "Opportunity – Modify All". Do this for each cluster of special perms (report export, certain field access, etc.).

Group into Role-Based Bundles: Create permission set groups that correspond to actual job roles – e.g. "Inside Sales Rep", "Field Sales Rep", "Sales Manager", "Sales Director". Add the relevant permission sets to each group (the ones that that role needs beyond the base profile). Use muting permission sets in the group if you need to *exclude* something that a component perm set includes.

Testing: Pick a few users from each existing profile and switch them to the new model in a sandbox or scratch org. Assign the base profile + appropriate group. Then test that they can do everything they could before (login as them or have them verify key tasks). Adjust the permission sets if something was missing.

Rollout: Once confident, update all users to the new profile and groups accordingly. It might be wise to do this in batches (e.g., one department at a time) and have support ready for any access issues to quickly assign missing permission sets.

Clean-Up: After migration, deactivate old profiles to prevent their use. Also, document the new structure so all admins know that e.g. "if you need to give a sales user extra rights, use these permission sets, don't clone profiles."

Continuous Review: Periodically review permission set assignments to ensure they still align with users' roles. The granularity of permission sets will make it easier to spot if someone has something they shouldn't (like why does a certain rep have the "Modify All Opportunities" perm set? Should they?).

This plan helps maintain least privilege and clarity by leveraging permission sets and groups for flexibility while keeping profiles simple."

This advice from ChatGPT aligns well with best practices. The idea is to gradually peel off the unique parts of profiles into permission sets.

One more cool thing: permission sets can now have user **expiration dates** as of recent releases. If you have contractors or temporary privileges (like giving someone extra access for a two-week project), you can assign a permission set and set it to expire on a date. This is fantastic for temporary least-privilege elevations – something that was very hard with profiles alone (you'd have to remember to remove them from a profile or change their profile later). We mention this because in enterprise security, the concept of "temporary access" with automatic removal is a big win for compliance (no more lingering access after it's needed).

Now, how do **permission sets and groups vs profiles** affect day-to-day user administration? One thing to note: with many permission sets, the user detail page can get crowded. Make use of the available "Permission Set Assignments" view and consider naming conventions to keep things clear (e.g., prefix all perm sets with their area, like "CRM – Export Reports" vs "CPQ – Approve Quotes" to distinguish domains). The clarity in naming helps when scanning a user's assigned sets.

It's also crucial to implement **governance** around who can create permission sets and groups. Too many cooks can create overlapping permission sets that do the same thing, causing confusion. A governed approach might designate a "Security Admin" or an Architect to approve new permission sets and ensure they fit into the overall scheme (versus someone creating a one-off permission set without considering reusability). Documenting which permission sets exist and their purpose in a central place (Confluence, a spreadsheet, or even using the Description field on the permission set) is very helpful for large orgs.

For experienced developers: consider using the Metadata API or SFDX to manage permission sets as part of version control. Profiles are often avoided in source control due to complexity, but permission sets are more amenable to being tracked as metadata. This way, changes to security config can go through code review in a Git repository, increasing transparency.

Before moving on, let's reflect with a quick Q&A style ChatGPT interaction that could clarify any lingering confusion between these approaches:

ChatGPT Prompt (Voice via Sora):

User (speaking): "I have a profile giving full access to an object, but I want most users to have read-only and only a few to have edit rights. What's the best way to do this declaratively?"

ChatGPT (responding in a friendly voice): "The best practice here is to use a combination of profile and permission set. Set the profile's permissions on that object to read-only (or even no access if some users shouldn't see it at all). This ensures everyone by default is limited. Then create a permission set that grants edit (and any other needed higher privileges on that object). Assign that permission set only to the specific users who require edit access. This way, you're following least privilege – nobody gets edit by default, and only those who truly need it have an extra permission set granting it. Over time, if someone no longer needs edit, you just remove that permission set from their user. This is much cleaner than cloning profiles for the sake of one permission difference."

The answer highlights how mixing profiles and perm sets can implement a tiered access scheme cleanly.

Pro Tip: *Embrace the "PermSets over Profiles" mindset, but don't forget to revisit old profiles.* If you migrate to using permission sets heavily, some permissions in the profiles might become unnecessary (because you moved them to perm sets). It's wise to gradually tighten the profiles by removing now-extraneous permissions – of course, test as you do so. The ultimate state might be profiles that are extremely restrictive (baseline only) and all actual access above that coming from perm sets. If you reach that nirvana, auditing permissions becomes far easier.

Now that we have a solid grasp of profiles vs permission sets and how leveraging permission set groups can simplify user management, we can confidently design a scalable permission architecture. With this in place, the next step is to ensure that even with proper object-level access, **sensitive data within those objects is protected** and that users only see the records they're supposed to. In other words, we move from "who can access Objects and fields" to "who can access specific Records and data values". That's where field-level security and record-level sharing come into play, which we'll cover in the next section.

We've sorted out how to manage *who gets what permissions*. Now let's focus on applying those permissions at a granular level: locking down fields and sharing records appropriately. In the next section, we'll explore advanced field-level and record-level security techniques to ensure that having access to an object doesn't mean seeing **all** the data within it.

IMPLEMENTING ADVANCED RECORD AND FIELD-LEVEL SECURITY

Even with object-level access properly configured through profiles and permission sets, we still have two crucial questions: **Can the user see or edit a particular field?** and **Can the user see a particular record?** Salesforce's security model gives us fine-grained control to answer those questions through field-level security and record-level sharing mechanisms. In this section, we'll dive into advanced techniques for protecting sensitive fields and controlling record visibility. This is where you enforce business rules like "Only HR can see Salary field" or "Partners should only see Accounts in their region" or "Managers can see their team's records but not others." We'll cover the full spectrum of record-sharing tools – criteria-based sharing rules, ownership-based sharing rules, manual sharing – and discuss how to use each effectively. We'll also discuss the role of the **"Grant Access Using Hierarchies"** setting and when you might turn that off for custom objects to tighten control. On the field side, we'll explore best practices for field-level security vs. other methods of hiding fields (and why field-level security is the real security). By the end of this section, you should be able to design a robust sharing model that ensures each user sees exactly the records (and fields) they should, and nothing more, aligning perfectly with least privilege.

Let's start with **field-level security** (FLS), often accessed via profiles and permission sets under "Field Permissions." Field-level security determines whether a user can see or edit the value in a given field on an object. This is independent of record ownership or sharing – even if a user has access to a record, if a field is hidden for them, it's as if that field is invisible (not on page layouts, not in reports results, not via API). For example, a profile might have "View All" on the Employee object, but you could mark the "Social Security Number" field as hidden for that profile – they'll see the employee records but the SSN field will be blank or not present for them. Field-level security is a **must-use** for sensitive personal or financial data. A classic mistake is assuming "if I remove the field from the page layout, users can't see it." Page layout is *UI* only – the field could still be accessible through reports, API, or other means if field-level security isn't also restricting it. **Always enforce through FLS**, not just layout, for true security.

FLS can be set on profiles or via permission sets (to give certain users access to fields others can't see). One approach is to have certain fields default to hidden for most profiles, and then create a permission set like "View Compensation Fields" to give HR or Managers the ability to see salary or bonus fields, for instance. This ties back to the prior section's advice: use permission sets for special cases so you don't have to clone profiles.

One nuance: some fields are "Required" at the database level (like a record's Name field) – those you obviously can't hide or else the object wouldn't function. But you can hide most custom fields and many standard fields. Also, FLS doesn't apply to formula fields in the sense that if a user can see the formula field, they implicitly see whatever the formula calculates (so if it uses a field they're not supposed to see, careful: formula might reveal info). Conversely, if they don't have access to a field a formula references, but *do* have access to the formula field, Salesforce will run the formula in system context, potentially exposing some info. Best practice: if a formula field includes sensitive info from another field, treat the formula field's visibility with the same sensitivity. (In sensitive cases, you might even mark the formula field as hidden and just use it for internal automation, not for user viewing.)

Now on to **record-level security**, often called the "Sharing Model". We touched on OWD and roles already, which are part of this. Let's recap and then extend: OWD sets baseline (e.g., "Accounts are Private by default"). Role hierarchy (if enabled for the object) opens up access upward. But what about access to peers or cross-functional? This is where **Sharing Rules** come in. There are two types of sharing rules: **Owner-based** and **Criteria-based.**

Owner-based Sharing Rules: These let you share records based on who owns them. For example, "Share all records owned by users in Role X with users in Role Y." Suppose your org has roles for each region's sales team. By default, a rep in Europe can't see a rep in Asia's opportunities (OWD private). But maybe you want all sales reps globally to share some leads. You could write a sharing rule: if owner's role = "Any Sales Rep" (or owner in role group "All Sales"), share with role "VP of Sales" or share with group "All Sales Reps" (yes, you can create a public group of all sales rep roles). However, owner-based rules are somewhat coarse – they share *everything* owned by certain users. They're good when teams own buckets of records that others need access to categorically.

Criteria-based Sharing Rules: These are more granular and based on field values in the record. For instance, "Share all **Account** records where Industry = 'Technology' with the **Tech Sales Team** public group." This doesn't care who owns the record; it cares about record data. Criteria-based rules are super useful for use cases like territory-like sharing (without using Territory Management feature), or sharing certain cases with another team (e.g., all high-priority cases share with the Escalation team). They're declarative: you choose the object, set a filter (criteria), and choose a target (roles, roles and subordinates, public groups, or queues) to share with, and what level of access (read or read/write). For example, "If Case Type = 'IT Support', share read/write with Role: IT Support Team Lead".

Manual sharing: This allows individual users to share individual records with others on an ad-hoc basis. If OWD is not public (private or read-only), users will see a "Share" button on records they own (or have certain access to), letting them grant access to specific people or groups. Manual sharing is often used for one-off needs, like a salesperson sharing a specific opportunity with a peer for collaboration. It's not a scalable strategy for large sets of data, but it's a nice user-driven flexibility. Administrators with proper permissions can also create manual shares via the UI or API if needed (for instance, if a single record needs special access outside of rules).

A note: manual sharing isn't available on all objects, especially some that have their own sharing mechanisms (like Campaigns use Campaign Member sharing, etc.) And if OWD is public read/write, obviously manual sharing isn't even needed (everyone already has access). Manual shares are stored in the "Share" tables (AccountShare, CustomObject__Share, etc.).

Salesforce also has specialized sharing constructs like **Account Teams** and **Case Teams.** Those are another declarative way for end-users to grant access: e.g., adding someone to an Account Team can give them read or edit on that account and optionally related opportunities. It's like manual sharing with structure and the ability to define default teams. We won't dive deep into teams here, but keep in mind they exist for collaborative scenarios. They operate within the sharing model (essentially when you add someone to a record's team, behind the scenes it creates a share for that user for that record, according to the team role's access settings).

Public Groups are your friend in sharing configurations. A public group is an admin-defined collection of users, roles, and/or roles & subordinates, and even other groups (they can contain groups). They are used to target sharing rules (and also for things like folder sharing for reports/dashboards, and queue membership). Rather than making a sharing rule per role, you often create a group that includes multiple roles or users, then share to that group. For example, group called "All Field Sales (Worldwide)" which includes roles: North America Sales, Europe Sales, Asia Sales. Then a single sharing rule can share records owned by Inside Sales (as a role) with group "All Field Sales" – meaning all field sales globally get to see those leads or accounts. This saves effort and is more maintainable (if a new region is added, update the group membership, not every rule).

Another advanced concept: **"Grant Access Using Hierarchies"** on sharing settings. By default, standard objects always have it on (you can't turn it off). For custom objects, you have a checkbox to allow or prevent roles hierarchy from granting access. If you uncheck it, that object's records won't roll up visibilities along the hierarchy. This is useful for something like a "Performance Review" custom object – maybe you want only the employee and HR to see it, not their manager (so you'd set

OWD private and turn off hierarchy access, then explicitly share with HR as needed via rules). It's a powerful option for specific cases where the hierarchical nature of roles doesn't match the data security requirements.

Let's outline a practical example to illustrate record-level sharing: **Suppose we have an object Project__c** (custom object). OWD is Private because projects are sensitive. We want: Project managers can see their projects (they are owners), their direct managers (in the role hierarchy) should also see them, and any project that is marked High_Risk__c = TRUE should also be visible to a special Risk Team. How do we do this?

OWD Private on Project__c ensures only owners (and their managers via roles) see projects by default.

"Grant Access Using Hierarchies" is on (by default for custom, it is on unless we uncheck). So managers (role hierarchy) get access to their subordinates' project records.

Create a public group "Risk Team" containing all users who are on that risk assessment team.

Create a **criteria-based sharing rule** on Project__c: criteria High_Risk__c = TRUE, share **Read/Write** access to group "Risk Team". Now whenever a project is flagged high risk, that team can access it as well (regardless of owner).

For one-off collaboration, if a project manager wants to share a project with another peer, they could use Manual Sharing (assuming we grant them access to the Share button via their profile – note: "All Internal Users" don't automatically get share button, but record owner does for their records).

Now, with that model, consider: if the risk team member opens a project record, Salesforce checks:

Does this user have Project__c in their profile/perm set access? (We must ensure yes, otherwise none of this matters – object access is prerequisite.)

Is there a share granting them access to this specific record? If they're not the owner and not in role hierarchy path, the sharing rule we made would grant it if High_Risk__c = TRUE and they are in the "Risk Team" group. If the project isn't high risk and they're not the owner or manager, they won't see it.

If criteria met, record is visible with read/write as specified; if not, "record not found" for them.

One more advanced concept is **Apex Managed Sharing** – writing code to insert share records programmatically for scenarios too complex for declarative rules. Because this chapter focuses on declarative, we won't dive into that, but know that for every share need, first try to use OWD/roles/rules. Only resort to code if absolutely necessary (like sharing based on complex calculations or triggers). And even then, consider if the requirement can be adjusted to fit a declarative approach, for maintainability.

Now, let's talk about *using ChatGPT to troubleshoot or plan record level security*. Often, an admin might encounter a situation: "User X can't see record Y, why?" There are multiple possible reasons. ChatGPT can act like a wise colleague walking you through possibilities.

ChatGPT Prompt (Text):

User: "A sales rep says they cannot see an Opportunity record that they should have access to. What steps should I take to troubleshoot this in Salesforce's sharing settings?"

ChatGPT: "Start by verifying the basics: Is the Opportunity's **Owner** that sales rep or someone below them in the role hierarchy? If yes and they can't see it, something's off (because owners can always see their records) – double-check they're looking at the right record. If the owner is someone else, then consider:

Check the Opportunity's **Org-Wide Default (OWD).** If it's Private or Public Read Only, the rep will need some form of sharing to see others' opps. If OWD is Public Read/Write and they can't see it, then it's likely something else (like field-level security making it appear missing, but they should at least see the record exists).

Assuming OWD is private, how could this rep normally see opportunities? Are they higher in the **Role Hierarchy** relative to the owner? If the rep and the opportunity owner are peers or in different branches, role hierarchy wouldn't grant access.

Are there any **Sharing Rules** for opportunities? For example, a rule that shares opps between certain roles or based on criteria (like share all opps of Industry X with Team Y). If such a rule exists, see if this record meets the criteria and if the rep is in the target group.

Could it have been **Manual Sharing**? Maybe the rep was individually shared on some records but not this one. You can check the Opportunity's Sharing list (in Salesforce Classic or via the Sharing button, or run a report on Opportunity shares) to see if there's an entry for that user or their group.

Confirm the user's **Profile/Permission Set** has at least read access on Opportunity object. If not, they wouldn't see any opportunities at all (so that likely isn't the case if they said 'should have access', they probably see others).

Lastly, consider if there's any weird scenario like an Apex sharing recalculation delay or territory management in play (if using Territory Management, ensure the rep is part of the territory for that opportunity's account).

In short, check ownership and role hierarchy, then sharing rules, then manual shares. Salesforce's "Why can't I see this record?" tool (in Lightning Experience, the user can click "Share" and sometimes see why they have or don't have access) could also guide you. Once you find the gap (e.g., maybe there was supposed to be a sharing rule that wasn't configured for that user's role), you can address it – possibly by adding a new sharing rule or adding the user to a group that's included in an existing rule."

This thorough answer covers the thought process. ChatGPT can also summarize sharing settings for an object if we feed it the configuration (perhaps via a JSON or CSV of sharing rules). Or for an explanation purpose: *"Explain to a new admin how criteria-based sharing differs from manual sharing"* – good for training contexts.

Next, let's mention **Territory Management** briefly. It's an alternative model (mostly for sales data) that allows record access by territory assignment rather than ownership. It's declarative to an extent but quite complex, and typically a large org feature. In territory management, an opportunity or account can be shared to users in the same territory. It's beyond scope to detail here, but if your org uses it, it essentially adds another layer of sharing rules automatically based on territory hierarchies. The best practice still is: only use it if your sharing needs truly revolve around geographic or segment territories that overlap in ways roles can't handle.

Now, on the field security side, a big "gotcha": **report and dashboard visibility.** Reports obey field-level security (they won't show data from fields a user doesn't have access to). So that's good. But be aware of things like *list views* and *search*: if a field is hidden, it won't show up in list view columns or search results for those users. Usually that's desired. But occasionally, an admin might hide a field and then wonder why users say "I can't search by X" – It's likely because they can't see X. This again underscores that hiding via FLS is robust. Conversely, if you need them to search or filter by a field but not actually see it on the record, that's tricky: they'd need visibility for search to match it typically. A compromise is sometimes using two fields (one for sensitive actual data hidden, one proxy that's less sensitive).

Criteria-based sharing has a limitation: you can't directly share to individual users (only roles, roles and subordinates, public groups, or territories/queues). So if you find yourself needing to share to one person specifically, consider making a public group for that person (maybe group called "John Doe Temporary Access" if needed). But usually, that signals a need for either manual share or maybe rethinking if that should be a broader rule.

One more advanced tip: Use the **"Sharing" button or report** to audit who has access to a given record. In Lightning, there's a "Sharing Hierarchy" view on records that shows which rules or shares give who access. It's very useful for understanding why a certain user can see a record.

It's also worth discussing **sharing recalculations:** when OWD or sharing rules change, Salesforce recalculates access in the background. This can be slow for large data volumes. Best practice: plan sharing architecture early, because flipping an object from Public to Private later in a million-record org can take time to rebuild all sharing tables. Salesforce does have the "Deferred Sharing Calculation" feature for very large orgs to group changes together to avoid performance hits.

Alright, time for a brief demonstration of design thinking via ChatGPT for a sharing scenario:

ChatGPT Prompt (Image Generation/Wireframe):

User: "Can you generate a simple diagram that shows the layers of Salesforce sharing (OWD, Role Hierarchy, Sharing Rules, Manual Sharing) as an inverted pyramid?"

ChatGPT: "Imagine an inverted pyramid divided into four layers:

- **Base Layer (Narrow tip at bottom):** Organization-Wide Defaults – this is the most restrictive baseline (who sees what by default).
- **Second Layer (above base):** Role Hierarchy – broadens access upward to managers.
- **Third Layer:** Sharing Rules – further opens access laterally based on criteria or ownership.
- **Top Layer (Widest):** Manual Sharing – individual record exceptions, broadest specific access.

The pyramid widens as you go up, indicating more access. An arrow along the side might say "Visibility increases" as we go from OWD (least visibility) upward to Manual Sharing (most specific but broad in effect for that record). The idea is to show that OWD is the foundation and additional mechanisms layer on more access."

Now, since we actually have an image from earlier research that matches this description, let's include it for clarity:

Salesforce's record sharing model is often visualized as an inverted pyramid. At the base (narrow end) are **Organization-Wide Defaults** – the most restrictive level setting baseline record access. Above that, the **Role Hierarchy** opens up visibility upwards to managers. Next, **Sharing Rules** (owner-based or criteria-based) allow wider lateral sharing based on roles or record criteria. At the top (broadest point), **Manual Sharing** provides one-off exceptions for specific records. The upward direction indicates how each layer progressively opens up more visibility to records.

As shown in the diagram, you always start by locking down with OWD, then ease restrictions as needed with the higher layers. If you mistakenly do the opposite (start open and try to restrict later), you'll find it much harder – Salesforce doesn't have a concept of "negative sharing" (except the new Restriction Rules which we'll get to next). So designing from least privilege outward is key.

Before wrapping up this section, let's highlight a **Pro Tip for experienced admins:** Use the **"View All" and "Modify All" permissions sparingly.** These are profile/perm set level perms that ignore sharing rules entirely ("View All" on an object lets the user see every record of that object, regardless of sharing; "Modify All" adds edit/delete on all records). System Administrators typically have "View All Data" which covers everything. But for standard users, granting "View All" or "Modify All" on an object basically overrides all your careful sharing rules. Sometimes it's needed for a specific power user or integration user, but be cautious. If, for example, a VP insists on seeing all Opportunities in the org, you might be tempted to give them "View All" on Opportunity. That works, but consider alternatives: could you instead use a sharing rule (like share all opps with Role = VP)? The difference is subtle, but using sharing rules keeps the visibility in the sharing model (which can be audited with sharing logs), whereas View All is a God-mode bypass. At least document why you gave anyone View/Modify All on an object. For compliance, you should regularly audit profiles/perm sets for those flags – they are necessary for admins and perhaps integration accounts, but not for most end users. A tip: Salesforce Security Health Check or Optimizer often flags profiles that have too many of these powers.

To conclude this section: mastering record and field-level security ensures that even if users have access to objects, they don't automatically get carte blanche to all data. We've layered our defenses: object access (profile/permset) AND record access (sharing) AND field visibility (FLS) all must allow something for a user to actually see or edit it. Think of it like multiple locks on a door – all must be unlocked for entry.

With these advanced sharing techniques, we can meet complex requirements declaratively. However, what if even these tools aren't enough to fine-tune access? Historically, the answer was either use Apex managed sharing or live with it. But Salesforce in recent releases provided a new declarative tool to *remove* access in ways sharing rules can't: **Restriction Rules.** In the next section, we're going to explore these modern security features – how they can trim down record visibility beyond the standard model, and other enhancements Salesforce has rolled out for savvy security admins.

We've now covered how to open up access where needed. Next, we'll look at the flip side: how to *restrict* access in special cases using new features like Restriction Rules, and how to leverage some of the latest security innovations Salesforce offers to tighten our security even further without writing code.

LEVERAGING RESTRICTION RULES AND MODERN SECURITY FEATURES

As Salesforce has evolved, so too have the tools available for fine-tuning security. In traditional Salesforce security, most mechanisms *add* access (you start locked down and then grant access via roles/sharing rules). But what if you need to **remove** or limit access beyond the norm? For instance, perhaps your OWD is relatively open for collaboration, but you have a subset of records that should be hidden from certain users. This is where **Restriction Rules** come in – a relatively new declarative feature that allows you to hide records from certain users under specified conditions, even if those users would normally have access. In this section, we'll dive into Restriction Rules: what they are, how to configure them, and when to use them (with cautionary tips, because with great power comes great responsibility!). We'll also highlight other modern security features and enhancements that might not have existed a few years ago, ensuring you're up to date on Salesforce's security capabilities as of 2025. This includes features like **Scoping Rules** (to refine search results), **Enhanced Permission Set targeting** (like user access policies that auto-assign permission sets based on criteria), and any relevant updates to sharing architecture or platform security that help you declaratively manage access in new ways.

Restriction Rules essentially allow you to write criteria that *remove* records from certain users' view. Think of it as the inverse of a sharing rule. Currently, restriction rules (at the time of writing) apply to custom objects, and certain standard objects like Events, Tasks, and perhaps a few others like Contacts in specific scenarios (especially with External users). The idea is: if your OWD for an object is more open (like Public Read/Write or Public Read Only), you might still have cases where you want to hide some records. For example, maybe you have an object that's generally public, but records with a certain checkbox should be hidden from all but a specific group. With a Private OWD, that's easy (don't share those). But if OWD is public, you historically had no way to *exclude* records – everyone could see everything. Companies faced this with things like internal cases vs. customer cases, or perhaps Accounts marked as "Secret" when normally accounts are shared.

How do restriction rules work? They are configured in Setup (similar to sharing rules UI, but separate section). You specify:

- **Object** (which object to restrict).
- **Who** it applies to: you choose a user criteria – either based on profile, or based on a field on the user (like a checkbox or maybe a custom permission assigned). Essentially, you define which users the rule affects. Often it's something like "Profile != System Administrator" meaning apply to all regular users but not admins.
- **Record Criteria:** a filter condition on records that should be hidden.

The effect is always to **prevent view** of those records for the specified users (if they would otherwise have access).

For example: *"On the Case object, for users with profile 'External Customer User', restrict records where Case.Type = 'Internal'."* If your OWD for Case is Public Read/Write (because normally customers and support see cases), this rule would make any case marked "Internal" invisible to external customer users, even though public access would have allowed it. It's a layer of subtraction.

Important: Restriction Rules *only remove read access*, they don't create something out of nothing. So if a user didn't have access to begin with, the rule doesn't matter for them (it doesn't grant or further restrict beyond none). And if a user has view access and meets the criteria, it will block that view. It doesn't currently restrict edit separately; if they can't read, they effectively can't edit either (no visibility, no action).

One must use restriction rules carefully. They can be a bit counterintuitive because Salesforce's core assumption was always "if you open via OWD, you can't selectively hide except with record types and page layouts" (which weren't true security). Now we have a true security mechanism to hide. So, why not just set OWD to private and use sharing? Sometimes business logic or performance demands a more open OWD with selective hiding. Perhaps you have 99% records that everyone should see, and 1% super sensitive – making OWD private and writing many sharing rules to replicate "everyone can see all non-sensitive records" would be arduous (like you'd need a rule per user or something huge). Instead, OWD open then a restriction rule for the 1% might be simpler. It's basically a way to handle exceptions.

A scenario: You have a **Contracts** custom object. Salespeople and Finance should see all Contracts, so you set OWD Public Read. However, some contracts are under NDA and only Finance should see them but not Sales. You could restrict: user criteria = Profile = Sales, record criteria = NDA = TRUE. Result: those NDA contracts are hidden from Sales profile users, even though OWD would normally give access. Finance still sees them because rule didn't target them.

It's powerful but be mindful: Overusing restriction rules can complicate understanding who can see what (as it's one more place to check when debugging access issues). Also, there are limits on how many restriction rules per object (currently I believe 2 per object and 5 conditions per rule, something like that). They also currently only work on *reads*. If a user tries to query via API or something, the system omits restricted records from results as if they don't exist.

One limitation: you cannot restrict admins (they see all). Also, certain high-level perms (View All Data) will bypass these. So this is for standard users; it's a business logic restriction, not an absolute (like encryption would be). Also, at present, you can't use OR conditions elegantly in a restriction rule – like one rule covers one scenario; if multiple, you may need multiple rules or craft the criteria carefully.

Scoping Rules – Another new feature – are not exactly security, but worth noting. Scoping Rules allow you to define filter conditions to "scope" what records users see by default in searches and lookups. Unlike restriction rules, scoping rules do **not** prevent access; they just hide things from initial view, while still allowing the user to find the record if they explicitly go to it or clear filters. For example, you might scope that a support rep's default case list only shows their region's cases, but they could still search for an out-of-region case if they needed to. It's more of a UI convenience and performance thing to reduce noise, not a security barrier. In any compliance discussion, note that scoping is not for data protection, only for usability.

Another modern feature: **User Access Policies** (Pilot/Beta feature) or even using Flow to auto-assign permission sets. Salesforce has been introducing ways to declaratively automate user access management. For instance, you might say "All users with Title = 'Manager' should automatically get the Manager Permissions permission set." Instead of an admin doing that by hand, you could have a flow or a user access policy do it on user create/update. This ensures consistency and is especially useful when combined with something like Restriction Rules. (E.g., mark a checkbox on user "Is Intern", have a restriction rule key off that to hide certain records, and also have a user access policy that assigns an "Intern Basics" permission set when that's true.)

Let's talk **Enhanced Folder Sharing** (for reports and dashboards) – not new, but often an overlooked part of declarative security. As of Lightning, report and dashboard folders can be shared with roles, groups, and permission sets (yes, you can share a folder with all users who have a certain perm set). This is handy for segmenting data access in analytics. For example, maybe you have a dashboard that should only be visible to Finance team: you create a public group "Finance Team", share the folder to that group. Ensuring the right folder sharing means sensitive reports aren't visible to those who shouldn't see them. Always align your folder sharing with your data sharing (if a report shows data that only certain roles should see, make sure only those roles have folder access).

Lightning Record Pages can also add nuance: you might have dynamic forms or visibility filters on components to hide certain fields or buttons unless criteria met. Remember, those are *UI-layer* controls, not security. For instance, you might hide a "Payout" field component unless profile = Finance. That's fine for UX, but also enforce with FLS behind the scenes. We mention this because modern Lightning allows many dynamic show/hide rules that can make it seem like security ("only managers see this component"), but a savvy user with API or report could bypass if not secured at field level.

One more modern tool: **Transaction Security (Enhancements)** – if you have Salesforce Shield, you can write policy rules that trigger on events, including potentially blocking or alerting certain actions. This is more about runtime monitoring (like "if someone exports >1000 records, alert security"). While not a direct access control, it's a modern feature for security monitoring and sometimes enforcement (you can block actions like logging in from disallowed countries, etc., with declarative conditions in Transaction Security Center). It's not exactly declarative config of object/field but it's declarative security policy, worth knowing if your org has Shield.

Now, how can ChatGPT assist with these modern features? Possibly by explaining these new concepts to team members or helping craft conditions.

Consider a conversation about Restriction Rules:

ChatGPT Prompt (Text):

User: "We have an object with OWD Public Read. How can I hide certain records from a subset of users without turning OWD to private?"

ChatGPT: "You can use a **Restriction Rule**. Since your org-wide default is Public Read, normally everyone sees all records. With a restriction rule, you can specify criteria to hide records for certain users. For instance, define a rule on that object where the user criteria might be a profile or a user field, and record criteria identifies the records to hide. Those users will act as if those records don't exist (they won't appear in lists or reports for them, and if they try to access by URL, they'll get an error). This lets you keep most records open while carving out exceptions. Just ensure you use it carefully – it's a newer feature and applies only to read access. Also, System Admins or anyone with View All on the object will still see everything regardless of the restriction rule."

Another example: Suppose we have a screenshot or export of existing restriction rules. We could feed that to ChatGPT to check logic or overlaps.

Pro Tip: Document and communicate any use of Restriction Rules to your admin team. They are powerful but not obvious – someone troubleshooting visibility might scratch their head, forgetting a restriction rule is in play. Consider naming conventions like prefixing rule names with "RESTRICT" clearly and maybe adding help text: e.g., "RR – Hide NDA Contracts from Sales profiles."

Additionally, keep an eye on Salesforce release notes – features like restriction rules have been gradually expanding to more objects. For instance, perhaps by now they include "Tasks" or "Activities" or more standard objects. Knowing where you can apply them is important. Currently they don't cover all (e.g., not on Opportunity yet as far as I recall). So if you need a similar effect on an object that doesn't support restriction rules, you'd fallback to Private OWD and selective sharing or maybe Apex sharing.

What about **profiles vs. permission sets in context of these new features?** Well, one interesting thing is you can target restriction rules by profile or by permission sets (there's something about user criteria including "User.PermissionSetAssignments contains X" I'd need to check if it's allowed; if not, you might use custom user field or permission set group). But if it allowed by permission set, you could, for example, restrict records for all users who do *not* have a certain permission (like an "Elevated Clearance" perm). Actually, currently, in UI, it allows user criteria using a field or profile or role, not directly permission set. But you could fudge by creating a checkbox field "Has Clearance" on user that you set via user access flows for those with a certain perm set, then restriction rule uses that.

Modern Login Security feature (just to note though identity in next section covers more): there are also platform events like "login flow" that can be configured declaratively to do extra checks at login, like ensure user accepted terms or validate something – not exactly a security feature in terms of access to records, but part of overall posture.

Summarizing modern access controls:

Restriction Rules: subtract access for certain records and users.

Scoping Rules: UI-only filtering for convenience.

Muting Permission Sets: subtract certain permissions from a group.

User Access Policies: auto-assign perms sets – ensures correct assignments without manual error.

Enhanced Sharing for Experience Cloud: if you have community/portal, there are special sharing sets and such but that's a separate deep topic.

Possibly mention **Field Access Wizard**: in Lightning, you can mass-edit field-level security for fields across profiles – a nice admin UI improvement (not new, but some don't know it's there; helps you quickly lock down new fields by updating multiple profiles at once).

Security Center (if on multiple orgs): A Salesforce product that aggregates security metrics across orgs, not relevant for one org config but good for multi-org governance.

Let's illustrate a use case for a *combination* of these features in a cohesive story:

Imagine a **healthcare scenario**: The object "Patient Record" is mostly open to all doctors (OWD Public Read), but celebrity patients should be restricted to only a small team. We could create a field "IsVIP = true" on patient. Use a restriction rule: for all standard doctors (user criterion e.g. profile = Doctor), hide records where IsVIP = true. Meanwhile, the VIP team users (perhaps profile = VIP Team or a custom field on user) are excluded from this rule so they continue to see those. The result: normal doctors won't even know those VIP patient records exist (out of sight), only the VIP team can see them. Additionally, we might implement an Event Monitoring rule: if anyone tries to access a VIP record (maybe an admin), we log that heavily for audit. That's layering security: restriction plus monitoring.

Now, ChatGPT could help verify such a setup:

ChatGPT Prompt (File Upload / Analysis): Admin provides the config: "Restriction Rule 1: UserCriteria: Profile = 'Doctor', RecordCriteria: IsVIP__c = TRUE".

User: "Given this restriction rule, who can see VIP patient records?"

ChatGPT: "With that rule, any user with the 'Doctor' profile will be prevented from seeing records where IsVIP = TRUE. Users with other profiles (e.g., an admin or maybe a 'VIP Team' profile if that exists) will not be affected by the rule and thus, assuming they have general access, they can see those VIP records. So presumably the VIP team either has a different profile or we've excluded them from this criterion. To be precise: All standard doctors (with Doctor profile) cannot see VIP patient records. The VIP team and administrators can see them (admins have View All so they're never restricted, and VIP team likely is a different profile not matching the rule's user criteria)."

That sounds correct.

It's worth noting that these newer features like restriction rules are part of Salesforce's move towards a more flexible, trust-based platform, often influenced by the "Zero Trust" approach in the industry (assume least privilege by default, even within an internal network).

Pro Tip: Before deploying a Restriction Rule or major sharing change, test the performance and user experience in a sandbox with sizable data if possible. Because if you have millions of records and a broad restriction rule, ensure it doesn't slow down list views or cause confusion. There are usually limits (like max 50 million records per restriction rule scenario recommended). Monitor the Sharing Rule Jobs page if applicable.

So, we've now covered how to add access (sharing rules) and how to remove access (restriction rules). Combined with robust FLS, we have unprecedented control. The remaining big pieces of security are about overall org security health and identity, which we'll tackle in the next sections. But first, let's summarize what we've gained in these first four sections: a deep understanding of how to manage access at every level and new tools to refine it. The next section will transition into ensuring our configuration aligns with security best practices and compliance through things like Health Check and other auditing tools.

We've fine-tuned the nuts and bolts of access control. Now it's time to step back and **audit our security posture**. In the following section, we'll explore how Salesforce's Security Health Check and other compliance features help ensure our declarative settings meet industry standards and regulatory requirements, keeping our org not only secure, but verifiably so.

ENSURING COMPLIANCE AND SECURITY HEALTH

Security isn't a "set it and forget it" task – it requires ongoing vigilance to ensure your configurations meet both Salesforce best practices and any industry/regulatory standards your organization must uphold. Thankfully, Salesforce provides tools to help administrators continuously monitor and improve their org's security posture. In this section, we focus on **compliance essentials** and how to leverage Salesforce's features like **Security Health Check**, **password policies**, and other settings to maintain a healthy org. We'll discuss what the Health Check score means and how to address its findings, cover important org-wide security settings (session security, password requirements, etc.), and explore ways to meet common regulations (like GDPR, HIPAA, SOX) using Salesforce's declarative tools. We'll also touch on features like **Field Audit Trail**, **Login History**, and **Event Monitoring** as they pertain to compliance. By the end, you should know how to not only secure your org but also demonstrate that security to auditors or stakeholders with concrete metrics and configurations.

Salesforce Security Health Check is a one-stop dashboard (found in Setup) that evaluates your org's security settings against a baseline (either the Salesforce-recommended baseline or a custom one your company can define, perhaps aligning to internal policy). It gives a percentage score. Don't be surprised if the first time you run it, the score is something like 45% – many orgs have room for improvement. The baseline is fairly strict (Salesforce's built-in one aligns to a high security standard). The Health Check examines things such as:

- Password Policy settings (min length, complexity, history, expiration days)
- Session Settings (session timeout length, IP range enforcement, whether sessions persist on login, etc.)
- User Authentication settings (like is MFA enabled, are security question for password resets allowed or not, etc.)
- Network Access configuration (trusted IP ranges usage)
- Clickjack protection settings for Visualforce/pages
- Whether TLS versions and certain security headers are enforced (for example, requiring HTTPS, strict transport security, etc.)
- Other org prefs like whether you allow embedding of SF in iframes (which could be a security risk).

It categorizes deviations as High-Risk, Medium-Risk, Low-Risk. A high-risk example: "Password expires in 90 days" is considered more secure than "password never expires" – if your profile settings say never expire, Health Check will flag that as high risk. Another: if your session timeout is very long (8 hours or never expire), that's high risk. Or if your org doesn't require secure HTTPS for certain communications.

When you open Health Check, it will list these settings and whether your org's value meets, partially meets, or does not meet the standard. You can often click an "Edit" or "Fix Risks" button next to some items to quickly tighten them up.

For example, if Health Check says "Minimum Password Length = 8 (Risk: High, Recommendation: 12)", you can go into Password Policies in Setup and change that to 12. Some items, like certain security headers, might require you to contact support or do something more involved if they're not enabled (but most are straightforward toggles).

Aiming for a 100% score is nice, but not always practical; some settings might be relaxed for user experience reasons. Instead, focus on high-risk items first. For instance, **multi-factor authentication (MFA):** Salesforce (as of early 2022) contractually requires customers to have MFA for internal users. If your org doesn't enforce MFA, that's a huge risk – an auditor would frown. Enabling MFA (via profile permissions or now via user session security levels) will be flagged if not done.

Password Policies: We should ensure they align with best practices (min length 12 or more, require mix of characters, don't allow a user's username in password, lockout after, say, 5 failures, etc.). Salesforce lets you set a baseline at org level, and you can override per profile if needed (though simpler to keep them all consistent).

Session Settings: Key ones include Session Timeout (shorter is safer – maybe 30 minutes of inactivity), whether to enforce logout on session timeout, use persistent login (probably disable for higher security), and whether users can stay logged in from trusted IPs without re-auth (maybe allow if behind VPN only).

One critical setting: "Lock sessions to the IP address from which they originated." Enabling this means if someone steals a session ID (like cookie) and tries to use it from another IP, it'll be invalid. This is good security but can cause issues for users who might change networks mid-session (e.g., switch from office wifi to cellular). Usually recommended to enable in high-security environments.

Another: "Require HttpOnly attribute" and "Require secure connections (HTTPS)" – these should be on to prevent cookies being accessible to scripts and to ensure no one can intercept plaintext traffic.

Network Access: Under Network Access, you can set trusted IP ranges for the org such that login from those doesn't get the login challenge (like verification code). If your company's IP range is there, employees logging from office might bypass some verifications – convenience vs security trade-off. Health Check might not directly score that, but if you allow logins from anywhere without any challenge, consider at least leaving login challenges on (which is default anyway if not in trusted range). Also, profile-level IP restrictions: if certain users should only log in from certain locations, that can be set in their profile. It's not flagged by Health Check per se, but is a good practice for sensitive profiles (like limit a financial data profile to office IPs only).

Compliance requirements vary: for **GDPR**, major focus is on data privacy – ensuring you know where personal data is and controlling access to it (we covered field-level security, encryption etc., which helps). Also, GDPR might involve using Salesforce features like **Individual object** (for managing consent) or data retention policies (perhaps using automation to delete or anonymize older personal data). That's beyond pure security, but related.

HIPAA (health info): would expect encryption of health fields (Platform Encryption), audit trails, and strong access controls (which we've done with FLS and sharing). Also maybe field history tracking on critical fields (so changes are logged).

SOX (financial controls): often means strict control of who can export data, who can modify key fields – so using permission sets to limit "Export Reports" permission to only finance, for example. Also change management: Salesforce's setup audit trail (logs changes in config) might be reviewed.

PCI DSS (payment card data): If you store credit card info in Salesforce (ideally avoid raw storage, but if needed), platform encryption is a must, and extremely limited FLS (maybe only a payments team profile can see those fields), plus masking where possible (Salesforce has a built-in tokenization for credit card fields in some contexts).

Auditing Tools:

Setup Audit Trail: tracks changes in configuration (last 6 months, downloadable). For compliance, it's good to review admin changes or have alerts when certain settings change (like if someone disabled MFA or changed a sharing setting).

Field History Tracking: You can track up to 20 fields per object (more if you have Field Audit Trail with Shield) for changes. This is good for compliance (e.g., track if someone changes an Opportunity's amount after closing, for anti-fraud).

Login History: Salesforce logs user logins (successful and failed) for the last 6 months (and more if you store via Splunk or Event Monitoring). Regularly reviewing login history can spot suspicious activity (like logins at odd hours or from countries you don't operate in).

Event Monitoring (Shield feature): provides detailed logs like reports export events, API usage, etc. Useful if you need to prove monitoring. For instance, you can see if someone exported a large report of contacts – which under privacy laws you might need to keep an eye on.

Salesforce Optimizer report is another tool: it looks at your org's config and highlights issues, some of which are security-related (like too many admins, or lots of users with Modify All Data). It's worth running occasionally.

Compliance certifications: If needed, mention that Salesforce itself has many (ISO, SOC, etc.) and offers **Shield Platform Encryption** which is certified for certain uses. But for your own org compliance, you use these features.

A nice feature: **My Domain** is required now (for Lightning and to use newer features). It also adds security by isolating your org's login URL and allowing you to set certain policies like preventing login from generic login if you want SSO only. Under My Domain settings, you can block login from login.salesforce.com forcing your custom domain (commonly done when you use SSO so that all internal users must go through corporate IdP).

We should also mention **Multi-Factor Authentication** (again) as a compliance point: Many regulations (and Salesforce's own contract) push for MFA. Enabling MFA can be done by permission set assignment of the "Multi-Factor Authentication for User Interface Logins" user permission (or by enabling a policy that requires it for all users). Starting Feb 2022, Salesforce will nag if you don't have it, but not block. But you really should implement it. The methods: Salesforce Authenticator app, Time-based one-time password (TOTP) apps, Yubikey/physical keys (WebAuthn) – these are all supported. MFA can be enforced at SSO level too (so if they login via corporate SSO which already did MFA, that counts).

Aligning with compliance often is about documentation and proving it. For example, if ISO 27001 audit asks "how do you ensure least privilege in Salesforce?", you can show them role hierarchy diagrams, permission set assignments, and the Health Check report for overall settings. If GDPR audit: show data classification on fields (Salesforce has a feature to tag fields as Personal, Sensitive, etc. – do use that, it's metadata that helps track where personal data lives). Also show how a user's data can be exported or deleted on request (maybe via the Data Export and anonymization processes).

Let's simulate an admin using ChatGPT to interpret a Health Check:

ChatGPT Prompt (Text):

User: "My Salesforce Health Check score is 65%. It flags 'Password Complexity' and 'Session Timeout' as high risk. What should I do?"

ChatGPT: "A score of 65% suggests some important security settings are below recommended levels. The high-risk items you mentioned:

Password Complexity: Likely your org isn't requiring a mix of letters, numbers, symbols, or minimum length is too low. Check your Password Policies (Setup > Profiles > Password Policies or Setup > Password Policies for the default) and ensure you require at least 1 letter, 1 number, 1 special character. Also, minimum length should be at least 8, preferably more like 12.

Session Timeout: This might be set too long. High risk indicates maybe it's set to 'Never expire' or several hours. Best practice is something like 30 minutes or 1 hour of inactivity. Shorten your session timeout setting in Session Settings.

After adjusting these, click the 'Recalculate' on Health Check. The score should improve and, more importantly, your org will be safer. Gradually address other items as well. For instance, if Health Check flags not using HTTP-Only cookies, enable that (it's a checkbox in Session Settings: 'Enable secure and HTTPOnly attributes'). Use the Health Check as a guide, but always test that tighter settings don't break user functionality (for example, very short session timeouts could frustrate users if they get logged out too quickly). Overall, aim for that green zone (approx 80%+)."

That advice is practical. ChatGPT can also help explain *why* a setting is important. Sometimes users want to know "why must we change passwords every 90 days, can't it be 365?" ChatGPT could give balanced perspective (security vs user friction, many are moving away from frequent changes in favor of MFA).

Automation for compliance: You can schedule a Health Check extract or use the Metadata API to monitor changes. Some companies integrate it into monitoring solutions. At the least, have a periodic (quarterly) security review where you check Health Check, user access, etc.

Sharing compliance duties: It's not just the Salesforce admin, but working with InfoSec team. Many InfoSec folks might not know Salesforce specifics; they'll appreciate if you can translate Salesforce config into their compliance checklist language. For example: "Requirement: All user access changes are reviewed by management monthly." In Salesforce, you might satisfy that by producing a user list or using an app that attests user's permissions sets are correct. Or "Ensure least privilege": you can show that each user's permission sets match their role, and none have excessive rights.

Data Mask / Anonymization for Sandboxes: worth noting – compliance often requires that non-production (sandbox) have sanitized data if production has sensitive info. Salesforce offers Data Mask (paid tool) to scramble or delete sensitive fields in sandboxes after copy. If not using that, at least manually delete things like plaintext passwords or personal data in sandboxes if developers don't need real data. This is part of a larger compliance picture.

Now, let's incorporate how ChatGPT could help with compliance documentation or analysis:

ChatGPT Prompt (Operator with File):

User: "I'm uploading our Security Health Check results (a CSV export). Summarize the critical issues for me and any quick wins."

Assume we gave it a CSV of settings vs baseline.

ChatGPT: "Summary of Critical Issues: The Health Check indicates 3 high-risk settings:

Password Expiration – Currently set to 'Never expires'. Recommended is 90 days or less. *Quick win:* Set a 90-day expiration to mitigate this.

Session Timeout Length – Currently 4 hours. Recommendation is 30 minutes. *Quick win:* Reduce session timeout to 30 min or 1 hour to significantly reduce risk of unattended sessions.

Allowed Login IP Ranges – Not defined for any profiles (which means logins are allowed from any IP by username+password alone). *Quick win:* Define IP ranges for at least privileged profiles (like admins) to add an extra layer of security, or ensure MFA is in place to mitigate open IP access.

Medium risks noted include not enforcing history tracking on certain objects and not requiring 2FA for API logins (which Salesforce currently doesn't natively support but can be mitigated via integration security). Overall, tackling the above three will raise your score and security posture significantly. Next steps: implement MFA (if not already), and address medium risks like disallowing HTTP connections (force HTTPS only)."

This kind of analysis can save an admin time and help in reporting to management in plain language.

Finally, emphasize that after all these measures, the org should not just be secure, but you should be able to **prove** it's secure. Running a Health Check before and after changes is one way to show improvement (like a metric: we went from 65% to 92%). That's often something you put in an IT security report or executive summary.

Pro Tip: *Integrate security checks into your deployment lifecycle.* For example, if you use CI/CD for Salesforce metadata, consider automating a scan of profiles/perm sets for risky permissions, or ensure new fields have classification and FLS set intentionally. There are tools that can parse Salesforce metadata to enforce security rules (e.g., a build fails if a profile unexpectedly gets Modify All on something or if a new field containing "SSN" in its API name isn't encrypted or restricted).

Now that we have our environment configured securely and an ongoing process to keep it that way, let's transition to one of the most critical aspects of security: ensuring the *people* who access our org are who they say they are, and they do so in a secure manner. That brings us to identity management and authentication, which we'll cover in the next section.

We've confirmed that our org's doors and windows are strong and well-locked. Now it's time to focus on the gate and keys themselves – in the next section, we'll explore declarative best practices for **Identity Management and Authentication**, from multi-factor auth to login networks and beyond.

IDENTITY MANAGEMENT AND AUTHENTICATION BEST PRACTICES

The strongest locks on data mean little if the wrong people can obtain the keys. Identity management – verifying that each user is who they claim to be and only that user – is the front line of security. In Salesforce, identity and access management includes everything from how users log in (and from where), to how we enforce multi-factor authentication, single sign-on, and session security. In this section, we'll cover best practices for managing user identities and authentication declaratively. This includes implementing **Multi-Factor Authentication (MFA)** for all users (a must in today's security climate), setting up login IP ranges and login hour restrictions for added security where appropriate, configuring session settings for optimal balance of security and usability, and leveraging tools like **Login Flows** or user policies for additional verification steps. We'll discuss how to integrate Salesforce with identity providers for SSO to centralize and strengthen authentication, and how to handle common scenarios like user provisioning and deactivation securely. By following these practices, you ensure that only authorized individuals get into your Salesforce org, and even if their credentials are stolen, attackers can't easily abuse them.

Let's start with **Multi-Factor Authentication (MFA)**. MFA adds a second verification step at login (something you have, like a code or a mobile app confirmation, in addition to something you know, like your password). Salesforce provides multiple MFA methods:

The **Salesforce Authenticator app** (mobile app that can do one-tap approve and even intelligent location-based approvals).

Time-based One-Time Password (TOTP) apps (Google Authenticator, Microsoft Authenticator, etc. where you scan a QR code from Salesforce and then the app generates 6-digit codes every 30 seconds for you to enter).

Physical security keys (U2F/WebAuthn devices like YubiKeys that you plug in or tap).

SMS (text) is actually supported as MFA in Salesforce but generally not recommended due to security concerns with SIM swapping; Salesforce prefers the above methods.

Enabling MFA in Salesforce can be done by a few methods:

By **user permission**: "Multi-Factor Authentication for UI Logins" permission. This can be assigned via permission set to profiles/users. If a user has this permission, upon next login they'll be forced to register an MFA method and use it going forward.

Or via the **Session Security Level** policy: You can require a "High Assurance" session for certain pages or for login. Salesforce now basically requires all internal logins be MFA so practically you give everyone that perm or enable an org-wide setting.

If using SSO (SAML) with an external Identity Provider (like Okta, Azure AD): If the IdP enforces MFA, you're covered without having to set the Salesforce perm (Salesforce trusts the IdP's MFA).

Why MFA? Because passwords alone are weak (users reuse them, phishers steal them). With MFA, even if a password leaks, an attacker likely can't log in without also having the user's phone or key. As mentioned, Salesforce contractually mandates it for direct logins. (They don't yet have a way to enforce API logins with MFA – instead, you use things like IP restrictions or VPN for API users, and the move towards token-based auth vs. storing user passwords).

Best practice: Enable MFA for all human users. The only exceptions might be integration users where you can't do MFA on API calls – for those, lock down by IP and principle of least privilege (and monitor closely). Many orgs have moved to SSO only: that is even better, as you can disable direct logins for internal users (set "Is Single Sign-On User" on their user record or block login page for them) and rely on corporate login which does MFA and maybe other checks.

Next, consider **Login IP Ranges:**

Org-wide Trusted IPs (in Network Access): If you add your company's office IP range here, logins from there skip the extra verification code challenge (but still need username/password and MFA if enabled). It doesn't restrict logins from elsewhere; it just marks some IPs as trusted.

Profile Login IP Ranges: These *restrict* where users with that profile can log in from. If not in range, login is blocked entirely (they get an error). This is powerful for say, limiting certain sensitive profiles to only work from office network or VPN addresses. But in modern remote work era, profile IP restrictions are less used except maybe for integration profiles or highly privileged profiles (like you might say SysAdmin profile can only log in from HQ's IPs – though real admins often need to log in from anywhere to fix things, so weigh carefully).

Alternatively, you could achieve similar restriction via SSO policies in your IdP (like only allow SSO login if from certain conditions).

Login Hours (also set per profile): You can restrict at what times login is allowed. For instance, you could prevent standard users from logging in outside of business hours. This is rarely used in practice (because sometimes people do need after-hours access or in different time zones). But some orgs use it for, say, call center profiles: they shouldn't access CRM outside their shift. It's a blunt instrument and can be bypassed if someone just leaves a session open before hours (it doesn't auto-kick them at the end time except they can't start new session). Still, it's available.

Session Security settings:

Session Timeout: as discussed, 30 minutes to 2 hours is typical. Shorter for sensitive orgs. If someone is active, they won't be logged out – it's idle time.

"Require re-login after Login-As" – if admins use the "Login as User" feature, this setting (if on) makes them re-auth as themselves after finishing to ensure no session hijack. Good to enable if you use that feature.

Session security levels: default and high assurance. You can mark certain pages/components to require high assurance (meaning user must have done MFA in this session). For example, maybe viewing a report with very sensitive data triggers a step-up MFA prompt if not already high assurance. Declaratively, you might set this up via App or via code. But typically, if MFA is at login, you're always high assurance from the get-go.

Single Sign-On (SSO) integration:

Using SSO with an external Identity Provider (IdP) like Okta, Azure, ADFS is strongly recommended in enterprises:

It centralizes credentials (users remember one password, and you can enforce strong policies there).

You can instantly disable a user in one place (IdP) to cut off access across many apps including Salesforce.

You can incorporate advanced security: IdP might do device trust checks, geo checks, risk-based authentication, etc., which then covers Salesforce access too.

Admins in Salesforce no longer manage passwords (less reset requests, less risk of stored passwords).

You can use standards like SAML or OIDC. Declaratively, setting up SSO is mostly configuration: exchanging certificates, configuring a Connected App (or the Salesforce settings for SAML).

For user provisioning, Salesforce can be integrated with systems to auto-create users via APIs or use Salesforce User Provisioning (SCIM) if supported by IdP to sync users.

From a declarative perspective, once SSO is set, you might check "Prevent login from login.salesforce.com" (so only IdP can auth users). But always keep a backdoor like one admin user not SSO, in case IdP is down and you need to login directly to Salesforce.

My Domain: enabling My Domain is required for SSO and also gives you the ability to do things like set a custom login policy. E.g., you can have MyDomain login page only allow certain methods (like remove username/password option if you only want them to go "Login with SSO").

Login Flows: A super interesting declarative feature (with a bit of flow builder). You can attach a Screen Flow to run right after a user logs in but before they get to app. This can be used for custom login steps – for example:

- Display a Terms of Service and require acceptance (common for compliance).
- Do a survey or message of the day.
- Do additional verification: For instance, if a user is logging in from a new device, maybe ask a security question (this would require building that logic though). Or check a user field "Training Completed" and if not, present a message and log them out.
- Login flows are powerful but you must design carefully so you don't lock everyone out by a broken flow. Always test with a non-admin user first with login flow (as admin typically bypasses by default or you can set it for specific profiles).
- User Management:
- Ensure you remove access promptly when people leave (deactivate user, maybe freeze first if needed and then deactivate). In compliance terms, having an Identity workflow ensuring HR or IT disables accounts in timely manner is key.
- Use **Queues** and **Public Groups** to manage group-based access rather than sharing with individuals whenever possible – easier to update when personnel changes.
- For external identities (communities/Experience Cloud), those users have similar settings – you likely require verification (like email confirmation). Also consider options like allowing self-service password reset but with secure practices (you can customize the email templates for user retrieval, etc., to make sure they aren't leaking info).

Device Trust:

If your org is on Salesforce Mobile or using the mobile app, consider security settings like PIN requirement for the app or disabling the ability to use some features if device isn't secured. Some of this is handled by Mobile Device Management outside SF, but note that Salesforce app itself can be configured to require a 4-8 digit pin or biometrics for access after initial login, adding a layer if someone steals a logged-in phone.

Monitoring:

We touched on login history. A wise practice is to set up alerts for certain events: e.g., if an admin logs in outside business hours or from foreign country, maybe review that. There are products and log analysis which can do that; or in Salesforce, you could use transaction security policies to detect, say, "multiple failed login attempts" or "login from blacklisted country" – some of that might require Apex or event monitoring though.

Profile vs. Permission Set for identity:

We normally set many identity things at profile (like IP ranges, login hours). But you can also assign "Bypass MFA" perm (for maybe integration user). Actually, better: use "API Only" profile for integration so they can't GUI login at all, and put IP restrictions on it. Many orgs create an "Integration User" profile which has API only and IP restricted.

Experience Cloud (Community) Users:

They have separate login mechanisms but similar principle – these often require login from anywhere (customers), but you might enforce extra verification for sensitive actions (like communities can use identity verification if doing something sensitive like viewing cases). Check things like "Require customers to verify their email before activating" – that's a setting.

One more angle: **Delegated Authentication** – rarely used now since SSO is better, but SF supports delegated auth where it calls out to a web service to validate passwords. Usually not needed in modern setups.

Now, let's incorporate ChatGPT's role as an assistant in identity management. Possibly:

- Asking it steps to set up MFA, or
- Troubleshooting login issues,
- Helping generate a communication plan to users (e.g., an email about upcoming MFA requirement).

For example:

ChatGPT Prompt (Text):

User: "We want to enforce MFA for all our Salesforce users next month. How should we roll this out to minimize disruption?"

ChatGPT: "Rolling out MFA is crucial for security. Here's a plan to do it smoothly:

Preparation (Now): Communicate to all users that MFA is coming. Explain what MFA is and why it's important. Provide a short guide or training on how to register the Salesforce Authenticator app or another verification method. Emphasize the date when it becomes mandatory.

Pilot: Enable MFA for a small group of tech-savvy or volunteer users first (like a pilot group or IT team). Make sure they can log in and address any unforeseen issues (e.g., issues with older devices and Authenticator).

Support: Set up a help channel (like a Slack room or email helpline) specifically for MFA issues, especially during the rollout week. Possibly prepare help desk staff to handle common questions.

Enforcement Day: Using a permission set, enable the MFA requirement permission for all users. Do this off-hours if possible. Users logging in will be prompted to set up MFA if they haven't already. Many may have set it up proactively from your communication.

Post-Rollout: Monitor login success/failure rates in Login History. Follow up with any users who haven't logged in with MFA (maybe they are inactive or on leave). Also, remind users that next time they change or lose devices, they need to update their MFA method (and provide instructions for that).

As a backup, ensure admins have spare authentication methods or the ability to disable MFA for a user temporarily if someone is completely locked out (you can remove the MFA perm temporarily to let them in and then re-enable).

With good communication and support, the disruption should be minimal and users will quickly adapt to the new login step."

This is sound advice and covers both technical and user relation aspects.

Voice interaction example (some folks might use ChatGPT voice while in the admin console):

ChatGPT Prompt (Voice via SORA):

User (speaking): "I have a user who keeps getting login challenge emails every time they log in. How can I reduce these?"

ChatGPT (voice): "It sounds like the user is logging in from IPs Salesforce doesn't recognize as trusted, triggering verification codes. There are a few approaches:

- If these IPs are actually company offices or VPN, add them to **Network Access trusted IP ranges.** That way, Salesforce won't challenge logins from those IPs with a verification email.
- If the user's IP varies (maybe they travel or work from coffee shops), encourage them to use the Salesforce Authenticator app or a verification method, so at least the challenge is quick (approve on phone rather than typing email code).
- Check the setting 'Browser timeout' – Salesforce by default remembers a browser for a user for some time (until cookie expires). If this user clears cookies or uses private browsing, that could cause frequent challenges. Ensuring they allow the cookie to persist will reduce prompts.

Ultimately, setting key office IPs as trusted is the best way to reduce these emails for multiple users while still keeping security."

These are correct suggestions (the cookies and trusted IP piece).

Pro Tip: For any highly privileged user (like system admin), consider requiring even stricter controls:

- Use a separate hardened machine or browser profile for admin tasks,
- Possibly dedicate an integration user for integrations so the main admin can have MFA.

- Regularly review who has the System Admin profile or Modify All Data permission. That list should be very small. Use permission sets to grant limited admin abilities (like a "User Manager" perm set for helpdesk to reset passwords instead of giving full admin).

User Access Reviews: Periodically review user access – an admin can generate a report: Users with last login date, profile, top permission sets. See if any active users haven't logged in (maybe they left and someone forgot to deactivate), or any with overly powerful profiles who no longer need them.

Expiring Inactive Users: You can set an Apex/Flow to auto-deactivate users after X days of inactivity if policy (some strict orgs do that). Not directly declarative but something to consider with flow + some scheduled trigger possibly.

We've covered that identity management is partly about features and partly about processes (like timely deprovisioning).

Finally, tie in that strong identity practices complement the earlier content: we secured data internally, now we secure entry points. The final layer after that is protecting data even if someone got it – which is encryption, our next section.

With robust identity and authentication measures in place, we've significantly reduced the chance of unauthorized access. Now we turn to one more crucial layer of defense: **encrypting data** so that even in the unlikely event of a breach, sensitive information remains protected. In the next section, we'll explore Salesforce's Platform Encryption and other data protection strategies.

SALESFORCE PLATFORM ENCRYPTION AND DATA PROTECTION

The ultimate safeguard for sensitive information is encryption – encoding data such that only authorized parties can read it. Salesforce provides powerful encryption capabilities to protect data at rest within the platform, ensuring that even if someone were to gain unauthorized database access or backups, the data would be meaningless without the keys. In this final section, we'll cover **Platform Encryption** (part of Salesforce Shield) and related data protection best practices. We'll explain how Salesforce's encryption works (it's largely transparent to end users and applications, but with some important limitations), when and how to implement it declaratively, and what considerations to take into account (performance, search, filtering, etc.). We'll also discuss strategies for handling encryption keys, such as **Bring Your Own Key (BYOK)**, and how encryption fits into compliance requirements (for example, meeting regulations that require encryption of personal data). Beyond encryption, we'll mention other data protection features like **Data Masking for sandboxes, backup strategies, and data retention policies.** By the end, you'll understand how to add a strong final layer of security around your Salesforce data – keeping it safe not just from unauthorized view in the app, but even at the database level.

Salesforce Shield Platform Encryption allows you to encrypt certain standard fields (like Contact LastName, etc.) and custom fields, files, and attachments at the infrastructure level. Unlike the classic "Encrypted" custom field type (which was a basic masking and reversible with a perm called "View Encrypted Data"), Platform Encryption uses strong encryption (AES-256) and is tied to a **key management** infrastructure. One big difference: with Shield encryption, there is no "View Encrypted Data" permission – users either have access to the field (via FLS/sharing) and see plaintext, or they don't have access and see nothing. There's no concept of some users seeing the encrypted gibberish and others seeing plaintext; encryption is mostly invisible if you have access. This is by design to ensure encryption doesn't accidentally become a "half-baked" control. It encrypts data at rest in Salesforce's databases and in files.

However, encryption isn't just a magic on/off switch – you enable it per field (or file object), and there are **trade-offs:**

Search & Sort: Encrypted data is not easily searchable or sortable. For example, if you encrypt the Contact Email field, users can't search by email or do a SOQL WHERE Email = 'test@test.com' unless using deterministic encryption and even then only exact matches. Also, you can't use encrypted fields in ORDER BY in SOQL or certain list views, because the platform can't sort on ciphertext meaningfully. Salesforce documentation lists which standard fields are encryptable and what functionality impact is (some support deterministic encryption which allows equality comparison but still not partial search or case-insensitive matching).

Filters and Formula: If you have workflows, formula fields, or validations referencing an encrypted field, those still work if deterministic, but anything requiring reading the actual value in Apex might not (unless you decrypt via platform, which is not available to Apex anyway). In general, encryption might limit using that field in some process builder criteria or flows, depending on context. Fortunately, things like email deliverability or uniqueness checks have adapted or have notes if encryption is on.

Performance: There's a slight performance overhead to encryption, especially on large data volume, as Salesforce has to decrypt/encrypt on the fly for each record display or edit. Usually negligible for users, but bulk operations or very large reports might slow a bit.

Encrypted data length: ciphertext can be longer than plaintext. Salesforce handles it under the hood, but some limits like data storage might slightly increase (e.g., encrypted text might count a few bytes more).

Backup: If you do data export, the extracted data from an encrypted field in a CSV is **still encrypted** unless you specifically have set up a process to decrypt it (which generally you shouldn't). Actually, with Shield encryption, even admins don't get a "decrypt" ability outside the app – the data is meant to always require the platform to decrypt. If you run an export via Data Loader, you'll get the plaintext (because it goes through API as your user with access). But if someone got hold of a database backup from Salesforce's servers, it's encrypted.

When to use Platform Encryption? Identify fields that contain highly sensitive info: government-issued IDs (SSN, EIN, etc.), health data, passwords or secrets, financial account numbers, etc. Also consider fields that regulations specifically say must be encrypted at rest. Don't encrypt everything arbitrarily – focus on those with material impact if leaked. Remember, field-level security already prevents unauthorized viewing within Salesforce's UI. Encryption is protecting against threats *below* the app layer (like someone with DB access, or if Salesforce as a service is compromised in a way that exposes raw data, which is very unlikely but some industries require that precaution).

Implementing Platform Encryption:

Declaratively, once you have Shield, you:

Generate an **org-specific key** (Salesforce can manage keys or you can upload your own – BYOK). The key is stored in a special keystore. There's a concept of key rotation (you can periodically rotate to a new key) and key destruction (in extreme cases, to render data permanently unrecoverable – e.g., a legal right-to-erasure could be done by destroying a key, though that nukes all data encrypted by it).

In Setup, you go to Platform Encryption and mark which fields to encrypt. For standard fields, there's checkboxes (if supported). For custom fields, it's an attribute on the field (you'd edit field and check "Encrypt" or create new field with "Encrypt" if you made it an encrypted text field – but careful, the older "Encrypted Custom Field" type is different from Shield encryption!). Actually, with Shield, you can encrypt standard fields without changing field type. For custom, you might need to change field type to encrypted? Actually, with Shield, you can encrypt custom fields of type text, long text, etc. by checking a box in Encryption Policy without altering the field type to "Encrypted Text" (which was for classic encryption).

Salesforce then encrypts existing data in background (for large data sets, it might take some time but usually transparent).

After that, users see no difference except perhaps if a user without field access tries to query via API, they get nothing as usual. With classic encryption, if they didn't have "View Encrypted Data", they'd see masked **** in UI. With Shield, no such concept: if you have field access, you see actual data normally.

Key Management:

By default, Salesforce generates and stores keys (and they have a master key that itself is in a Hardware Security Module etc. – they have a robust approach).

BYOK: Some companies want to manage their own keys. Salesforce allows you to supply your own key material from an external key management system. That way, if you revoke the key from Salesforce, all data is unrecoverable, giving you ultimate control. BYOK involves some setup using a command-line tool to import a key.

Key Rotation: You can rotate keys for good practice (say annually). Salesforce can re-encrypt data with new key behind scenes. They support up to 5 active keys historically (to allow phased rotation).

Always backup your key or seed if you do BYOK; losing it means losing data permanently because even Salesforce can't help then.

Encryption and Testing:

In sandbox, you might not enable encryption (since it's an add-on, might not be active in sandbox by default, but you can request it). But to test how your app behaves with encryption on, you should have at least one sandbox with encryption to see if any custom code breaks (like if code tries to do a SOQL ORDER BY on an encrypted field, it'll throw runtime error).

Some features improved: as of Spring '23, Salesforce allowed Einstein search to search encrypted data (with some limitations). They are trying to mitigate the trade-off.

Other Data Protection Tools:

Data Mask (mentioned): For sandboxes, to scramble personal data after refresh.

Backup & Restore: Salesforce now has a native backup service (paid) – but even if using that or external backups, ensure backups are stored encrypted too and access-controlled.

Data Retention: Use tools like Time-Based Workflow or flow to auto-delete or anonymize records that are no longer needed (if policy says delete closed cases after 7 years, etc.). Or use the built-in "Auto-archive" on Field Audit Trail to move old history out of main tables.

Monitoring: If you have event monitoring, watch for reports or queries that retrieve a lot of encrypted field data – though if an attacker is inside, that might still show up, at least you'd see unusual access patterns.

Combining encryption with other security: If a field is encrypted and a user lacks field permission – well, they can't see it anyway. If they have permission, they see it normally. The encryption's benefit is for all those scenarios outside of normal app usage, like backups or physical storage. It is also a checkbox for compliance: many regimes just say "encryption at rest must be enabled for sensitive data".

One might ask: **"Should we encrypt fields that are already masked via FLS for most users?"** Possibly yes if the data is sensitive enough (like salary info, SSN). FLS protects in-app, encryption protects if the database were accessed or if Salesforce employees had to access data for support (with Shield, even Salesforce support can't see encrypted data in raw form without your key). It also adds another layer: even an admin with access, if they ever exported data and saved a file on disk, that file would have plaintext unless they separately encrypt it – encryption within Salesforce doesn't protect once exported by a legit user. So you still need policies about exporting data. But at rest in Salesforce's cloud, it's safe.

Example scenario:

A financial services company uses Salesforce to store client financial records. They enforce encryption on any fields containing bank account numbers, tax IDs, etc. They use BYOK so that if there's a legal order or a breach, they could revoke the key to protect clients. They also restrict who can export reports containing those fields (possibly by not granting "Export Reports" perm widely). They schedule key rotations yearly. In their compliance documents, they proudly state "Client sensitive fields are encrypted using AES-256 encryption via Salesforce Shield; keys are rotated annually and managed with BYOK, and data is thus protected both in transit and at rest, meeting regulatory requirements for encryption."

ChatGPT as a helper on encryption:

Maybe ask it things like "Which fields should I encrypt?" or to explain a weird error related to encryption.

For instance:

ChatGPT Prompt (Text):

User: "We turned on Platform Encryption for the Case Description field. Now our support agents say they can't search cases by keyword. Why did that happen?"

ChatGPT: "When you encrypt a field like Case Description, it affects search functionality. Encrypted fields are not indexed for full-text search, so agents can't search by words in that field anymore. This is a known trade-off with Shield Platform Encryption – it ensures data is secure, but you lose the ability to do partial or full-text searches on that encrypted text. Agents can still search by other criteria (like case number, subject if not encrypted, etc.), but not the Description text content. To mitigate this, some organizations leave less sensitive fields unencrypted if search on them is critical. Another approach could be using deterministic encryption, but that only supports exact matches, not full text. You'll need to decide if the security benefit for Case Description is worth the loss of search, or consider alternative solutions like a proxy field or external search tool that has access to decrypted data (though that reintroduces risk)."

This explanation helps the admin team decide next steps (maybe they remove encryption from that field if search is vital and accept risk or see if they can train agents to not include highly sensitive info in description so it doesn't need encryption).

Another scenario: generating a policy document via ChatGPT:

ChatGPT Prompt (Typing):

User: "Draft a brief policy statement for our security manual on the use of Salesforce Platform Encryption for confidential data."

ChatGPT: "Policy on Salesforce Data Encryption: All confidential or sensitive personal data stored in Salesforce shall be protected using Salesforce Shield Platform Encryption. The Information Security team, in collaboration with Salesforce administrators, will identify fields requiring encryption (e.g., government-issued IDs, financial account numbers, health information) and enable platform encryption for those fields. Encryption keys will be managed in accordance with corporate key management procedures, including annual rotation and secure storage (BYOK is utilized to maintain control of encryption keys). This encryption ensures that data at rest within Salesforce's databases is unreadable to unauthorized parties, adding an extra layer of defense beyond standard access controls. Users with authorized access to the data will not experience significant changes in functionality, but certain operations (such as search or sorting on encrypted fields) may be restricted by design. Under no circumstances should encryption be disabled for a designated confidential field without approval from the Information Security Officer. This policy helps ensure compliance with relevant data protection regulations and safeguards client and company sensitive information even in worst-case scenarios of data leakage."

This kind of output could be directly useful for documentation.

Pro Tip: Don't forget about encrypting **Files and Attachments** if you store sensitive documents in Salesforce. Shield can encrypt file data (the actual content) and even search terms, but note that if you encrypt files, you lose the ability to preview them in some cases (older versions had that limitation – it might be addressed but historically encrypted files couldn't be previewed, they had to be downloaded and decrypted on the fly). Check current state; likely still some limitations. Only encrypt files if those files themselves contain sensitive info and you're comfortable with any preview limitations.

Also, **Communities:** If sharing data with customers/partners, be mindful of what you encrypt because if encryption prevents search or something, it affects them too.

Now let's bring it all together in closing the chapter:

We've gone through layers: config, sharing, modern restrictions, compliance, identity, encryption. A quick wrap-up would underscore how these all interrelate and the philosophy of balancing security and usability (for example, not to over-encrypt or over-restrict that users can't do their jobs – find that sweet spot, often through iterative adjustments and user feedback, all while maintaining core security principles).

SUMMARY

In this chapter, we explored Salesforce declarative security from end to end – building a **layered defense** that protects data at every level. We began with fundamentals: using **org-wide defaults** to lock down record access by default, then thoughtfully opening access via roles, profiles, permission sets, and sharing rules. We learned that profiles should be kept minimal (just the baseline permissions), with **permission sets and permission set groups** granting granular access in a flexible, scalable way – a modern best practice to avoid profile proliferation. We reviewed **field-level security** to hide sensitive fields from those who shouldn't see them, and **record-level sharing** to ensure users only see the records they're meant to. Through criteria-based sharing and manual sharing, we can accommodate complex business scenarios without any code.

We then looked at new tools like **Restriction Rules**, which let us go a step further by removing access to certain records in orgs that are otherwise open – adding a powerful "last mile" control for special cases. Alongside, we touched on **modern features** such as permission set muting, scoping rules for UI convenience, and user access policies to automate assignments. These advancements underscore Salesforce's commitment to fine-grained, admin-friendly security.

To maintain a strong security posture, we saw how to leverage **Security Health Check** and other compliance tools. By regularly reviewing the Health Check score and addressing high-risk settings (password policies, session settings, etc.), we keep our org aligned with best practices and regulatory requirements. We highlighted the importance of **MFA** – ensuring every user's login is doubly verified – and saw how declarative settings can enforce that without burdening users unnecessarily. We also emphasized **identity management**: from login IP ranges and hours to single sign-on integration, it's critical to control entry points to our org. With login flows and careful profile restrictions, we can add extra verification steps for those who need it, further aligning our Salesforce access with the principle of least privilege.

Finally, we added a robust **data protection** layer with platform encryption. By encrypting sensitive fields (and files) at rest, we guard against threats beyond the application layer. We discussed how to implement encryption thoughtfully – targeting the right data and managing keys responsibly – to meet compliance mandates and give our stakeholders confidence that their data is safe, even in extreme scenarios. We also considered the trade-offs (like search limitations) and ways to mitigate them, ensuring security enhancements do not unduly hamper business users.

Throughout the chapter, we included real-world examples and **ChatGPT callouts** demonstrating how AI can assist in security configuration – whether it's troubleshooting a user's access issue, generating a quick tip for cleaning up profiles, or outlining a rollout plan for MFA. These interactions showcase how modern admins can collaborate with tools like ChatGPT to save time and gain clarity, effectively having a smart co-pilot for security tasks.

Key Takeaways:

Layered Security Model: Use the combination of OWD, roles, profiles, permission sets, and sharing rules to enforce least privilege. Start with the most restrictive settings and grant access as exceptions, not the other way around.

Profiles vs Permission Sets: Simplify profiles to baseline access and favor permission sets (grouped into permission set groups) for adding permissions. This yields flexibility and easier administration as your org grows.

Field and Record Controls: Protect sensitive fields with Field-Level Security and use sharing rules (owner-based, criteria-based) to share records thoughtfully. Employ manual sharing sparingly for one-offs. Remember, page layouts are not a security measure – always enforce via profiles/perm sets for fields.

Newer Security Features: Apply Restriction Rules if you need to hide subsets of records in otherwise open data sets. Use muting permission sets to fine-tune grouped permissions. Keep up with Salesforce's evolving security features to continuously refine access control.

Ongoing Compliance: Regularly run Security Health Check and address any high-risk or medium-risk items. Ensure your org meets or exceeds your company's security policies (password rotation, session timeout, etc.). Leverage audit logs, field history, and event monitoring to keep an eye on security-critical activities.

Strong Identity Management: Implement MFA for all users – this single step dramatically reduces account compromise risk. Use login IP ranges or SSO policies to constrain where logins can occur for especially sensitive roles. Integrate Salesforce with corporate SSO for centralized control, and turn off direct logins when appropriate to enforce use of the more secure SSO path.

Data Encryption: Enable Platform Encryption for fields containing highly sensitive data to add encryption-at-rest. Plan encryption with awareness of its impact on app functionality (search, sort) and adjust processes accordingly. Manage encryption keys diligently, and use BYOK if your policy demands maximum control. Combine encryption with other measures (like not over-sharing those fields and monitoring access) for defense-in-depth.

Balance Security and Usability: Perhaps the most important lesson is finding the right balance. With the tools at your disposal, you can tailor security precisely – but always consider the end-user experience. Aim for a secure org that **enables** business productivity safely, rather than one that stifles it. This might mean using dynamic forms or assistance tools to guide users rather than simply denying access, or investing in user training about security features like MFA to smooth adoption.

By following these best practices, you've essentially built a **fortress of declarative security** around your Salesforce data – one that is robust yet adaptable. You can now confidently meet demands from IT auditors, client security questionnaires, and executives who ask "Is our data safe in Salesforce?" – because you have the settings and evidence to say, "Yes, and here's why."

This concludes our deep dive into security and access control. With a solid security foundation in place, you're ready to tackle the next challenge: leveraging Salesforce's powerful **declarative automation** tools to streamline and enhance business processes (all while respecting the guardrails we've set up). In the next chapter, we will shift focus from protecting what we have to optimizing how things get done – exploring how tools like Flow, Approval Processes, and Orchestration can be used (in tandem with ChatGPT's guidance) to build efficient, intelligent automations within our secure Salesforce environment. Get ready to unlock the next level of Salesforce's "clicks, not code" capabilities and see how automation can drive productivity in a well-architected, secure org.

7

DECLARATIVE PROCESS AUTOMATION (FLOW, ORCHESTRATION, AND BEYOND)

Salesforce Flow stands at the heart of declarative process automation, empowering Salesforce professionals to automate complex business processes with "clicks, not code." In this chapter, we will learn how **Flow** – and its extended family of tools like Flow Orchestrator – enable rapid, flexible automation that drives business agility and efficiency. Experienced developers and admins know that a well-crafted flow can replace dozens of lines of Apex or retired tools like Workflow Rules and Process Builder. We will explore the variety of flow types Salesforce offers, from user-guided Screen Flows to behind-the-scenes record-triggered automations, illustrating how each contributes to streamlining operations.

The chapter begins by distinguishing the different **Salesforce Flow types** and when to use each. You'll learn how to choose the right tool for the job, whether it's a Screen Flow for a guided user wizard or a Scheduled Flow for overnight data updates. Next, we dive into advanced flow design principles, demonstrating how to structure flows for clarity and reuse. As flows grow in number and complexity, adhering to best practices ensures they remain maintainable and robust.

We'll then shine a light on common pitfalls that can plague declarative automation. Even with powerful no-code tools, mistakes like hard-coded

IDs, overly complex branching, or missing error handling can lead to fragile processes. By recognizing these issues early, you can steer clear of them or fix them before they cause trouble. This naturally leads into a design pattern gaining popularity: the **"One Flow per Object"** approach. We'll explain how consolidating automations for a given object can simplify management and reduce unintended conflicts, drawing on real-world guidance.

Modern Salesforce orgs are also contending with legacy automation built in now-sunset tools. We'll provide a practical guide to **migrating Workflow Rules and Process Builder to Flow**, helping you develop a migration strategy that preserves existing business logic while taking advantage of Flow's richer capabilities. This includes tips on using Salesforce's Migrate to Flow tool and reimagining automations for efficiency post-migration.

Beyond individual flows, Salesforce now offers **Flow Orchestrator** to handle multi-step, multi-user processes that were once difficult to manage declaratively. You'll learn how Orchestrator can string together multiple flows and human work steps into cohesive end-to-end processes – for example, an onboarding sequence that involves HR, IT, and Finance in turn. We'll cover Orchestrator's key concepts (stages, steps, and work items) and see how it elevates declarative automation to a new level of complexity management.

Finally, we'll wrap up with best practices for **deploying, testing, and maintaining** your flows over time. Building a great flow is only part of the story – you also need strategies for safe deployment, rigorous testing (including Salesforce's new Flow testing features), version control, monitoring, and continuous improvement. By the end of this chapter, you'll have a 360° view of declarative process automation in Salesforce: the tools at your disposal, how to design flows expertly, how to avoid known pitfalls, and how to ensure your automations remain reliable as your Salesforce org evolves.

EXPLORING SALESFORCE FLOW TYPES

Salesforce provides several types of flows, each suited to different automation scenarios. Choosing the correct flow type is the first step to building effective declarative solutions. The main flow types we'll explore are:

Screen Flows: Flows that include screens for user interaction. These run only when initiated by a user (for example, via a button or Lightning page) and guide the user through a series of screens and inputs. Use Screen Flows to create wizards or guided data entry processes; for instance, a service agent stepping through a guided troubleshooting script or a sales rep completing a multi-step lead qualification form. Screen Flows improve user experience by collecting or displaying information with context – they are the declarative way to build custom UI without code. Users can invoke screen flows from many entry points – a Lightning record page, a quick action button, an Experience Cloud site, or even within a Flow Orchestration. This flexibility means you can surface tailored flow experiences wherever needed in your application.

Record-Triggered Flows: Flows that launch automatically when a record is created, updated, or deleted. These run in the background (no user interface) and are often configured to run either **before-save** (to update fields very efficiently) or **after-save** (to perform actions like create related records, send emails, etc.). A before-save flow (also called a **Fast Field Update**) makes changes to the record almost instantly as part of the save operation, whereas an after-save flow can perform additional actions (such as creating related records or sending notifications) after the record is saved. Record-triggered flows are analogous to database triggers – they respond immediately to data changes. For example, you might use a record-triggered flow to automatically assign a new Task when an Opportunity is marked "Closed Won," or to update an Account's status when all its related Cases are closed. Record-triggered flows have become the workhorse for enforcing business rules in real time.

Scheduled Flows: Flows that run at a specific time or on a recurring schedule (daily, weekly, etc.). These are ideal for batch jobs and routine maintenance tasks. A scheduled flow might run every night at 2 AM to flag stale opportunities, or every Monday morning to send weekly summary reports to sales teams. When you configure a scheduled flow, you can specify criteria so it processes only certain records (for example, opportunities with a Close Date in the past week). Like record-triggered flows, scheduled flows run in the background without user input. They can perform bulk processing – for example, closing out all cases that haven't been updated in 30 days – making them useful for housekeeping and data quality enforcement.

Autolaunched Flows: Flows that have no trigger on their own – they are launched by other means (such as from Apex code, via the REST API, from another flow, or even from a custom Lightning component or button). Autolaunched flows run in the background and do not support screens or user interaction. Think of autolaunched flows as reusable "functions" in your org's automation toolkit: they can accept input parameters, execute complex logic (even calling subflows or Apex), and return outputs to the process that invoked them. These flows are often used as subflows (invoked by other flows) or to encapsulate automation that might be kicked off by various events outside the UI. For example, you might create an autolaunched flow that calculates a "Lead Score" and call it from multiple places (a record-triggered flow on Lead creation, a Screen Flow that a user runs on-demand, and even an Apex trigger). Platform Event–triggered flows are another kind of autolaunched flow – they launch when a platform event message is received, allowing you to respond to external event streams in real time.

Each flow type has its role. If a business process requires user input or guided steps, a Screen Flow provides that interactive experience. If the process needs to run invisibly in the background in response to data changes, a Record-Triggered Flow is usually the best choice. For logic that must run at specific times or intervals, Scheduled Flows fill the need. And for everything else – especially reusable or on-demand logic – Autolaunched Flows offer maximum flexibility.

ChatGPT Prompt (Typing): "I need to automate updating related records whenever an Opportunity is closed, and sometimes I want to let a user manually trigger that update via a button. What flow types should I use?"

ChatGPT: "Use a **Record-Triggered Flow** for the automatic updates when an Opportunity is closed (after-save). This will run in the background whenever the trigger conditions are met. For the manual trigger via a button, create a **Screen Flow** or **Autolaunched Flow** that can be invoked from the button. The screen flow could provide confirmation or options to the user, or you could use an autolaunched flow if no user interaction is needed. Then, embed that flow in a Lightning action or button. In summary: record-triggered flow for the automatic part, and a screen/auto-launched flow for the on-demand button."

Selecting the appropriate flow type at the outset helps ensure your automation is efficient and maintainable. In practice, you will often use a combination of flow types in an org. For example, a Screen Flow might invoke an Autolaunched Flow (subflow) to perform calculations behind the scenes, or a Record-Triggered Flow might create a Task and also launch an autolaunched flow for additional processing. Understanding the strengths of each type will guide you to the right approach as requirements evolve.

Now that we've covered the spectrum of flow types and their use cases, let's explore how to design flows effectively. In the next section, we'll look at principles and techniques to build flows that are clear, modular, and easy to manage as your business processes grow in complexity.

ADVANCED FLOW DESIGN PRINCIPLES

Building flows is not just about making them work – it's about making them clear, efficient, and easy to maintain. As your org's automation grows, well-designed flows save time and prevent headaches. Here are some advanced design principles to apply when creating Salesforce flows:

Document and Name Everything Clearly: Just like well-written code, a well-designed flow should be self-explanatory. Always fill in your flow's description with its purpose, the objects it touches, and any assumptions. Use consistent naming conventions for flow elements and variables – for example, prefix text variables with "txt" or "var", label decision outcomes like "NoOpenCases" or "NeedsApproval", etc. Clear labels help you (and the next person) understand the flow at a glance. Adding brief descriptions to steps (via the "Notes" field on elements) can provide helpful context. If you come back to a flow six months later, you shouldn't scratch your head wondering what "Screen 4" or "var-

Temp" was for – good labels and notes will save your future self a lot of time.

Modularize Using Subflows: Avoid gigantic, monolithic flows that try to do everything. Instead, break complex processes into smaller subflows that can be called as needed. Subflows let you re-use common logic rather than duplicating it in multiple flows. For example, if several processes require sending a "Welcome" email, build one **Send Welcome Email** autolaunched flow and call it from various parent flows. This keeps each flow focused and easier to read. Modular design also means when a requirement changes, you can update one subflow in one place rather than hunting down redundant logic in many flows. Think in terms of "Lego blocks" – create small flows for discrete tasks (like calculating a discount, assigning ownership, etc.) and assemble them into larger processes via subflow elements. If you find yourself scrolling endlessly in one massive flow canvas, that's a sign the flow should be refactored into subflows.

Structure Your Logic with Simplicity: Flows can implement complex business rules, but you should strive to keep the structure as simple as possible. Use **Decision** elements to branch logic cleanly, but don't cram too many conditions into one decision if it hurts readability. It can be clearer to use multiple decision blocks in series, each with a specific yes/no question, than one mega-decision with 10 outcomes. Also, avoid hard-coding values inside flow logic. For example, don't embed a specific record Id or a fixed discount percentage in a decision – that's difficult to spot and update later. Instead, use **Custom Labels** or **Custom Metadata** to store business constants, or use formula resources to calculate conditions. (One technique is to create a formula resource that encapsulates a complex condition, then simply have a Decision check if that formula is true – this keeps the decision logic visually clean.) This way, if a business rule changes (say, the discount threshold moves from 10% to 15%), you can adjust a label or metadata value without editing the flow. The goal is to make the flow logic as declarative and adaptable as possible.

Design for Bulk and Performance: Salesforce flows execute under the same governor limits as Apex code, so efficient design matters. Always consider how your flow will handle large volumes of records. Use

Collection variables to process lists of records in a single operation, and avoid performing SOQL queries or DML **inside a loop**. For example, if your flow needs to update 100 related records, it's far better to gather them in a collection and update them all at once with a single **Update Records** element, than to loop 100 times performing one update per record. The latter approach could quickly run into limits (and is slower). Likewise, use **Get Records** outside of loops with proper filters to retrieve only the data you need. Efficient flow design – sometimes called "bulkification," borrowing the Apex term – ensures your automation will scale as data volumes grow. Salesforce's recent enhancements like **Fast Field Updates** (before-save flows) improve performance for certain operations , so leverage them when appropriate. In summary, treat your flow logic with the same performance mindset as code: do things in bulk and avoid unnecessary operations. Keep in mind Salesforce's limits (e.g. a single transaction can execute at most 150 DML statements and 100 SOQL queries). A flow that updates records inside a loop could hit those limits quickly. Designing with bulk operations prevents hitting such limits and keeps your flows running fast.

ChatGPT Prompt (File Upload): "I'm uploading a screenshot of my Flow (ChatGPT history is off for privacy). Can you review its design and suggest improvements?"

ChatGPT: "Sure – I've received your flow diagram. (I see you've disabled chat save, which is a smart step to keep this design confidential.) Looking at the flow, I notice a few areas for improvement. First, I see you've hard-coded an Id in one of the decision criteria – it's generally better to avoid hard-coded IDs. Instead, consider using a **Get Records** to query that record by a unique field, or store the Id in a Custom Metadata type if it's a config value. Second, your flow has two separate paths that both send a similar email – that's a great opportunity to use a subflow. You could create one email-sending autolaunched flow and call it from both places, which would simplify maintenance. I also recommend renaming "Decision 3" to something meaningful (perhaps "Check Invoice Status") so that anyone reading the flow can understand its purpose. Overall, the flow's structure is solid – with a few tweaks (externalizing that Id, reusing a subflow for emails, and improving some labels) it will be much easier to manage long-term."

Pro Tip: Use subflows to handle any logic that requires higher privileges. For example, if a Screen Flow running in user context needs to update a record the user can't normally edit, you can have it invoke an Autolaunched subflow that runs in system context to perform that update. This pattern keeps your main flow user-friendly while a helper flow does the "heavy lifting" with elevated permissions behind the scenes.

By applying these design principles, you'll make your flows more robust and easier to work with. A well-designed flow should feel like a clear flowchart of the business logic, not a tangled web. When another admin or developer opens your flow, they should quickly grasp what it does. In the next section, we'll look at common pitfalls and mistakes in declarative automation – essentially, what **not** to do – and how to fix or avoid those issues to keep your Salesforce processes running smoothly.

AVOIDING PITFALLS IN DECLARATIVE AUTOMATION

Even with the best intentions, it's easy to make mistakes when building declarative automations. Let's look at a few common pitfalls that can undermine your flows, and how to remedy them:

Hard-Coded IDs or Values: One of the most notorious mistakes is hard-coding record IDs, user IDs, or magic values directly in a flow. For example, referencing a specific Opportunity Id in a Decision element or using a fixed discount percentage in multiple places. Hard-coded references make your flow brittle – if you refresh a sandbox or the business logic changes, the flow breaks or behaves incorrectly. **Remedy:** Replace hard-coded IDs with queries (**Get Records**) that find the needed record by criteria (e.g. a unique name or attribute), or store constants in Custom Metadata/Custom Settings so admins can update values without modifying the flow. This way, your flow adapts across orgs and over time. Salesforce explicitly advises against hard-coding IDs in flows , and using dynamic references ensures your automation won't unexpectedly fail due to an Id mismatch.

Over-Complex or Unmanageable Flows: As requirements evolve, some admins keep piling onto a single flow until it becomes a tangled mess, or they create new flows for every little variation without a strategy. The result can be flows that are huge, hard to follow, or multiple flows on one object triggering in unpredictable order. Overly complex flows are not only difficult to maintain – they can also hit limits or behave in unintended ways. You might notice a flow doing way more than one logical job, with dozens of elements and branches (or conversely, five different flows all firing on the same object update). **Remedy:** Simplify and organize. If a flow is doing too much, break it into subflows (as discussed) or consider the **"One Flow per Object"** design pattern (coming up in the next section) to consolidate logic. If you have many flows on one object, consider using Salesforce's trigger order feature to control execution sequence, or better yet, merge logic where it makes sense. The key is to avoid having automation logic scattered in a way that no one person can understand the full picture. A cleaner, modular approach will save you from headaches down the line.

Ignoring Bulk Limits and Performance: Another pitfall is assuming flows will only run for single records or small datasets. A flow might work in testing for one record, but then fail or run slowly when 200 records trigger it at once. Signs of this pitfall include loops that do a SOQL query or update each time through (causing limit errors) , or flows that process large record sets without any batching. **Remedy:** Always build flows with bulk operations in mind. As noted earlier, use collections to handle multiple records in one go, and move any data retrieval or updates outside of loops. Test your record-triggered flows by performing mass updates (e.g., import 100 records) to see how they hold up. If something in your flow could potentially operate on many records, assume someday it will – and design accordingly. In practice, this means combining actions (like doing one update for 100 records, not 100 updates for 1 record each). By avoiding inefficient designs, you prevent runtime exceptions and ensure your automation scales. Salesforce's governor limits are strict, but well-built flows can handle large volumes within those limits if you follow best practices.

Inadequate Error Handling: Perhaps the most overlooked aspect of flow design is error handling. By default, if a flow hits an unhandled error (like a null value or a validation rule failure during a record update), it simply fails and may display a very generic error message to the end user (or no message at all if running in the background). This leads to frustrated users and admins unaware of issues. **Remedy:** Always plan for errors by using **Fault** connectors on critical elements (like record updates, creates, etc.). A fault path can, for example, display a friendly error message on a screen flow ("Oops, something went wrong with updating the record. Please contact support.") instead of the dreaded generic Salesforce error. For autolaunched flows, fault paths can log an error record or send an email notification to admins with details (using an Email Alert action). By catching exceptions, you ensure your flows fail gracefully – or even better, attempt recovery or alternative actions when something goes wrong. It's also good practice to test your flows with bad data or missing data to see where they might break, so you can add checks or fault handlers in those places. In short, don't ignore errors – handle them deliberately so that a flow failure doesn't turn into a business outage.

ChatGPT Prompt (Speaking): "(Voice) Uh, my screen flow just shows 'An unhandled fault has occurred' – what does that mean and how do I fix it?"

ChatGPT: "It sounds like your flow encountered an error that wasn't caught by a fault path. In Salesforce, when a flow fails without a configured error handler, users see an 'unhandled fault' message, which isn't very informative. To fix this, you should add **Fault** connectors to your flow elements that might fail (like record updates or creates). For example, connect a fault path from the problematic element to a Screen that shows a user-friendly error message (explaining what went wrong or what to do next). Also, consider logging the error – you can do this by creating a record in a custom "Flow Error Log" object, or sending an email alert to an admin. This way, instead of the generic message, users get guidance, and you get notified with details when something goes wrong. Essentially, 'unhandled fault' means the flow hit an exception with no instructions on what to do, so by adding fault handling, you'll resolve the issue and make the flow more robust."

By steering clear of these pitfalls, you make your declarative automations far more reliable and easier to maintain. It's worth periodically reviewing your flows with these "gotchas" in mind – often a few small adjustments (like removing a hard-coded value or adding an error handler) can significantly improve a flow's durability. Next, we'll discuss a specific design pattern that addresses the complexity of managing multiple flows: the **One Flow per Object** approach, which can greatly simplify your automation landscape.

IMPLEMENTING "ONE FLOW PER OBJECT" DESIGN PATTERN

Managing automation becomes especially tricky when multiple flows operate on the same object. The "One Flow per Object" design pattern addresses this by consolidating all record-triggered automation for an object into a single flow (per context – typically one before-save flow and/or one after-save flow for that object). The idea is to have a central flow that acts as a hub for all that object's logic, rather than many disparate flows running independently.

Why One Flow per Object? This approach brings several benefits. First, it avoids conflicts and uncertainty around flow execution order. If you have five flows all firing after an Account update, the execution sequence can be hard to control (and before the Flow Trigger Explorer existed, it was essentially arbitrary). With one flow, you eliminate that issue – all the logic is in one place and it executes in a predictable, unified sequence. Second, a single flow can often do the work more efficiently than multiple separate flows. For example, instead of each flow querying for the record or related data independently, one flow can query once and use the results across various branches. This reduces redundant operations and helps prevent hitting governor limits. Salesforce architects have noted that consolidating automation into one flow per object tends to reduce the total number of queries and DML operations, making it less likely to encounter limits or race conditions. Third, having one flow to rule them all makes it much easier to understand and maintain your business logic. If an object's processes are all in one flow, you know exactly where to look to update or debug them. There's no risk of forgetting about a "rogue" flow that also runs on, say, Contact creation – everything is centralized.

How to Implement: Start by reviewing existing flows on the object. Identify what each does and see if they can be logically combined.

Often, you'll find flows that were built at different times for similar triggers. In a one-flow-per-object model, you can merge their logic by using Decision elements or subflows within one master flow. A common strategy is to use one **before-save** flow for all quick field updates on the object (for example, formula-like calculations or setting default field values), and one **after-save** flow for everything else (creating related records, sending emails, invoking subflows, etc.). Within that after-save flow, use Decision branches to handle the different scenarios that multiple flows used to handle. For example, one branch could handle "If Opportunity is Closed Won, then do X," another branch "If Stage changed from Prospecting to Qualification, do Y," and so on – all within the same flow. Each branch can then invoke subflows or perform actions as needed. This segmentation within one flow replaces having separate flows for each scenario.

When implementing this pattern, make sure to thoroughly test that the combined logic still behaves as expected in all cases. It can be helpful to use subflow calls for pieces of logic to keep the main flow from becoming too monolithic. Essentially, your main object flow becomes a dispatcher that routes to subflows or different sections based on conditions, but all under one roof. Document the flow well (as it becomes a critical piece of automation) – include in the description which business processes it covers and any important notes for future maintainers. Once in place, this single-flow approach greatly simplifies ongoing maintenance: need to add a new action on Case close? You'll add a branch in the one Case after-save flow, rather than creating yet another flow.

Real-World Example: Suppose your org had three flows on the **Opportunity** object: one calculates and updates a custom "Health Score" field when certain fields change, another creates a follow-up Task when an Opportunity is closed, and another updates related Account info when an Opportunity's Amount goes above a threshold. In the one-flow model, these would all be unified. You might build a single Opportunity after-save flow that: (1) checks if relevant fields changed and updates the Health Score (perhaps in a before-save subflow for efficiency), (2) after save, checks if Stage changed to Closed Won and then creates a Task, and (3) also after save, checks if Amount > threshold and calls a subflow to update the Account. All of that logic lives in one flow, with three Decision outcomes. In the end, you have one flow to manage instead of three, and it performs the actions in a controlled sequence within one transaction.

ChatGPT Prompt: "I have several record-triggered flows on the Contact object doing different things (assigning owners, sending alerts, etc.). Would it be better to combine them into one flow?"

ChatGPT: "In many cases, yes. Combining them into a single Contact flow can simplify management and reduce redundancy. You'd use Decision elements to handle the different conditions that each separate flow was handling. This ensures all your Contact logic runs in one place sequentially, rather than in separate flows where order is harder to control. Before moving to one flow, verify that the combined logic won't hit any limits (one flow handling all might do more work at once). Usually, though, Salesforce flows are optimized enough that one flow can handle multiple tasks efficiently. You'll want to clearly document each branch of the flow so it's obvious which scenario (owner assignment, alert sending, etc.) is being handled. The result is easier troubleshooting – when a Contact is updated, you check one flow's execution rather than juggling multiple flows. Just be sure to thoroughly test the unified flow. If it's well-designed, the one-flow-per-object approach can definitely pay off in streamlined maintenance."

Pro Tip: Salesforce now allows you to set a specific **Trigger Order** on record-triggered flows. If for some reason you do keep multiple flows on the same object, use this feature to control the order in which they execute (e.g., assign one flow Order 1, another Order 2). This helps avoid ambiguity. However, be cautious – relying on trigger order can add complexity. Often, consolidating flows (or at least minimizing how many you have per object) is a cleaner solution in the long run.

Adopting the "One Flow per Object" pattern can greatly enhance the manageability of your org's automation. It forces you to be intentional about an object's business logic and reduces the sprawl of automation. Administrators who have migrated dozens of Workflow Rules into a single object flow often report that it's much easier to adjust criteria or add new actions without losing track of what impacts what. In the next section, we'll tackle the challenge of legacy automation – specifically, how to migrate old Workflow Rules and Process Builder processes into the modern Flow framework in a systematic way.

MIGRATING WORKFLOW RULES AND PROCESS BUILDER TO FLOW

With Salesforce Flow now the centerpiece of automation, migrating legacy Workflow Rules (WFR) and Process Builder (PB) processes into Flow has become a priority. Salesforce has officially announced that Workflow Rules and Process Builder will no longer be supported after December 31, 2025 , so customers are urged to transition their automations to Flow Builder well before that date. How should you approach this migration in practice?

1. **Assess Your Existing Automation:** Begin by taking inventory of all active Workflow Rules and Process Builder processes in your org. Document what each does – the object, the trigger criteria, and the outcome (field update, email alert, task creation, etc.). This assessment will help you identify overlaps or easy wins. For example, you might discover you have three workflow rules on Case that all send slightly different email alerts for related conditions; those could potentially be combined into one flow. Also identify any processes that are outdated due to changed requirements – you can plan to retire those rather than migrate them.

2. **Plan the Migration Strategy:** Not every legacy automation needs a one-to-one recreation in Flow. Often, multiple workflow rules on the same object can be consolidated into a single record-triggered flow with Decision branches (as we discussed in the previous section). Likewise, a Process Builder with many criteria nodes might translate into a single flow with multiple Decision elements. Decide whether to merge or separate logic based on what makes sense. Salesforce recommends adopting a "rebuild and enhance" mindset – meaning, use the migration as a chance to improve the design (e.g., use before-save flows for what used to be workflow field updates, combine related automations into one flow, and apply best practices like subflows and fault paths). Identify high-impact or frequently-triggered automations and prioritize those in case you need to migrate in phases.

3. Use the Migrate to Flow Tool (Where Applicable): Salesforce provides a Migrate to Flow tool that can automatically convert Workflow Rules into flows (and as of Spring '23, it can convert certain Process Builders as well). This tool is accessible in Setup and is great for simple rules. Using it is straightforward: select a workflow rule and click **Migrate to Flow** – Salesforce will generate an equivalent flow for you, set to **Inactive**. For example, a workflow rule with criteria "Status = Closed" and action "Field Update: Priority = High" will become a record-triggered flow on that object with the same entry criteria and an Update Records element to set Priority to High. After using the tool, review the resulting flow in Flow Builder. Salesforce's conversion is usually accurate for the basic logic, but you'll want to check things like: Did it set the flow to run **after-save** or **before-save** appropriately? Are picklist values or formula criteria translated correctly? The tool saves time, but it's not foolproof – especially for Process Builders with multiple branches or scheduled actions. Those might require some manual tweaking or a partial rebuild.

4. Rebuild Complex Processes Manually: For more complex Process Builder processes (for instance, those with many decision branches, scheduled actions, or those that invoke Apex), it might be better to rebuild them from scratch in Flow. Use the PB as a blueprint: recreate each criteria node as a Decision in a new flow, and each action as the equivalent Flow element (Field Updates become Update Records, Email Alerts become core Send Email actions, outbound messages might become HTTP Callouts with Flow integrations, etc.). The good news is Flow can do everything Process Builder could do – and more. In fact, some things that were clunky in PB (like looping through related records or handling record deletions) are easier in Flow Builder. While rebuilding, apply improvements: for example, combine two criteria that result in similar outcomes into one Decision with OR logic, or replace hard-coded values in the PB with variables or constants in the flow. Ensure that your new flows are optimized (e.g., if the old PB updated 10 child records in a loop, your new flow can update a collection of child records in one go). Take advantage of Flow's advanced features (like Loop and Collection operations) to make the new solution more efficient than the old.

5. **Test Thoroughly in a Sandbox:** Before turning anything off in production, test your new flows in a safe environment. Run scenarios for each migrated rule or process: does the flow fire at the right time and perform the same (or better) outcomes? It can be helpful to do side-by-side comparisons: for example, if you have not yet deactivated the old automation in a sandbox, trigger both the old and new automations (on separate but similar records) and verify the results match. Pay special attention to side effects: because flows run in a single transaction by default, an old PB that did something in a scheduled action might now happen immediately in a flow, which could be fine but should be noted. Also test bulk behaviors – where a workflow might not have considered bulk updates, your flow will handle batches of up to 200 records, so simulate a mass update to ensure the flow handles it without errors or hitting limits.

6. **Deploy and Monitor:** When you're confident everything works, deploy the new flows to production. It's wise to schedule this during a low-usage period. The key is to **activate the new flows and deactivate (or delete) the old Workflow Rules and Process Builders in quick succession** so you don't double-execute logic. One approach is to first deactivate the Workflow Rules or PB processes (so they stop firing), then activate the new flows. Alternatively, you might deploy the flows in advance and leave them inactive until cutover time (the Migrate to Flow tool by default leaves converted flows inactive). Once switched over, closely monitor the system. Watch for any flow error emails or unusual behavior in the days following. Engage users to report any discrepancies. It's normal to do a little fine-tuning in production after a big migration – maybe a flow needs an added null check or an updated assignment that wasn't apparent in sandbox testing. As a safety net, keep backup copies of your old Workflow Rules and Process Builders (you can leave them deactivated or export their metadata) until you're fully confident in the new flows.

ChatGPT Prompt: "We have about 20 Workflow Rules and a few Process Builders in Sales Cloud – what's the best way to migrate them to Flows?"

ChatGPT: "Start by grouping them by object and function. Identify which ones can be combined into a single flow. Salesforce's Migrate to Flow tool can handle the simple one-for-one conversions (for example, straightforward field updates or email alerts). For the more complex automations (especially those Process Builders with multiple branches), plan to rebuild them manually in Flow – which is actually an opportunity to streamline the logic. Use before-save flows for anything that was just doing field updates (because those run much faster, similar to how workflow rules did). Use after-save flows with Decision elements to handle what used to be multiple workflow rules or PB criteria. Migrate in phases if necessary: maybe start with a couple of rules, test them in a sandbox, then deploy to production. Always deactivate the old rule or process once the new flow is active, so you don't get duplicate actions. And communicate with users if needed – the goal is that nothing changes for them in terms of functionality, just that it's powered by Flow now. By migrating methodically, you'll end up with cleaner, more powerful flows and you'll be off workflow rules/process builders well before the end-of-support deadline."

Pro Tip: Don't perform a migration "heads down" without reevaluating your automations. Use this chance to clean up and improve. Often you'll discover redundant or outdated rules – consider retiring them instead of rebuilding. Also involve business stakeholders for UAT when possible, to ensure the new flows meet their needs. Migrating to Flow isn't just a technical task; it's an opportunity to realign your automation with current business processes and best practices.

Transitioning legacy automation to Flow may seem daunting, but it usually pays off in simplified maintenance and better performance. Admins who have done it often find that what used to require dozens of workflow rules can be handled by a handful of consolidated flows. Once this migration is complete, you'll be fully leveraging Salesforce's modern automation toolkit. Next, we'll explore Salesforce Flow Orchestrator – a tool that extends Flow's capabilities to coordinate multi-step, multi-user processes that go beyond what a single flow can easily handle.

LEVERAGING SALESFORCE FLOW ORCHESTRATOR

As powerful as individual flows are, some business processes are too complex for a single flow to manage easily – especially those involving multiple people, long time frames, or coordinated parallel tasks. **Salesforce Flow Orchestrator** is designed to handle these scenarios by orchestrating multiple flows and user interactions into a cohesive, guided process. Think of Flow Orchestrator as the conductor of a multi-part automation symphony: it doesn't replace Flow, but rather organizes and sequences multiple flows (and manual steps) across users and over time.

What Orchestrator Does: Salesforce defines Flow Orchestrator as a tool to "orchestrate multi-user, multi-step, and multi-stage flows". In practice, this means you can have a business process that spans several stages and assign different steps to different users or teams, all coordinated by the Orchestrator. For example, imagine an employee onboarding process: Stage 1 might have HR fill out a new hire form (Screen Flow assigned to an HR user), Stage 2 might have IT perform equipment setup (background automation plus an interactive checklist for an IT technician), Stage 3 might wait 30 days and then ask the new hire's manager to complete a probation review (another user-assigned step). Flow Orchestrator can manage this entire sequence – ensuring each step happens in order, handing off tasks to the right person at the right time, and even running steps in parallel when possible.

Key Components: Flow Orchestration introduces a few new concepts on top of regular flows:

Orchestration: This is the container for your end-to-end process. You create an Orchestration (either auto-launched or record-triggered) that kicks off the whole sequence. The Orchestration is what you deploy and activate (similar to activating a flow).

Stages: A Stage is like a phase in the process. Stages group one or more **Steps**, and stages execute sequentially (you can even separate stages with a Decision element to support conditional branching).

Steps: Each Step represents a unit of work. There are two types: **Interactive Steps** (assigned to a user or group, requiring human action via a Screen Flow) and **Autolaunched Steps** (which execute an auto-launched flow behind the scenes without user input). For an Interactive Step, you'll specify *who* it's assigned to (a specific user, queue, role, etc.), *which* flow should be run, and *where* the flow should appear (for example, on a record page via the Work Guide component). For an Autolaunched Step, you simply point to an autolaunched flow (often a subflow you've created for a specific task) that should execute at that point.

Work Items and Work Guide: When an Orchestrator reaches an interactive step, it generates a Work Item – essentially a task record that indicates a user needs to complete something. The assigned user is notified (e.g., via a Salesforce notification or email) and can find the task in the **Flow Orchestration Work Guide** component in Salesforce. The Work Guide is a Lightning component you add to record or Home pages; it shows users their active Orchestrator tasks and lets them launch the associated Screen Flows to complete those steps. As each Work Item is completed (or marked done), Orchestrator transitions the process to the next step or stage.

Use Cases for Orchestrator: Any process that involves multiple handoffs or actors over time is a candidate. For example:

Complex Approvals: While Salesforce has standard Approval Processes, Orchestrator can handle more elaborate multi-person approval scenarios. Imagine a contract that needs legal review, then finance approval, then final sign-off by a VP – possibly with a loop for edits if any approver rejects. Orchestrator can manage that sequence, assign each approval step to the right person, and even loop back if changes are required.

Case Escalation and Resolution: A high-priority support case might require a Tier 1 agent to collect information, then hand off to a Tier 2 expert for diagnosis, and finally involve a manager to approve a solution or perhaps reach out to the customer. With Orchestrator, each of these can be a stage with its own interactive or background steps, ensuring the case goes through all necessary phases. Some steps could even run in parallel (perhaps notifying a customer success manager while the technical investigation is underway).

Employee Onboarding/Offboarding: As mentioned, onboarding often spans HR, IT, Facilities, and the new hire's manager. Orchestrator is perfect for launching an onboarding checklist for each department – some steps can run in parallel (IT and Facilities setting up equipment simultaneously), and others in sequence (HR must complete paperwork before IT accounts are provisioned, etc.), with Orchestrator coordinating and waiting where appropriate.

Multi-Department Service Requests: For instance, a new product launch might require Marketing to create collateral, Sales to update pricing and communicate to prospects, Support to update the knowledge base, and Product to monitor early feedback – all as part of one "Launch" process. Orchestrator could coordinate tasks for each department (some sequential, some parallel) and ensure that, say, marketing materials are approved before sales communications go out.

In short, Orchestrator shines when a business process isn't just a single linear flow, but a coordinated dance between multiple players and automated actions. It brings transparency (via the Work Guide's task list) and ensures nothing falls through the cracks during long-running processes (Orchestrator can wait days, weeks, or months between stages if needed).

It's worth noting that Flow Orchestrator is an add-on in Salesforce (at the time of writing, it may require specific licenses or incur usage-based costs). For example, Salesforce might charge per orchestration run or per Work Item created. Always check with Salesforce pricing and your account executive about Orchestrator usage terms. However, don't let that deter you from leveraging it for critical processes – many organizations find the efficiency gains and error reduction in complex processes well worth it.

How to Build an Orchestration: You create an Orchestration using the Flow Builder (when you click "New Flow," you'll see options for **Record-Triggered Orchestration** or **Autolaunched Orchestration**). After choosing one, you define the Stages of your process and the Steps within each stage. For each Step, decide if it's interactive (assign to a user with a Screen Flow) or automatic (run an autolaunched flow). You will typically need to have the Screen Flows and autolaunched subflows built ahead of time so you can select them for each Step. A best practice is to keep those flows simple and focused on the task at hand (e.g., an "Approval Screen Flow" that just captures approve/reject and comments, or an "IT Provisioning Flow" that creates user accounts). This makes it easier to manage and reuse them. You can arrange Steps in a stage to run sequentially or in parallel (parallel steps will all be assigned at once and Orchestrator waits until all are completed before moving on). You can also add a Decision element between stages to decide which stage to go to next, based on outputs of prior steps (for example, if an approval is rejected, jump to a Rework stage; if approved, skip straight to a final stage). Once your orchestration is built and activated, be sure to add the **Work Guide** component to relevant Lightning pages so users can see their assigned tasks and launch flows to complete them.

ChatGPT Prompt: "We're planning a new hire onboarding process involving HR, IT, and Finance. Can Flow Orchestrator help, and how would we set it up?"

ChatGPT: "Yes, Flow Orchestrator is an ideal fit for that scenario. You would create an Orchestration with stages for each department's part of the process. For example, Stage 1: HR paperwork (assign an interactive step to an HR rep to complete a Screen Flow for HR tasks), Stage 2: IT setup (maybe two steps in parallel – one autolaunched step to create system accounts, and another interactive step assigned to an IT user for a hardware setup checklist), Stage 3: Finance setup (interactive step for the Finance team to set up payroll and benefits). You would specify who each step is assigned to – e.g., the HR business partner, an IT queue, the Finance manager – and which flows they need to run. The Orchestrator would then automatically create tasks for each step and notify the assignees. As they complete each task (using the Work Guide component in Salesforce to view and launch their steps), the Orchestrator moves the process to the next stage. If any stage needs to wait (say Finance shouldn't start until the employee's first day), you can configure a wait in the Orchestration. In the end, you get a coordinated onboarding process without writing any code. The key is to break the process into clear stages and steps, and have supporting flows for each step. Once that's set up, Orchestrator takes care of the sequencing and assignments."

Pro Tip: When deploying Orchestrations with interactive steps, add the **Flow Orchestration Work Guide** component to relevant Lightning pages (such as a Case page or a custom Onboarding record page). This gives users a clear to-do list of their orchestration tasks and guides them on what to do next. It greatly improves the user experience by providing context and easy access to the flows they need to run for each step.

Flow Orchestrator is a relatively new addition to the platform, but it unlocks scenarios that were previously very difficult to automate without custom code or clunky workarounds. By intelligently chaining together flows and user tasks, it enables true end-to-end process automation. As you begin using Orchestrator, start with a pilot use case (perhaps one of the examples above) to get familiar with stages and steps. Once mastered, you'll likely find many complex processes can be streamlined using this capability. In our final section, we'll discuss best practices for deploying and maintaining your flows (and orchestrations) over time, including testing, monitoring, and iterative improvement.

FLOW DEPLOYMENT, TESTING, AND MAINTENANCE

Building flows is only half the battle – you also need a strategy for safely deploying them and keeping them running smoothly. This section outlines best practices for releasing changes to flows, testing automation, and maintaining your flows over the long term.

Deploying Flows Safely: It's strongly recommended to build and test your flows in a sandbox (or scratch org) before touching production. Flows can be deployed using change sets or metadata deployment tools just like code. When deploying a new flow, you have a choice: you can deploy it as **Inactive** and then activate it in production manually, or deploy it already **Active** (if using a metadata API deployment with the flow marked active). A cautious approach is to deploy inactive, perform a quick smoke test in production by running it as an admin user, then activate it for all users. Also, remember that flows have versions – when you modify an active flow and click "Save As," you create a new version. The new version won't run until you activate it, and the old version remains available. This is great for rollback (you can always re-activate a prior version if needed). However, be mindful of the 50-version limit per flow. If you approach 50 versions, you'll need to delete some older versions to make room (or consider cloning the flow into a fresh flow if you want to preserve a history of changes). It's good practice to periodically clean up inactive flow versions that are no longer needed. Name your flows and versions clearly to indicate their purpose or changes (many teams include version numbers or dates in flow descriptions for tracking). Finally, plan deployments for times when you can monitor the outcome – deploying a critical flow right before a major business event (like end-of-quarter) isn't ideal. Plan changes when you have a buffer to quickly address any issues if something isn't right.

Testing and Quality Assurance: Because flows execute business logic, testing them is essential. Use Flow Builder's **Debug** feature extensively while building – it lets you run the flow as if under real conditions (you can provide input values and even simulate running as a different user profile). For record-triggered flows, Debug can simulate a record create/update and show you each step and outcome. Beyond manual debugging, Salesforce introduced a **Flow Testing** framework for record-triggered flows. This allows you to create declarative tests for your flows: you define input scenarios and expected outputs, and Flow Builder can run these tests on demand or during deployments (much like Apex unit tests). Embrace this capability! For example, you might create a test that "if a Case is escalated, the flow posts to Chatter and sets a follow-up task" and have the platform verify that behavior automatically whenever the flow changes. In addition to flow-specific testing, include flows in your broader testing plans. During User Acceptance Testing (UAT), have end users exercise scenarios that trigger your flows to ensure everything works as intended and is user-friendly. Catching an issue in a flow before it hits production is just as important as catching a code bug early. If your org has a formal release management process, treat flow changes with the same rigor as code changes: get a second pair of eyes to review complex flows, and consider storing flow definitions in source control to track changes over time.

Runtime Monitoring and Error Handling: Once flows (and orchestrations) are live in production, keep an eye on them. One easy proactive step is to configure **error notifications**. By default, if a flow interview fails without being caught by a fault path, Salesforce sends an error email to the last user who modified the flow. It's a good idea to specify a dedicated recipient (like an integration user or an admin group) for flow error emails so they don't get overlooked. Better yet, incorporate error handling in the flow itself: use Fault connectors to handle exceptions and notify the right people. For instance, a fault path on an Update Records element could send a custom email (via an Email Alert action) to the admin team with the error details, or create a record in a custom "Flow Error Log" object. Monitor the **Paused and Failed Flow Interviews** list in Setup – this shows any flow instances that are waiting (paused flows, perhaps for an Orchestrator step or a scheduled path) or that hit an error and stopped. Regularly reviewing this can alert you to issues that users might not report. Additionally, leverage Salesforce's logging: you can turn on Debug Logs for a user and reproduce a scenario – the debug log will include flow execution steps (look for lines with FLOW_START, FLOW_ELEMENT_BEGIN, etc.). This is invaluable for troubleshooting complex flows or orchestrations in conjunction with other logic. Some advanced teams use external logging frameworks (for example, the open-source Nebula Logger can be invoked from flows to log messages to a custom object) to capture detailed flow runtime info for analysis. In summary, don't "set and forget" your flows – monitor them like you would any critical part of the system, so you can address small issues before they become big ones.

Ongoing Maintenance and Improvement: Salesforce delivers enhancements to Flow with each release. Stay informed via release notes, Salesforce blogs, and the community. You may find new features that allow you to simplify flows that were built with workarounds. For example, the **Collection Filter** and **Collection Sort** elements introduced in Spring '22 eliminated the need for some loop logic that older flows had to use. Make it a habit to review your existing flows periodically. Are there flows that are no longer used or needed? If so, deactivate or remove them to reduce clutter and potential interference. Are there steps that every flow seems to do (like set a "Last Touched Date" on a record)? Perhaps centralize those in one subflow or a trigger flow to ensure consistency. Also, maintain documentation for your flows – whether it's inline (description fields) or an external document. Knowing *why* a flow was built (the business context) is extremely helpful when you or someone else needs to update it later. Consider a quarterly or biannual "flow audit" where you scan through flows, check for things like numerous versions, complex branches, or rising error counts, and then decide if refactoring or tuning is needed. As the business changes, your flows should adapt – don't be afraid to refactor a flow (or break it into simpler flows) if it starts handling too many divergent tasks. And always remember to leverage the declarative tools Salesforce provides: for example, if you find an old flow doing something that a newer feature (like Dynamic Related Lists or a Validation Rule) could handle more simply, evaluate if a flow is still the best solution for that piece of logic.

ChatGPT Prompt (Operator): "Use Operator to log in to my Salesforce org and check for any inactive flows that have no active versions."

ChatGPT (Operator): "Logging into Salesforce and scanning your Flow list... I found 2 flows that are inactive and have no active versions: **Lead Mass Update (Inactive)** and **Old Contract Flow (Inactive)**. These appear to be unused. I recommend archiving or deleting them if they're truly no longer needed, to keep your org tidy."

Pro Tip: Don't let your flow roster and versions sprawl indefinitely. If you find a flow has accumulated dozens of versions over time, take that as an opportunity to review and possibly refactor it. You might consolidate logic or start a fresh flow for clarity. Salesforce allows a maximum of 50 versions per flow , so as a matter of good housekeeping, trim old versions when you can. Maintaining a lean set of well-documented flows will make it much easier for you (and your successors) to manage automation in the long run.

In summary, treat your declarative automation with the same care as programmatic code: deploy methodically, test thoroughly, monitor actively, and iterate for improvements. Flows and orchestrations are living parts of your Salesforce org – give them attention and they will serve your business reliably. With the strategies from this chapter, you can ensure your Salesforce automations remain robust, scalable, and responsive to your organization's needs over time. The next chapter will transition from automation to user experience, where we'll explore building intuitive UIs and dynamic pages on the Salesforce platform – a perfect complement to the powerful backend processes you now have in place.

SUMMARY

In this chapter, we explored the full spectrum of declarative process automation on Salesforce and learned how to do it the *right* way. We began by distinguishing the types of Salesforce flows – from Screen Flows that guide users step by step, to record-triggered and scheduled flows that run behind the scenes, to autolaunched flows invoked by other processes. With real-world scenarios, you saw when to use each flow type and how choosing the correct tool sets the foundation for success. We then explained advanced flow design principles: crafting flows that are modular, well-structured, and maintainable. You learned to use subflows to break down complex processes, to avoid hard-coding and other anti-patterns, and to build flows that can handle scale (bulk data) gracefully. We highlighted common pitfalls – like hard-coded IDs, overly complex collections of flows, and missing error handling – and showed how to avoid or remediate them (for example, utilizing Decision logic wisely and adding Fault paths for resilience).

One major theme was **simplification**: the "One Flow per Object" pattern demonstrated how consolidating automations can improve reliability and clarity. By managing an object's logic in one flow, you reduce ambiguity in execution order and make it easier to adjust business rules in one place. This concept, along with guidance on migrating legacy Workflow Rules and Process Builders into modern flows, positions your Salesforce org for the future. Migrating those older automations not only ensures supportability as Salesforce retires Workflow Rules and PB, but also gave us a chance to rethink and optimize processes – often combining several old rules into a single efficient flow. We outlined a migration approach (inventory, plan, convert with the tool, rebuild complex logic, test, deploy) so even large orgs can systematically transition to Flow, reaping benefits in consistency and power (for instance, using before-save flows for speedy field updates and after-save flows for everything else).

Salesforce Flow Orchestrator extended our toolkit beyond single flows. By learning how to orchestrate multi-step, multi-user workflows, you are now equipped to automate processes that span departments and time. We saw how Orchestrator's stages and steps can assign work to the right people in the right order – automating complex scenarios like multi-tier approvals and employee onboarding with clicks instead of custom code. This opens up new possibilities to streamline business operations that previously might have required external project management or manual coordination. Combined with core Flow features, Orchestrator gives you the ability to design both the micro-flows (individual task automation) and the macro process (the end-to-end journey) on the Salesforce platform.

Throughout this chapter, we also demonstrated how ChatGPT can be a valuable companion for Salesforce professionals. We saw examples of using ChatGPT to review flow designs, generate best-practice suggestions, troubleshoot errors, and even brainstorm orchestrations. Whether through text or voice – or leveraging advanced modes like Operator and Sora – AI tools can augment your workflow, from documenting a flow to automating parts of the deployment process. Embracing these tools can accelerate learning and help ensure you're following best practices (for instance, having ChatGPT double-check for hard-coded values or missing fault paths in your design).

To keep your declarative automations running smoothly, you learned the importance of rigorous testing and proactive maintenance. Treat flows with the same discipline as code: use sandboxes for development, create test scenarios (especially now with Flow Test features), and monitor your flows in production with error logs and performance reviews. By doing so, you catch issues early and continuously improve your automations as business needs evolve. Salesforce's point-and-click automation is incredibly powerful, but with great power comes the responsibility to govern and refine it – a theme we encountered in topics like version control, documentation, and flow governance.

In conclusion, declarative process automation in Salesforce – when applied using the best practices we've covered – can deliver robust, scalable solutions with a fraction of the effort of writing code. You can build everything from simple field updates to sophisticated multi-user processes in a declarative way, increasing agility and aligning IT solutions closely with business experts. You now have a well-stocked toolbox: multiple flow types for different needs, design patterns to keep flows clean, strategies to avoid pitfalls, methods to migrate legacy automations, and even Orchestrator for big-picture processes. With this knowledge, you can implement automation that is not only effective and efficient, but also maintainable and adaptable for the long haul.

Next, we turn our attention to the **user experience** side of the equation. Automating processes is crucial, but so is providing users with an intuitive and efficient interface. In the following chapter, we'll explore how to build dynamic, user-friendly Salesforce pages and UIs (using tools like Lightning App Builder and Dynamic Forms) to complement the automated processes you have put in place. With a combination of strong back-end automation and thoughtful front-end design, you'll be well on your way to delivering a polished, enterprise-grade Salesforce experience.

8

USER EXPERIENCE AND UI CONFIGURATION (APP BUILDER & DYNAMIC FORMS)

A Salesforce solution lives or dies by its user experience. No matter how powerful the automation behind the scenes, an unintuitive or sluggish interface will drive users away. This chapter focuses on building an **intuitive and high-performing user interface** using Salesforce's declarative tools. When users find the system easy, fast, and tailored to their needs, **adoption soars** – and with it, the success of your implementation.

We begin with the **Lightning App Builder**, the canvas for creating custom Lightning pages. With App Builder, you can arrange standard and custom components into modular layouts without writing a single line of code. An effective Lightning page isn't just a collage of fields; it's a thoughtfully designed interface that guides the user's attention. We'll discuss how to use tabs, accordions, and other layout elements to organize information so that pages load quickly and users can find what they need at a glance. The goal is to craft pages that feel **clean and responsive**, whether in Sales Cloud, Service Console, a partner Experience Cloud site, or on a mobile phone.

Next, we explore **Dynamic Forms**, Salesforce's game-changing feature for building tailored record pages field by field. Dynamic Forms let you **display or hide sections and fields based on criteria** like the user's profile, the record type, or even specific field values. This means each user sees exactly the information they need, when they need it. You'll learn how Dynamic Forms can eliminate the proliferation of page layouts – historically a headache for admins – by consolidating variations into one smart page. We'll walk through examples of using visibility rules to simplify complex pages, improving usability for internal users and partners alike.

Of course, a great user experience isn't just about what data is shown – it's also about what actions are available. We'll cover **Dynamic Actions**, which control which Quick Action buttons appear for a user in a given context. By configuring Dynamic Actions, you can streamline the interface so that, for instance, a support agent sees an "Escalate Case" button only when a case meets certain criteria, or a partner user in a portal sees a simplified action set compared to an internal user. Fewer irrelevant buttons mean less visual clutter and fewer mistakes, leading to more efficient workflows.

Performance and responsiveness are recurring themes throughout this chapter. Users expect pages to load in a snap, so we'll discuss strategies to **optimize Lightning page performance**. This includes limiting the number of fields and related lists that load by default, prioritizing high-value components, and using tools like the Salesforce Optimizer and the App Builder's built-in Analyzer to identify what might be slowing your pages down. Declarative configuration gives us many knobs to turn – component visibility, page structure, data density – and we'll detail how to tweak these for speed without sacrificing functionality.

Underpinning all of these tips are fundamental **user-centric design principles**. Even in a declarative world, concepts like consistency, simplicity, and accessibility guide us to better outcomes. We'll weave in principles of **good UX design** – such as keeping layouts consistent across pages, designing for mobile as well as desktop, and ensuring our configurations meet accessibility standards. Features like Salesforce Path (with **Guidance for Success** prompts) will be highlighted as ways to provide in-app guidance, reinforcing process adherence in a user-friendly way.

By the end of this chapter, you'll have the tools and best practices to create Lightning pages that are not only powerful and dynamic, but also **clean, fast, and delightful to use.** An optimized UI means users can get their work done with fewer clicks and less confusion, which ultimately boosts productivity and satisfaction. Now, let's dive in by opening the Lightning App Builder and exploring how to lay the groundwork for an optimal user experience.

LEVERAGING LIGHTNING APP BUILDER FOR OPTIMAL USER EXPERIENCES

Designing a great user experience in Salesforce starts with the Lightning App Builder – the hub for assembling **Lightning record pages** and other custom pages. Lightning App Builder allows you to drag, drop, and configure out-of-the-box or custom components to create pages tailored to different business needs. In this section, we'll cover how to use App Builder to create **modular, responsive layouts** that ensure users see important information quickly without being overwhelmed. We'll also look at tips for component placement, alignment, and load optimization across various contexts (Sales Cloud, Service Console, Experience Cloud, and mobile).

At its core, a Lightning page is made up of regions defined by a **page template.** Choosing the right template is your first design decision. For example, a sales record page might use a **three-region layout** (header and two columns) to take advantage of wide desktop screens, whereas a community page for partners might favor a single-column layout (for simplicity and mobile friendliness). No matter the template, a key best practice is to keep the layout **modular:** break information into logical sections or tabs rather than one long scroll. This modular approach not only makes the UI cleaner, but also improves performance by loading data in chunks.

Using Tabs and Accordions: One of the simplest ways to modularize a Lightning page is by using the Tabs component (or Accordion, which is similar but vertically stacked). For example, instead of a contact page that shows Details, Related Lists, Activities, and Charts all at once in a long scroll, you could create tabs labeled **"Details," "Related,"** "Activi-

ties," and **"Analytics."** The Details tab (as the default) might show key fields and maybe one or two crucial related lists, ensuring a snappy initial load. Other information is tucked into secondary tabs, loaded only when the user clicks them. This technique keeps the first screen lightweight and focused. As a bonus, tab names set expectations and help users navigate ("I know where to find what I need"). Salesforce even lazily loads tab contents, meaning if "Related" is not viewed, the queries for those related lists don't run, saving time on initial page load.

Real-World Example – Account Page: Imagine designing an Account record page for a sales team. You might start with a **Highlights Panel** at the top showing the Account Name and a few key fields like Industry, Rating, and Phone – just the information a rep needs at a glance. Directly beneath, an **Activities** or **News** component could surface upcoming tasks or recent news about the account (keeping the rep informed). Using App Builder, you place these in the top region for visibility. Now, to avoid information overload, you add a Tabs component in the main region. The default tab is **"Overview,"** containing a strategically chosen subset of fields (perhaps a compact layout or a Dynamic Form section). Other tabs could include **"Contacts & Opportunities"** (with related lists for key contacts and open opportunities), **"Cases"** (showing support cases for the account if relevant), and **"Details"** (for less frequently needed fields like full address, description, etc.). By structuring the page this way, the sales rep lands on an Overview that loads quickly and shows the most important info. If they need more, it's one click away on a tab.

ChatGPT Prompt (Typing):

User: "I'm building a Lightning record page for our Account object. What are some best practices to ensure the page is user-friendly and loads fast?"

ChatGPT: "Great question! First, use a **tabbed layout** to organize information – for example, an Overview tab with key fields and additional tabs for related lists, analytics, etc. This way, only the Overview loads initially, making it faster. Keep the **Highlights Panel** at the top with 3-5 critical fields (like Account Name, Industry, Phone) so users see important info at a glance. Avoid cramming every field on the page; instead, leverage **Dynamic Forms** to show less-used fields only when needed (if your org has that enabled). Also, include the **Related List Quick Links** component – it gives users a handy menu of related records without loading all those lists at once. Finally, test the page in different screen sizes (desktop and mobile) to ensure its responsive – Lightning will adjust, but you may want to hide or reorder some components for smaller screens. In short: keep it clean, chunked into tabs, and focused on the user's primary needs."

Notice in the ChatGPT advice above the mention of **Related List Quick Links.** This standard component can be a lifesaver for usability: it displays a horizontal list of related list names at the top of the page (with record counts), acting like a table of contents for related records. Users can click a link (e.g., "Opportunities (5)") and jump straight to that related list section. Including this on a record page (often right under the Highlights Panel) is a **pro tip** for long record pages – it prevents endless scrolling by giving a quick navigation aid. It also loads very fast since it's just a summary of links, not the full related lists themselves.

Different Salesforce products might warrant different design choices in App Builder:

Sales Cloud (Standard Lightning Experience): Sales users often care about quick context and pipeline info. Use components like **Path** (to show Opportunity stages) at the top for opportunities, and maybe a **Report Chart** to highlight a KPI (e.g., pipeline dollar sum) on an account or opportunity page. Keep the layout **two-column** for desktop, with the left for primary info (fields, related lists) and a narrower right column for secondary info (like guidance text or additional insights). Ensure that on narrower screens the components stack in a logical order (App Builder lets you configure the order for the phone form factor, or you can mark some components "desktop only" or "phone only" using visibility filters).

Service Console: In a console app, users often have multiple panels open. Lightning App Builder supports **console page templates** that include pinned regions. For example, a Case page in a console might have a left column pinned with core case details (so those never scroll away), while the right side is a tabbed component for related information (contact info, case history, knowledge base suggestions, etc.). The App Builder allows you to designate a region as "pinned" for consoles, meaning it stays visible while agents flip through subtabs. Use this to keep crucial info (like case status, customer name, SLA info) always on screen. Console users value efficiency, so also consider using the **Utility Bar** (configured in the app, not page, but part of UX) for things like macros or quick text – though that's beyond App Builder, it complements your page design.

Experience Cloud (Partner or Customer Communities): When building pages for external users, simplicity is paramount. Community users might be less trained and have more basic needs. In App Builder, you can create pages specifically for Experience Cloud and even assign them to audience criteria. For example, you might design a pared-down version of a case detail page for customers logging support tickets – only a few essential fields and a feed, perhaps. Note that not all internal Lightning components work in Experience Cloud, but many do. Always test the page in the community context. A good practice is to **hide internal-only components** from external profiles using component visibility filters (for instance, a section that only internal support agents should see can be set to visible only if User.Profile = "Support Agent"). We'll discuss Dynamic Forms and visibility in the next section, but the concept applies in App Builder broadly.

Mobile (Salesforce mobile app): The beauty of Lightning pages is that, by default, they work on mobile too. However, a page that's perfect on desktop might be cumbersome on a phone. App Builder offers a **"Smartphone" preview** and the ability to set component visibility by device. Use these! You might decide to create a separate page layout for mobile users if the use case diverges greatly (Lightning pages can be assigned by form factor – one for desktop, another for phone – in the Activation settings). Often, though, you can use one page and simply hide a few components on mobile. For example, a wide report chart or a dense related list might be better omitted on a phone for performance and readability. Mark those components with a filter like "Form Factor != Phone". The result is a cleaner mobile experience with only the essentials. **Consistency** is key here: mobile users should still recognize the page, just optimized. For instance, keep the Highlights Panel and key fields, but maybe collapse some secondary sections.

As you add components, consider **alignment and spacing.** Lightning does a pretty good job automatically, but you have control over some things like using blank space components or adjusting component order to make the page balanced. A common pattern is to use a **two-column layout** where the left column holds the main record details (which might themselves be split into subtabs or sections) and the right column holds supporting information (like Related Lists or Related Quick Links, possibly a sidebar component such as Recommendations or an Einstein component if enabled). This mirrors the natural reading order (left to right) and keeps primary data in the "hot zone" of the page.

Pro Tip: Maintain a **consistent layout structure across similar objects** to help users learn the interface. For example, if on Account pages the first tab is "Overview" with key info, do the same on Contact and Opportunity pages. If the Highlights Panel shows 5 fields in a certain order on one page, use that same order on others when those fields exist (e.g., always show Name, then Phone, then Email in the highlights for any core object). Consistency reduces the mental effort for users – once they learn one page, they can guess where things are on another.

Before moving on, always **preview and test** your new Lightning page. Use the App Builder's preview (you can even simulate as different profiles or record types). Better yet, do a user acceptance test: have a few end users try a sample record with the new layout and watch where they click or get confused. They might say "I expected to find X in the first tab" or "This section is not useful for me." Use that feedback to iterate. The declarative nature of App Builder means you can tweak pages quickly. Don't be afraid to adjust tab orders, add helpful labels, or remove components that aren't adding value.

With a solid Lightning page structure in place using App Builder, you've set the stage for an optimized UI. Next, we'll supercharge this setup with **Dynamic Forms**, which will allow our pages to truly adapt to the context – showing the right fields to the right users at the right time. This dynamic field capability will build on the foundation of the Lightning page for an even more tailored experience.

MASTERING DYNAMIC FORMS FOR TAILORED RECORD PAGES

One of the most significant enhancements to Salesforce UI configuration in recent years is **Dynamic Forms.** Dynamic Forms enable you to break the traditional "Details" component of a record page into individual field sections that you can control with precision. Instead of a one-size-fits-all page layout for all users, you can now **display or hide specific fields and sections based on who is viewing the record or what data it contains.** In this section, we'll guide you through using Dynamic Forms to create tailored record pages, reducing the need for multiple page layouts and record types. We'll also cover practical examples of field-level visibility rules, and discuss use cases for both internal users (employees) and external users (partners or customers via Experience Cloud).

Enabling Dynamic Forms: If you have a Lightning record page (for a custom object or a supported standard object), enabling Dynamic Forms is straightforward. In the Lightning App Builder, when editing a record page, select the record detail component (the traditional fields section). If Dynamic Forms are available for that object, you'll see an "Upgrade Now" or "Upgrade to Dynamic Forms" button. Clicking this will convert your "Details" into a collection of Field Section components. Alternately, you can start from scratch: in the App Builder's component panel, there will be a **"Fields" tab** alongside standard components, where you can drag individual **Field Section** components onto the page. Each Field Section is essentially a container where you can add fields (from that object) as you please, much like editing a classic page layout but in a more flexible way.

Once Dynamic Forms are enabled, you gain a new superpower: **conditional visibility** at the field or section level. Every Field or Field Section component has a "Set Component Visibility" property. This is where the magic happens – you can add filter criteria so that the component only displays when those conditions are true. The criteria can be based on the record (fields on the record), the device (e.g., form factor = Desktop or Phone), or the user's context (their profile, role, or a permission via User fields).

Consider a scenario: You have a custom object for "Project" that is used by both internal employees and external partners. Internal users should see internal budget fields and internal status notes, while partners should not. With Dynamic Forms, you can create an **"Internal Details"** field section containing fields like Internal Budget and Margin, and set its visibility to show only when User.Type = Employee (or when the user's profile is one of your internal profiles). In the same page, you could have a "Partner Details" section with fields relevant to partners, visible only to partner profiles. Both internal and external users access the *same* Lightning page URL, but Dynamic Forms ensures each sees a different view tailored to them – all without writing code or maintaining separate page layouts. This greatly simplifies administration: if you need to add a new field for all users, you add it once to the dynamic page (perhaps in a commonly visible section). No need to touch multiple layouts.

Another common use case is simplifying pages based on record stage or type. For example, on a Case record, you might want to show a field "Resolution Details" only when a case is Closed, since while working on an open case that field is irrelevant. Using Dynamic Forms, you can put **Resolution Details** (and maybe "Closure Category" etc.) in a section and set the filter: *Case Status equals Closed*. The result: agents don't see those fields until they actually close a case, decluttering the UI during the active lifecycle of the case. Or think of an Opportunity: you could have a section of "Legal Approval Info" that only appears if the Opportunity's Amount > $100,000 *and* a custom "Discount > 20%" checkbox is true – effectively guiding the user that big deals with heavy discounts require legal info, while not bothering them with those fields on smaller deals.

Dynamic Forms shine especially in reducing the need for multiple **record types or page layouts** that were previously used just to handle minor differences. For instance, many orgs had separate page layouts for, say, "Domestic Account" vs "International Account" just because one had a "VAT Number" field visible. With Dynamic Forms, one Lightning page can cover both: put "VAT Number" field on the page and make it visible only if Account.BillingCountry is not USA (or whatever your condition is). Fewer page layouts means fewer assignments to manage and fewer places to remember to add new fields in the future. This is a huge win for maintainability.

Let's walk through setting up a Dynamic Form step by step, then dive into use cases:

Upgrade to Dynamic Forms: Open an existing Lightning record page in App Builder. Suppose we have a **Custom Object** called "Inspection__c". We click the Record Detail component and choose *Upgrade*. App Builder will create a default Field Section for each section it finds on the current layout (e.g., it might make a "Detail" section with all fields as one block). This is a starting point; you can then add, remove, or rearrange fields freely.

Create Field Sections: Maybe we decide to split the fields into logical groups. We drag a new **Field Section** component onto the page (say, in the left column, Overview tab). We name it "Main Info". Inside that section, we add important fields like Inspection Date, Location, Status, etc. Then we add another Field Section and label it "Additional Info" with less critical fields. Field Sections can have 1 or 2 columns inside them, just like classic page layouts; choose what fits the content. Perhaps "Main Info" is 2-column for compactness, while "Additional Info" is one-column for readability.

Apply Visibility Rules: Now the fun part – we decide that "Additional Info" should only show for certain users or conditions. For example, if some fields are sensitive, we might set "Additional Info" section to only show if the user's profile is "Inspector Manager" or if a checkbox "Show Advanced Fields" is true on the record. We click *Add Filter* in the component's visibility properties: we can pick **User.Profile = Inspector Manager** as one rule; if we want multiple criteria, we can add more and decide if all must be true or any (AND/OR logic). Perhaps also *OR* User.-Role = "Inspection Admin". The UI builder will show a little eye icon on the component indicating it has conditional visibility. We can test it by using the Preview as different profiles.

Remove Redundant Layouts: After building a dynamic page, we might realize a separate page layout or record type is no longer needed. For example, if we previously had a record type "Advanced Inspection" with its own layout for managers, we could potentially retire it and use a single record type with dynamic sections. However, tread carefully here – changes to record types or layouts can impact other things (like picklist values and default values). Always communicate with stakeholders before consolidating. But definitely, for any *new* requirements, favor dynamic forms over proliferating layouts when possible.

To illustrate the benefit, let's consider an internal vs. external use case in detail:

Internal vs. External: A **Case** object used in a customer community. Internally, support agents track detailed information (e.g., internal resolution notes, escalation steps). External customers raising the case should see only the basic status and communications. We can create one Lightning Case page that serves both. How? On that page, include all sections needed. Perhaps a section "Internal Notes" (with fields visible to support profile only), and another section "Customer Info" (with fields like Contact Name, which maybe the customer can also see). In the Experience Cloud site, because the external user is logged in with a high-level profile like "Customer Community User," the Internal Notes section will auto-hide (visibility rule: user profile is not Customer Community). The customer sees a clean, simple case layout. The support agent, opening the same case inside Salesforce, sees everything including the private notes. Without dynamic forms, you'd need two different page layouts and maybe even two record pages (one for community, one internal). With dynamic forms, one page adapts to both audiences. **Note:** As of now, standard Experience Cloud record detail components don't fully support dynamic forms – an external user might still see the classic layout if using standard components. A workaround is to embed the dynamic Lightning page in the community via the *Record Detail* component set to "Use publisher layout" = false. This gets technical, and Salesforce may improve support in future releases. Always test with an external user account to ensure the dynamic filters behave as expected for that context.

Another scenario is reducing record types used solely for minor field differences. Suppose we have a **Custom Object "Policy"** with two types: Personal and Business. Historically, we might have had two record types and two page layouts because Business policies had a few extra fields for company info. With Dynamic Forms, we can potentially merge into one record type "Policy" (if business logic allows) and use field visibility: e.g., a field "Company Name" visible only if Policy.Type = Business. Fewer record types can simplify processes like approvals or automations that otherwise had to branch per type.

Dynamic Forms also **improve page load times** in some cases. If heavy or numerous fields are hidden based on criteria, they don't all load every time. For example, if a section with 20 formula fields (some possibly slow) is only shown when needed, viewing a record that doesn't meet the criteria avoids loading those formula computations at all. However, be mindful: Salesforce will still enforce Field Level Security and required field logic. If a field is required (by field definition or page layout), ensure your dynamic strategy accounts for that. For instance, if a field is required for saving, a user who never sees it (because you hid it) could be blocked from saving if they somehow trigger that condition. As a rule, required fields should generally be always visible to the user or have a default value to allow saving. Dynamic Forms won't override object-level required fields or field security; those act as a floor. So use dynamic visibility to hide *optional or conditionally needed* fields, not ones that are always required for saving.

Let's see Dynamic Forms in action with a hypothetical admin task:

ChatGPT Prompt (File Upload):

User: uploads a CSV file named "FieldVisibilityRequirements.csv"

The CSV contains two columns: Field Name, Visible To (with values like "Agents only", "All users", "Managers if Priority=High", etc.)

User: "ChatGPT, here is a list of fields and who should see them. How should I configure Dynamic Forms for these requirements?"

ChatGPT: "Let's interpret your requirements from the file. For example, Customer_Email is marked 'Agents only', meaning only internal support agents should see it. In Dynamic Forms, I'd create a field section (or add the field to an existing section) and set its visibility filter to show only when the user's profile is one of your internal agent profiles (or perhaps User.Type = Employee). Next, Discount_Code is marked 'Managers if Priority=High'. That one likely belongs in a section visible when **Case Priority = High** AND **User.Role = Support Manager** (for instance). You can accomplish that by adding two filter conditions to the field or its section: one based on the Case.Priority field value, and one based on the user's role. Make sure to choose the **"All conditions must be true"** option so both must match. For fields marked 'All users', you can place them in the default section with no visibility filters (visible to everyone by default). Essentially, you'll use Dynamic Forms' filters to translate each rule: profile-based, role-based, record field-based, or any combination. One more tip: organize fields into sections by who needs them. It might simplify to have an "Agent-Only Details" section and a "Manager Approval" section, etc., each with a corresponding visibility rule, rather than individual rules on every field."

ChatGPT Prompt (Privacy Tip):

User: "The CSV I uploaded has some real customer data examples. Is it safe to upload such files to ChatGPT?"

ChatGPT: "It's best to be cautious. When using any AI service, avoid sharing sensitive personal data unless it's scrubbed or anonymized. If the data is confidential, you should remove or obfuscate identifying details before uploading. OpenAI's privacy policy states that input data may be used to improve the model (unless you opt out or use a business/enterprise arrangement). So for brainstorming configurations, use dummy data or just share the structure (like field names and intended audiences) rather than actual customer info. In short: treat AI like you would any external consultant – don't give it secrets it doesn't need."

As the AI suggested, remember to **sanitize sensitive information** when seeking help. In a real admin scenario, you might export field metadata or a requirements spreadsheet and discuss that with ChatGPT. It's a great way to generate a game plan (the AI could draft a list of visibility rules or even pseudocode for criteria-based formulas), but always uphold your company's data privacy standards.

Now, having configured Dynamic Forms, take a step back and admire what's been achieved declaratively: a single Lightning page that adapts to different users and scenarios in real time. You didn't have to duplicate pages for each profile or overload one page with everything for everyone. Maintenance becomes easier – one page to update instead of many – and users enjoy a cleaner interface.

Before moving on, note that Dynamic Forms currently support **most standard objects (Account, Opportunity, Case, etc.) and custom objects**, but there are a few exceptions. Objects like Activities (Tasks/Events) and some others don't yet support Dynamic Forms as of this writing. Always check Salesforce release notes for updates. If an object isn't supported, you'll simply not see the "Upgrade" option or the Fields tab in App Builder for that object, meaning you'll stick to classic page layouts for those for now.

With dynamic field visibility under our belt, the next logical step is dynamic control of actions. Just as we tailored which data appears to whom, we can tailor which buttons and quick actions a user sees on a record page. This ensures that at any given time, users are presented only with the actions that make sense, further streamlining their experience. In the next section, we'll explore **Dynamic Actions** and how they enhance user efficiency by decluttering the button bar on Lightning pages.

ENHANCING USER EFFICIENCY WITH DYNAMIC ACTIONS

Every Salesforce user is familiar with the buttons (actions) at the top of a record page – **Edit, Delete, New Task, Email**, and custom actions that perform specific tasks. In classic Salesforce, and even early Lightning, which buttons showed up was mostly controlled by the page layout and global layout settings, with little context sensitivity. Enter **Dynamic Actions:** a feature that lets you manage and filter these quick actions directly in Lightning App Builder, much like Dynamic Forms for fields. In this section, we'll cover how to configure Dynamic Actions to show contextually relevant action buttons on your record pages. We'll go through examples of setting up filters so that actions appear on certain devices, for certain record conditions, or for certain users. We'll also consider use cases across desktop vs. mobile, standard vs. console apps, and even Experience Cloud scenarios.

Enabling and Using Dynamic Actions: On a Lightning record page in App Builder, clicking on the Highlights Panel (the component that holds the record's actions and key fields) will reveal if Dynamic Actions are available for that object. For many standard objects (like Account, Contact, Opportunity, Case) and all custom objects, you'll have the option to *enable Dynamic Actions*. If you do, you essentially decouple the quick actions from the traditional page layout's action section. Instead, you'll manage them in App Builder. Practically, you'll see an "Add Action" menu in the properties panel where you can choose which actions to show. This includes standard actions (Save, Edit, Delete, etc.), Quick Actions (object-specific or global), and productivity actions (Log a Call, New Task, etc.).

The real power, however, lies in applying **visibility rules to each action.** Just like fields, you can set criteria so an action only shows when certain conditions are met. For example, consider the **"Change Owner"** button on a Case. You might decide it should only be available to users with a Supervisor role *or* when the case status is New (maybe after that, ownership changes go through an approval, so the button isn't relevant). With Dynamic Actions, you can click "Change Owner" in the App Builder and set: *Visible when User.Role = Support Supervisor OR $Profile.Name = "Support Manager"* (for the role criterion) **AND** *Case Status = New* (for the record criterion). The UI lets you add these conditions in a point-and-click manner, very similar to how you did for Dynamic Forms.

What's the effect? End users will only see the **Change Owner** button in the highlights panel if they meet those conditions. Otherwise, it's hidden from their interface – out of sight, out of mind. This prevents less-authorized users from even attempting actions they shouldn't, and it simplifies the choices for everyone else.

Let's go through a few practical examples of dynamic actions that improve user efficiency:

Profile-based Actions: An "Approve Discount" custom action (perhaps triggering a Flow or Approval Process) should only show for managers. Instead of creating a separate page layout for managers with that button, just make it a dynamic action visible to the "Sales Manager" profile. Managers see the button on opportunities that need their approval, reps do not.

Record-based Actions: A "Close Opportunity" action should appear only when an Opportunity is in the later stages (Proposal/Price Quote or

Negotiation, for example). You can set the filter to *Stage = Negotiation OR Stage = Proposal*. At earlier stages, the "Close" action (which might be a quick action that sets Stage to Closed Won or Closed Lost via a screen flow) is hidden to avoid confusion. By the time it shows up, the user is ready to use it.

Device-based Actions: Perhaps you have a "Scan Business Card" action that launches a mobile-only app or uses device camera – you'd want that button on mobile, but it wouldn't make sense on desktop. Dynamic Actions allows you to set visibility by form factor. You could mark that action *visible when Form Factor = Mobile*. Conversely, a "Print to PDF" action might be desktop-only (since printing from the mobile app is not typical); set it to *visible when Form Factor = Desktop*. This way, the highlights bar on each device shows only relevant options.

Experience Cloud considerations: Suppose you expose some record pages to partner users. Dynamic Actions can respect the user's profile or license type, so you could hide certain internal actions from external users easily. For example, the standard "New Task" action might be something you only allow internal users to use (partners may not log tasks in your org). You set that action to not show if the user's profile = Partner Community User (by adding an inverse condition or a direct one for internal profiles). Now partners won't even see a greyed-out button – it's just not there, providing a cleaner UI. Keep in mind external users also might have a different set of actions (perhaps a "Request Support" action), which you can conversely hide from internal users if it's redundant for them.

Setting up dynamic actions is intuitive. In App Builder:

Enable Dynamic Actions (if not already). This typically gives you the choice for Desktop, and separately for Mobile, on some objects. For instance, you might see checkboxes "Enable dynamic actions for Desktop" and one for Mobile. You can configure them separately or together.

Once enabled, the list of actions becomes editable. Add any actions you need (there's a search to pick from standard and quick actions).

For each action, click the **filter icon (eye)** to add visibility rules. These can stack with AND/OR logic similar to fields. There is a known limit of

10 filters per action, but if you exceed that, consider simplifying by using a formula field as a flag (e.g., create a formula field "ShowManagerActions__c" that computes true/false for all complex logic, then just have one filter "ShowManagerActions__c = true").

Save and activate the page.

One thing to note: after enabling dynamic actions, **the page layout's assigned actions are no longer controlling the Lightning page** (for that form factor). It's often wise to remove or tidy up the page layout's actions to avoid confusion, especially if some users still use Classic or if you have some profiles not on the dynamic page version. In Lightning, once dynamic actions are on, Salesforce will ignore the page layout's action list for that Lightning page. But say you haven't activated the page for all profiles yet – some might still see the old layout. Keep consistency by updating the layout or eventually moving everyone to the new dynamic action-enabled page.

Let's see how ChatGPT might assist with a dynamic action scenario:

ChatGPT Prompt (Voice Mode):

User (speaking): "Hi ChatGPT. I want certain action buttons on the Opportunity page to show only for specific stages. Can you guide me?"

ChatGPT (responding in a friendly voice): "Sure! You can definitely do that with Salesforce Dynamic Actions. Let's say you want the **'Send Contract'** action to appear only when an Opportunity is in the Proposal stage. In Lightning App Builder, edit your Opportunity page. Click on the Highlights Panel, then find or add the 'Send Contract' action. Next, set a visibility rule: choose the Opportunity's Stage field, equals, Proposal/Price Quote (or whatever the exact value is). Save it. You could also add more criteria, for example if you only want it for opportunities above a certain amount, include Amount >= 50,000 as another condition. Once configured, users will see the button only in those situations. Don't forget to hit Activate to apply the page. Dynamic Actions are a great way to keep your UI contextual – so your users only see actions when they're relevant!"

In the above scenario, the user spoke a question and ChatGPT responded with the steps. This highlights how an admin could use voice (perhaps through a mobile phone or a voice-enabled interface) to get quick "how-to" advice while configuring the system. As Salesforce professionals, we often multitask – imagine tweaking a config while asking a voice assistant for guidance, keeping your hands free to click along. Pretty neat!

Dynamic Actions aren't just about hiding buttons; they're about **surfacing the right options at the right time.** Users will be more efficient when they don't have to wade through a dozen buttons to find the one action they need. Also, by reducing visible actions, you guide users toward the intended path. For instance, a service rep in a certain stage of a case might see only "Escalate" and "Request More Info" because those are the only two logical actions at that point – much clearer than seeing Edit, Change Owner, Change Status, Escalate, Close Case, etc., some of which might be wrong to use at that time.

From a maintenance perspective, Dynamic Actions centralize the logic in the Lightning page, which can be easier than remembering to edit multiple page layouts. However, keep documentation of these rules (perhaps in the description of the page or an internal wiki) – because if someone else edits the page later, they need to understand why a given button has a filter. Fortunately, App Builder makes the presence of filters visible with an icon on each action, so it's fairly clear.

A quick check on **Mobile vs Desktop**: Salesforce lets you configure dynamic actions separately for the mobile app. On some objects, enabling dynamic actions for mobile is distinct. If you do so, you might choose a slimmed down set of actions for phone users (since the small screen fits fewer buttons without extra taps). For example, on a mobile contact page, you might show just "Call" and "Log a Call" actions, whereas on desktop, you show "Edit", "Email", "New Task", etc., in addition. Think about what a user on the go truly needs versus at their desk.

What about **Console**? In a console, actions typically still appear in the highlights panel of a primary tab, so dynamic actions apply the same way. But console users often have keyboard shortcuts or other utilities, so actions are used a bit differently. It's still worth tailoring them – e.g., maybe hide "Edit" for console users if they have inline editing and prefer not to open the edit dialog.

To round out this section, let's consider how dynamic actions, when combined with dynamic forms, provide a highly personalized UI: The fields a user sees are tailored, and the actions they see are tailored. A sales rep and a sales manager might be looking at the same record and see slightly different layouts and buttons, each suited to their role – all done declaratively. And when that context changes (say a record moves to a new stage), the UI adapts in real time: new fields might appear (via Dynamic Forms), and some buttons might disappear or activate (via Dynamic Actions). It creates a sense that the system is "smart" and aware of the user's needs, which is the hallmark of a good user experience.

We've now configured what data is shown and what actions are available in a contextual way. The next challenge is ensuring that this dynamic, rich interface remains **performant** and responsive. After all, loading a bunch of conditional sections and actions won't help if the page takes too long to display. In the following section, we'll shift focus to performance tuning and responsive design techniques to make sure our beautifully customized pages are also blazing fast and adaptable to various devices.

DESIGNING FOR PERFORMANCE AND RESPONSIVENESS

A Salesforce page that's perfectly designed functionally can still fail if it's slow or clunky. Users have little patience for pages that take ages to load or that freeze their browser. Therefore, an essential part of UI configuration is ensuring that pages are **performant (fast to load)** and **responsive (adapting to different devices and screen sizes smoothly)**. In this section, we'll discuss techniques to streamline Lightning record pages for speed. We'll cover limiting what loads by default (using tabs and visibility, which we touched on), controlling the volume of data on a page, prioritizing high-value information, and using Salesforce's tools like Optimizer and the Lightning Page Analyzer. We'll also address some real-world cases where performance tuning made a measurable difference.

First, let's clarify what factors influence Lightning page performance:

Page complexity: The number of components on the page, the number of fields displayed, and the number of related lists or chart components all contribute to load time. Each standard component (like Related Lists, Related List Quick Links, Activities, etc.) may internally execute one or more server calls or queries.

Data volume in components: A related list showing 5 records loads faster than one showing 50 (especially if each related record has many fields or formula calculations). A component showing "All Open Opportunities" will load slower in an account with 1,000 open opps versus one with 5.

Client device and network: A user's browser and hardware matter (an older computer or slow internet will magnify any slowness). Also, large pages affect mobile devices more strongly due to potentially lower processing power.

Browser performance: Lightning is a client-side heavy application. It does a lot in the browser (JavaScript rendering). So, having too many dynamic components can strain the browser's memory and CPU as it processes the page.

Given those factors, here are best practices to design for performance:

1. Limit Fields Displayed in the Default View: Do you really need to show *all* 50 fields of an object on initial load? Likely not. We already applied this principle with Dynamic Forms: only key fields upfront, with others hidden until needed. As a rule of thumb, try to keep the number of fields in the **initial visible region** of the page as low as reasonably possible – perhaps 10-15 key fields. If the record has tons of fields, group them into collapsible sections or separate tabs. This not only helps users focus, but it reduces the rendering time. (Note: formula fields or fields with heavy calculations can particularly slow things down; if you have many such fields, consider if all must be on the page).

2. Break Up the Page with Tabs (and accordions): We mentioned this for usability, but it's equally a performance tactic. Salesforce has confirmed that only the components in the active tab load initially. By putting less-frequently-used components in secondary tabs, you delay their loading until (and unless) the user navigates there. For example, an

"All Related" tab that contains a component listing dozens of related lists can be hidden on load. Users who need that info can click it (and wait a second as it loads then), but others aren't penalized on the initial page load. Design your default tab to contain the **essential** information only. Everything else gets a place, but not front and center.

3. **Place Heavy Components Off the Default Tab:** Some components are inherently heavier. The standard **Related Lists** component is one – if you put "Related Lists" on the main tab and the record has, say, 10 related lists configured, the page will query for all 10 lists' data immediately. Instead, consider using **Related List – Single** components to expose only the top 1-3 related lists that are most important, and put the rest of the related lists in a secondary tab. For instance, on a Contact page, you might show only the **Cases** related list (if that's crucial for a support agent) on the main tab, and move other lists like Campaigns or Orders to another tab. The idea is to **minimize the number of queries and data** pulled during the initial paint of the page.

Similarly, interactive components like **Report Charts** or **List Views** embedded on the page can be heavy. If you include a Report Chart (say, pipeline chart on an Account), that's an extra query + rendering. It might be worth it for the value it provides, but ensure it's necessary and possibly put it on a side or secondary area if it's not mission-critical to see immediately.

4. **Control Related Lists Count and Size:** If you have to display many related lists, follow Salesforce's guidance: *no more than four related list components per tab* for optimal performance. And within each related list, showing 5 rows is faster than showing 10 or 15. You can configure each related list component to show a certain number of records by default – lower that if possible (users can always click "View All" to see more). Another trick: use **List View** components with filtered data instead of massive related lists. For example, if an Account has thousands of cases, loading all cases in the related list could be slow. Instead, maybe embed a List View component that shows only "Open Cases for this Account" (with a filter in the list view criteria). That might significantly cut down what loads, and again users can navigate to full list if needed.

5. Use Analyze Tools: Salesforce provides an **"Analyze" button** in the Lightning App Builder (at the top bar when editing a record page). This tool will simulate page load and give you a report with a performance score and specific suggestions. It might say things like "This page has 8 related lists, consider reducing to 3 per tab" or "The Details component has 40 fields, consider using Dynamic Forms to reduce initial load." Pay attention to these recommendations – they come from Salesforce's own analysis of what slows pages down. Also, the **Salesforce Optimizer** (which is a broader org-wide tool) can flag "Slow Lightning Pages" if any are detected, often pointing out similar issues (too many components, etc.). As an admin, running Optimizer periodically and reviewing the Lightning Experience section can alert you to pages that need tuning.

6. Optimize for Mobile Separately: We touched on hiding components on phone form factor. This is crucial for performance too. A page that is fine on desktop might crawl on a phone if it has many sections. Even though Salesforce auto-collapses tabs into a single-column scroll on mobile, all those components might load if they're on the default tab (since on phone, tabs often become an accordion). Use the form factor filters to actually exclude non-essential elements from mobile entirely. For example, maybe skip that embedded report chart on mobile – it's not as useful on a small screen, and it incurs load. Perhaps skip showing 5 related lists; instead show just one or two key ones for mobile. Keep in mind an interesting benchmark that Salesforce has shared: Having more than ~25 fields in the detail section or more than ~5 related lists visible can noticeably impact **mobile performance.** So use that as a guideline to trim mobile pages.

7. Remove Duplicate or Unneeded Components: It sounds obvious, but it's worth auditing your pages to see if anything redundant is there. Sometimes an admin might accidentally have two instances of the same related list, or a component that is largely unused (maybe a Rich Text area left blank as a spacer, etc.). Every component call is overhead. Keep pages lean. A classic mistake is leaving both "Related Lists" and also individual Related List components for the same object on a page – you only need one approach. Or showing the same information twice (perhaps a formula field that's also essentially conveyed by Path or highlights). Tighten it up.

8. Consider Data Volume and Archival: If a particular page (say, an Account page) is slow because that account has thousands of related records (activities, cases, etc.), consider whether old data can be archived. Salesforce recommends archiving or deleting stale data not just for storage reasons but for performance. For example, closing tasks after a year or moving old cases to an archive custom object could lighten the load when viewing an account. This is more of a data management strategy, but it plays into user experience when those edge cases of giant records come up.

Let's illustrate the impact of performance tuning with a brief story: A customer service department complained that the Case page in their org was loading very slowly – often 5-6 seconds before an agent could start reading the case details. The admin used the Lightning Page Analyzer and found that the page had **10 components loading on startup**, including 6 related lists (some showing 10 records each) and a large details section with 35 fields (including several formula fields pulling info from related objects). Armed with best practices, the admin redesigned the page: They moved four of the related lists to a new tab "Other Related Records," leaving only two key related lists (Contacts on the case, and Recent Cases from that Account) on the main tab. They split the details into two tabs: a primary "Case Info" tab with the 10 most important fields, and a secondary "All Details" tab with the rest. They also enabled dynamic forms to hide a "SLA Violation Reason" field unless the case actually violated SLA (so that formula only calculates and shows in those instances). After these changes, the initial load had fewer queries and less data. The Case page load time dropped to ~2 seconds. Agents immediately noticed the improvement – "Cases pop right up now!" – which in a call center environment can directly improve productivity and customer satisfaction.

ChatGPT Prompt (Operator Mode):

User: "Operator, find official Salesforce guidance on how many related lists per page is recommended for performance."

ChatGPT's Autonomous Agent (Operator): (The Operator agent opens Salesforce Help documentation behind the scenes.)

ChatGPT (Operator): "According to Salesforce's best practice documentation, it's recommended to **display no more than 4 related lists on a single page** (or per tab) to avoid slow load times. In fact, one Salesforce resource suggests reducing it to 3 per tab for optimal speed . It also advises placing related lists in separate tabs if you have many of them, so they don't all load at once."

The Operator went and fetched the info so you didn't have to search manually.

In the above imaginary interaction, we used an "Operator" mode of ChatGPT to automatically retrieve specific info (like a particular number or limit from documentation). This shows how an admin could quickly confirm a performance guideline while working on a page. Instead of Googling, they ask the AI to get the detail. The agent reads, say, Salesforce's help article on Lightning performance and reports the relevant piece (in this case, confirming that keeping related lists to a minimum is important).

Finally, let's talk **responsiveness**. Lightning pages by nature are responsive (they reflow for different screen sizes), but as a designer you should still test and tweak. We mentioned using component visibility to hide things on phone. Also consider the order of components: On a desktop two-column layout, the left column content will appear above the right column content when stacked on a narrow screen. So ensure that in a phone view, that ordering still makes sense. If not, you might rearrange components or use the "Mobile & Lightning Experience Actions" section to specify a different set of actions for mobile (in case you aren't using dynamic actions). The goal is that whether a user is on a big monitor, a tablet, or a smartphone, the page is usable and not awkward.

One more tool: **Lightning Usage App** (in Setup). This app can show you performance metrics like average page load time by profile, or the slowest pages in your org, etc. It's more about adoption and usage, but it does have a Performance section if enabled. Keep an eye on those stats if your users are complaining about slowness – it might point you to a specific page or user segment that needs attention (e.g., maybe your field sales folks on mobile are consistently having 8-second loads – a sign to optimize those pages for mobile).

In summary, designing for performance is about **showing the user what they need and nothing more, at least not until they ask for it**. We have the tools (tabs, dynamic visibility, single related list components) to apply that principle. By combining these with Salesforce's analyzers and a bit of attention to detail, you can deliver Lightning pages that feel light and snappy. With a fast foundation in place, we can now turn to broader user experience principles. In the next section, we will integrate classic UX design concepts – consistency, intuitive design, mobile-first thinking, and accessibility – into our Salesforce declarative configuration approach, ensuring that our solutions are not just fast and functional, but also user-friendly in the best sense.

INTEGRATING USER-CENTRIC DESIGN PRINCIPLES

Thus far, we've focused on specific Salesforce tools and tactics (App Builder layouts, Dynamic Forms, Dynamic Actions, performance tweaks). Now it's time to zoom out and consider overarching **user-centric design principles** and how to apply them declaratively. Experienced Salesforce professionals know that successful solutions require more than just meeting functional requirements – they must be *easy to use*, *consistent*, and *adaptable* to users' needs. In this section, we'll discuss key UX (user experience) principles like consistency, intuitive navigation, mobile responsiveness, and accessibility, and illustrate how to embed these principles into your Salesforce configuration. We'll also see how ChatGPT can assist in the design process – from brainstorming user flow improvements to even generating quick UI mockups or wireframes for validation.

Consistency: Consistency in UI design means users can predict where to find things and how to perform actions based on past experience. In Salesforce terms, this can be as simple as using standard components and sticking to a uniform page layout structure across objects. For example, if every record page in your app has the Highlights Panel at top, details on the left, related info on the right, and tabs separating out secondary info, a user moving from, say, an Opportunity to a Case will feel at home. They won't have to hunt around for the "Related" lists or the "Notes" component – you've placed them consistently. Consistency also applies to naming and terminology: if one page has a tab called "Details" and another calls it "Main Info," that's a subtle inconsistency that could confuse. Strive to use the same labels for similar content (and align with Salesforce standard naming where possible – users are used to "Details" and "Related" as tab names by convention).

From a declarative perspective, achieving consistency might involve creating a **template Lightning page** for similar objects. While you can't literally templatize and reuse a page config across objects, you can certainly configure one and then manually replicate its structure for others. Some companies create an internal "UI style guide" – e.g., "All major objects will have a tabbed page with these standard tabs..." – and then configure accordingly. As a senior admin or architect, you might be responsible for reviewing pages built by different teams and aligning them.

Intuitive Navigation & Flow: An intuitive design means the user can accomplish their tasks with minimal clicks and without feeling lost. In Salesforce, think about the end-to-end user flow: If a sales rep needs to update an Opportunity and create a follow-up Task, how easy is it? Do they have to scroll and find the Task related list, click New, fill out, save, then navigate back? Perhaps instead you provide a **quick action** on the opportunity page: "Log Follow-up Task" that pops a mini form (global action) to create the task without leaving the page. That's using declarative tools (quick actions) to streamline a user flow.

Intuitive flow also means guiding the user's eye to what's important. Use **visual hierarchy**: important stuff higher on the page or highlighted. In Lightning, the Highlights Panel already does this for key fields. You can also use components like **Path** (coming next section) which graphically show stage progression – that's intuitive because it matches a real process. Another example: on a Contact page, if the primary goal is to call the contact, putting the phone number right at the top (and maybe even using a formula field to make it a clickable link or using the CTI component if you have one) makes the process obvious.

Mobile Responsiveness: We have already covered performance for mobile, but responsiveness is broader – it's ensuring the design works on various devices and screen sizes. This is certainly a user-centric principle nowadays since many Salesforce users (especially sales folks, field service, executives) use phones or tablets. How to integrate this principle declaratively? Use the **preview as mobile** feature when building pages. Check components like dashboards or rich text on a narrow view – do they become illegible? You might decide not to show a wide dashboard on mobile and instead show a textual summary (perhaps via a formula or a simple report chart).

Salesforce has introduced the concept of **Dynamic Interactions** and flexipages which can adapt layout for different form factors, but even without advanced features, our earlier step of conditionally showing/hiding components for phone vs desktop is key. Also, consider leveraging Salesforce's **mobile-specific features**: for example, the mobile app has the **"Mobile Only" navigation and actions**. Ensure that if you have important quick actions, you've added them to the Salesforce Mobile Navigation or global publisher layout as needed so mobile users can access them easily (if not using dynamic actions for mobile). A user-centric approach is to design **mobile-first** for critical use cases: start by asking, "If I were doing this on my phone, what would I absolutely need and what could I omit or do differently?" This often results in a very clean design that then translates well to desktop too (where you have space to maybe add some of the nice-to-haves back).

Accessibility and Inclusive Design: Even as you design, keep in mind the diversity of your user base. Some may have visual impairments, hearing impairments, motor issues, or simply different tech literacy. An accessible design helps everyone. For instance, high contrast color schemes aren't just for the visually impaired; they help in bright sunlight or on lower-quality monitors. Clear, easy-to-read text and obvious interactive elements benefit older users and new users alike. When configuring pages, use standard components when possible because Salesforce built them with accessibility in mind (e.g., standard Lightning buttons and fields have proper keyboard focus and screen reader labels). If you use custom Lightning Web Components from the app exchange or built in-house, check that they meet accessibility guidelines (like using lightning base components under the hood or providing alt text for images). We'll dig deeper into specific tactics later.

Guidance and Feedback: A user-centric design provides feedback and help at the right moments. Salesforce provides tools like **Guidance for Success** (with Path) and even **In-App Guidance** (little prompt bubbles you can configure in Setup). Declaratively, you can add a **Rich Text component** on a Lightning page as a banner with instructions (e.g., "Don't forget to fill in all required fields before saving!" or a more friendly message like "Tip: Be sure to associate an Opportunity to this case if it's a sales-related inquiry."). Used sparingly, these can be effective. Also, ensure that important messages (like validation errors or approval notices) are visible – you might configure the page to include the standard **Highlights Panel notifications** (which show temporary green success or red error messages at top after actions) – fortunately this is built in.

Let's bring ChatGPT into the picture as a design buddy. Sometimes explaining a design idea or visualizing it can be tough. This is where ChatGPT's new multimodal abilities might help a Salesforce designer. Imagine you want to propose a new layout for an object to your team – you could manually draw it on a whiteboard, or you could ask ChatGPT to help create a simple wireframe.

ChatGPT Prompt (Image/Wireframe Generation):

User: "ChatGPT, can you generate a simple wireframe for a Lead record page that follows best practices? The page should have an Overview tab with key fields, a Related tab, and a Path at the top for the Lead Status."

ChatGPT: *"Certainly! Here is a rough wireframe for the Lead page:*\n* (Imagine a visual: at the top, a Path component showing stages: New -> Contacted -> Qualified -> Converted. Below that, a two-column layout. Left column has a section 'Lead Information' with fields: Name, Company, Phone, Email, Lead Source... Right column has 'Additional Details' with fields: Industry, Number of Employees, etc. At the top of left column, a Highlights Panel showing Lead Name, Company, and a couple of key fields. There are tabs below: Overview (selected), Related (which would contain related lists like Activities or Campaigns).)*n*This design keeps the most-used fields front and center and separates supplementary info into another section or tab. The Path guides the user on lead status progression. The Related tab can host things like open activities or campaign history, not loading them until clicked." ChatGPT might provide an ASCII or symbolic representation of this wireframe or even use a tool to generate a simple image. In our text, we can describe it, as above.

In this hypothetical exchange, ChatGPT helped articulate a design. While it's not providing a polished Figma mockup, even a simple ASCII sketch or described layout can help communicate ideas to stakeholders or clarify your own thinking. You might even ask ChatGPT for alternative layouts: "What if I need a third column for a sidebar – how would that look?" and iterate on the design conversation.

Pro Tip: Create a **pilot page** and gather feedback. User-centric design is iterative. If you're rolling out a new dynamic page or UI overhaul, identify a few power users and let them try it first. Gather their feedback on what's intuitive or not. For instance, they might say "I expected the 'New Task' button to be on the main page, not under Related tab" – which might prompt you to surface that action using dynamic actions on the main tab for convenience. Because declarative changes are relatively quick to implement, you can afford to tweak and refine after an initial rollout.

Another facet of user-centric thinking is catering to different roles' needs. Let's compare:

Sales Reps: Likely care about minimal data entry and maximum insight. So for them, you want things like key fields (deal size, close date) easily editable, perhaps inline, and helpful info like account insights or past interactions visible. A sales rep on mobile might benefit from a **global action** that creates a follow-up event in one tap after a meeting – configure one for them.

Service Agents: Value having all context at their fingertips in one screen (hence console designs). For them, make sure the case page pulls in related knowledge articles or similar cases (there are components for that like "Recommended Articles" if you have Knowledge). Keep navigation minimal – e.g., open sub-records in subtabs rather than replacing the main tab (this is more about console settings, but part of UX).

Field Technicians: Might mainly use a mobile app (Salesforce Field Service, etc.). If configuring those UIs (which have their own declarative setup), focus on large buttons (easy taps), offline availability, etc. Even though Field Service mobile has a separate app, the principles carry over: don't overload with unnecessary info in their views.

Community Users: Often are less frequent users, so the UI must be *extremely straightforward*. For an experience cloud portal, consider using a **Guided Flow or Wizard** for complex tasks rather than exposing a raw record page. For instance, instead of a partner editing an Opportunity via the standard page (which might be overwhelming), maybe provide a simplified screen flow that updates key fields with guidance. Declaratively, that's embedding a Flow component on the page or using the Experience Builder to create a custom page.

We should also mention **internationalization/localization** as part of inclusive design if applicable: if your user base spans languages, ensure your custom labels (like tab names or field labels you created) are translated via Translation Workbench. An intuitive UI speaks the user's language – literally.

Salesforce's **Lightning Design System (SLDS)** is the baseline for consistency and accessibility. When we stick to declarative configuration, we automatically use SLDS components (they come with the platform). If you ever consider adding custom front-end (Aura/LWC), maintaining the SLDS styling ensures the custom bits look and behave like the rest of Salesforce – important for consistency. But in a declarative-first approach, that's usually not an issue: we use what Salesforce provides.

To summarize this section: always put yourself in the user's shoes. A configuration choice might seem logical to us as builders, but how will it feel at the end of a long workday for the user actually clicking the buttons? Build pages that **minimize friction** – fewer clicks, less reading, more doing. Build pages that **tell a story** – the information on them is organized in the natural order the user would need it. And build pages that **respect the user's context** – whether that's their device, their role, or their level of ability. By grounding our declarative work in these user-centric principles, we ensure that the technology truly empowers users rather than confuses or hinders them.

We've touched a bit on guiding users (with hints, Path, etc.) in this section. Next, we're going to dive deeper into one of Salesforce's marquee features for guiding users through processes: **Salesforce Path and Guidance for Success.** This is a great example of a built-in tool that embodies several of the principles we discussed (visual guidance, consistency for process, and contextual help). Let's explore how to leverage Path and Guidance to lead users on the right track during key business processes.

FACILITATING GUIDED USER EXPERIENCES WITH SALESFORCE PATH AND GUIDANCE FOR SUCCESS

One of the most visually distinctive and user-friendly elements introduced in Lightning Experience is the **Path** component, often seen at the top of records like Leads, Opportunities, and Cases. Path, combined with **Guidance for Success**, is a powerful way to **guide users visually and contextually through a business process.** It provides at-a-glance information about where a record is in a process and can offer tailored coaching tips or key fields at each step. In this section, we'll discuss how to configure Salesforce Path and use Guidance for Success to create guided experiences. We'll walk through examples for sales (Opportunity Stage path), service (Case Status path), and even custom processes. We'll also give tips on writing effective guidance text and using this feature alongside other dynamic UI tools.

What is Salesforce Path? Path is a Lightning component that represents the values of a picklist field (usually a Stage, Status, or other lifecycle field) in a track across the top of a record page. Each value (stage) appears as a chevron (an arrow shape) in sequence, highlighting the current value. For example, on an Opportunity, stages like Prospecting, Qualification, Proposal, etc., will show up as a horizontal sequence of chevrons, with the current stage highlighted in a different color. This gives users a clear sense of progression. Path can also display up to five *key fields* and a custom *Guidance for Success* message for the current stage.

Enabling Path: To set up a Path, you go to **Setup -> Path Settings** (make sure Path is enabled in your org, which is usually just a one-time step). Then you create a new Path specifying:

- **Object:** e.g., Opportunity.
- **Picklist Field:** e.g., Stage (it will list picklists on that object; you choose the one representing the process).
- Optionally, you choose a **record type** if you want the path to apply to only a specific record type (often not needed unless different record types have different processes).
- You then configure the stages:

- For each stage (picklist value), you can select up to 5 fields as **Key Fields.** These are fields you want the user to focus on at that stage. They will appear in the Path's side panel on the record page when that stage is active, kind of saying "Make sure these fields are filled in or reviewed now."

- You also can provide **Guidance for Success** text for each stage. This is a rich text area where you can write notes, tips, or instructions relevant to that stage.

Once configured, you **activate** the Path. Then on the Lightning record page, you need to **add the Path component** (if it's not there already). Typically, you place Path near the top of the page (often full width region). When a user views a record, they'll see the Path bar and can even click on the chevrons to mark current stage as complete (and if celebration is enabled for a stage like "Closed Won," confetti might rain down – a fun touch).

Using Guidance for Success effectively: The Guidance for Success is your chance to coach the user at each step. For example:

- On a **Lead** object, at the "Qualify" stage, Guidance might remind the sales rep: "Ask the lead about their budget and timeline. Make sure to fill in the Budget Confirmed field if they have funding. Provide a brief introduction of our company's value – see marketing one-pager attached. If lead is not interested, you can mark Unqualified."

- On an **Opportunity** at "Proposal" stage: "Ensure the proposal document has been sent and linked. Key fields to update: Proposed Amount and Close Date. Common gotcha: double-check any custom requirements with legal."
- On a **Case** at "Escalated" status: "You have escalated this case. Communicate to the customer that a senior agent is now involved. Attach any relevant logs or files before handing off. Key fields: Escalation Reason, Severity (adjust if needed)."

The text can include bullet points, numbered steps, or even simple formatting (bold, italics) to improve readability. Keep it concise – ideally one or two short paragraphs or a few bullets. Users should be able to scan it quickly. It's not meant to be a full SOP document, just the highlights or tips.

Also, adjust tone appropriately: For internal employees, you might be a bit informal or use company lingo ("Remember to mention the *New Customer Discount* if applicable!"). For external-facing guidance (if you ever exposed a path to customers in a community, which is less common), you'd word it differently. But typically Path is for internal users.

Key Fields and Path: The key fields you designate serve as a reminder of what data should be completed at that stage. When a path stage is current, the user sees those fields (and can even inline edit them right from the Path panel if I recall correctly). For example, on an Opportunity at "Negotiation" stage, you might show "Legal Approved? (checkbox)" and "Revised Terms Sent? (date)" as key fields to prompt the rep to complete those before moving to Closed stage. It's a gentle nudge built into the UI.

Let's go through setting up a Path for a **Custom Object** to illustrate a different scenario: Suppose you have a custom object Onboarding__c that tracks onboarding of a new client. It has a Status picklist: Onboarding Scheduled, In Progress, Completed, On Hold. You want to guide the user (an onboarding specialist) through this. Using Path, you configure:

- Key fields for "Onboarding Scheduled" might be "Kickoff Meeting Date" and "Onboarding Owner".
- Guidance: "Schedule a kickoff meeting with the client. Make sure to assign an Onboarding Owner (yourself or a colleague) who will be the main point of contact."
- For "In Progress": Key fields could be "Percent Complete" or "Next Milestone". Guidance: "Work through the onboarding checklist with the client. Update the Percent Complete as you hit milestones. If any blocker arises, communicate via Chatter and consider putting the onboarding On Hold."
- For "Completed": maybe no key fields (or just a final "Completion Date"). Guidance: "Congrats! Ensure all deliverables are sent. Gather feedback from the client. Close out any open tasks."
- For "On Hold": Key field "Hold Reason", Guidance: "Document why onboarding is paused and any next steps or follow-up date."

Once that Path is activated and added to the Onboarding page, your onboarding specialists have a clear visual queue and instructions at each stage of the client onboarding. This could reduce mistakes (like forgetting to capture a piece of info early on) and improve consistency across different specialists.

Now, how does Path interact with other dynamic features? Actually, quite well. Path is essentially read-only display (except marking complete). It doesn't enforce anything by itself (that's what validation rules or required fields are for). But you can use Dynamic Forms in tandem – e.g., you could show/hide certain sections of the page based on stage, in addition to Path's guidance. Some orgs even create a sort of "Stage-specific section" that becomes visible when a record is in a certain stage (so that, for instance, in Qualification stage, a section with qualification questions appears). Path's Guidance might say "Fill out the Qualification Questions section below", and Dynamic Forms ensures that section appears only at that stage. This might be overkill if you have Path key fields though – you could just make those key fields appear in Path rather than another section.

Remember also that Path can trigger a celebration for a particular stage (e.g., Closed Won Opportunity). It's a small thing but can boost user morale. You can enable it and pick how often to show confetti (every time vs sometimes). For sales teams, that visual cue of success can be fun.

Let's bring ChatGPT in to help craft some Guidance text, since writing helpful, concise guidance can be tricky:

ChatGPT Prompt (Typing):

User: "I'm setting up Guidance for Success on our Opportunity Path. Any suggestions for what to include at the 'Closed Lost' stage to help our sales reps learn from it?"

ChatGPT: "For a Closed Lost stage, guidance should encourage reflection and documentation. For example: '**Closed Lost – Learn and Document:** Sadly, we didn't win this time. Please select a **Lost Reason** (was it pricing, competitor, timing?) and add a brief note in **Lessons Learned** field. Understanding why we lost helps us improve for the next opportunity. If appropriate, set a follow-up Task or re-open as a lead for the future.' This kind of guidance prompts reps to properly close out the deal and gather insights."

User: "How about Closed Won?"

ChatGPT: "For Closed Won, be celebratory but still instructive: '**Closed Won – Great Job!** Make sure all information is up-to-date: confirm the **Final Amount** and **Close Date** are accurate. Set the customer status to **Customer** in the Account record. Don't forget to thank the team and maybe ring that sales gong! Also, consider if there's an opportunity for a customer testimonial or referral.' This balances a congratulatory tone with actionable next steps."

The above exchange shows how an AI can help you brainstorm effective content for user guidance, ensuring you cover useful points without being too long.

Pro Tip: Keep Path guidance updated. Business processes evolve, and if a stage's meaning or best practices change, update the guidance text to match. Also, involve the end users (like actual sales reps or support agents) in crafting the guidance. They often know best what tips would have helped them as newbies. A senior rep's insights included in guidance can effectively mentor all newer reps.

Another feature related to Path to mention: **Celebrations** (I touched on it) and also **Marking status as complete.** On some objects like leads and opps, users can click "Mark as Complete" on Path which basically advances the stage for them and triggers confetti if it's the final stage. Ensure users know about this – many figure it out intuitively, but some might ignore Path entirely if not introduced. Training or on-screen prompts can encourage them to use it. Once they do, they typically love the quick view.

Beyond Path, Salesforce has **"Next Best Action"** (now called Einstein Next Best Action) which can pop up recommendations. That's more advanced and often involves AI/flow. But conceptually it's also guiding the user (e.g., "Offer 10% discount now" suggestion). Mentioning it is optional; in declarative config context, it's an intersection of AI and UI. If you have it enabled, ensure its placed prominently and not conflicting with Path or other guides.

One more use case: **Custom Paths** on objects like **Volunteer Registration** or **Grant Application**. Any process where there's a defined set of statuses can benefit from Path. It adds a level of polish and clarity. Even if you don't use Guidance text heavily, the visual path alone helps users see progress. Humans generally like to know "where am I in this process, how much is left?" – Path answers that at a glance.

Finally, consider if **In-App Guidance** (little pop-up prompts you can create in Salesforce) might complement Path. For instance, the first time a new user enters a particular stage, you could have an in-app prompt saying "Notice the tips in the Guidance section above – they'll help you complete this stage." That might be unnecessary if your guidance is obvious, but something to keep in mind as an onboarding tool for new staff.

With Path and Guidance configured, users are not left to wander through stages blindly; they have a built-in coach on every record. This reduces the need to memorize procedures or reference external documents – the system itself provides the context-aware help. As we conclude this section, recall how far we've come: our UI now adapts in content (fields via Dynamic Forms), in available actions (Dynamic Actions), in structure (responsive design), and in process guidance (Path). The last piece of the puzzle is ensuring **every user** can effectively use this UI – in other words, making it accessible and compliant with standards. In the final section, we'll focus on UI accessibility and compliance, so all users, regardless of ability, can benefit from the great experiences we're building.

ENSURING UI ACCESSIBILITY AND COMPLIANCE

In the rush to deliver features and optimize performance, it's easy to overlook something fundamental: **Can every intended user actually use this interface?** Accessibility isn't just a nice-to-have or a checklist item for compliance; it's an essential aspect of user experience. A truly declarative *best practice* is to design with accessibility in mind from the start, rather than retrofitting later. This section covers how to evaluate and improve the accessibility of your Lightning pages and configurative UI, ensuring you meet standards like WCAG (Web Content Accessibility Guidelines) and other relevant regulations (such as Section 508 in the US). We'll discuss practical tips for keyboard navigation, color contrast, screen reader compatibility, and overall inclusive design strategies. By making our UIs accessible, we not only accommodate users with disabilities but often improve usability for everyone.

Salesforce's Built-in Accessibility: The good news is that the Lightning Design System and Salesforce's core components are designed to be accessible out-of-the-box. This means if you stick to standard Lightning page components and fields, you inherit a lot of accessibility features:

- Proper semantic HTML under the hood (for example, form fields have labels tied to inputs, lists have proper roles, etc.).
- Keyboard navigation support (e.g., you can tab through buttons and links, use arrow keys in menus or Path, etc.).
- ARIA labels and roles for screen readers (Salesforce includes these in their components).

- Responsive design (so on different zoom levels or screen sizes, things adjust rather than overflow or cut off).
- However, there are still things to watch out for in your declarative configuration:
- Color choices in things like Rich Text components or formula image fields.
- The way you structure text or instructions in Guidance or custom components.
- Any images or icons you add (ensuring they have alt text).
- Ensuring that hiding/showing content (like dynamic forms) doesn't confuse screen reader users (if content appears conditionally, is it obvious? Typically yes, since when it appears, it's just part of the DOM and the user can find it by navigating, but just be mindful).

Let's break down key areas:

1. **Color Contrast:** Make sure text is readable against its background. Salesforce's default color scheme is generally fine (dark text on light background, etc.). But if you've customized anything – for instance, maybe you colored some text in a Rich Text component to make a warning stand out – ensure sufficient contrast. A common standard is a contrast ratio of at least 4.5:1 for normal text (WCAG AA). For example, **yellow text on a white background** is nearly impossible to read (poor contrast). If you need an accent, use bold or icons rather than low-contrast colors. If you use conditional highlight formulas (some folks make formula fields that show a colored indicator like a green dot or red dot emoji), provide a text alternative for color-blind users. E.g., instead of just a green circle • for "Active", maybe also include the word "Active" or a checkmark symbol. Or use a formula that outputs "Active (green)" vs "Inactive (red)" – though redundant for some, it helps those who can't see the color.

2. **Keyboard Navigation:** Power users and users with motor disabilities alike often rely on keyboard navigation. Test your pages by pressing Tab and Shift+Tab to navigate through interactive elements. You should be able to reach buttons, links, and inputs in a logical order. Salesforce arranges this well usually. If you use custom Lightning components or

HTML (less likely in declarative config), ensure the tab order is set correctly (tabindex order should follow the visual/logical order). Also check that when a dialog opens (like a quick action modal), focus is trapped within it (Lightning modals do this by default).

Another keyboard consideration: some dynamic elements like accordions or drop-downs should be usable via keyboard (arrow keys to move, etc.). Standard ones are. If you find any part of your page where a keyboard-only user might get stuck, that's a flag. For instance, if you have an embedded Google map (just hypothetical) that might not be easily navigable by keyboard, you'd need to provide an alternative way to get that info (like an address field they can read).

3. Screen Reader Compatibility: A screen reader user "hears" the page structure. They rely on proper labels, headings, and region roles. Lightning pages have landmarks (like the navigation bar, main content, etc.) built-in, which is good. But pay attention to:

- Field Labels: Always use the field's actual API label when placing it (the platform does, unless you override with a custom label via formula or something). Don't blank out labels or use misleading ones.

- Section Headers: When you add a section in a dynamic form, give it a meaningful label (e.g., "Financial Information") – screen readers can use these as navigation points.

- Alt Text for Images: If you embed an image in a Rich Text component (like a company logo or a diagram), add an alt text or caption in the rich text so it's conveyed. If an image is purely decorative, you might say "Image: decorative swirl" or simply omit if it's not important.

- Icons/Emojis in text: Some admins like to prefix field labels or tab names with emojis (to add visual flair). E.g., "📞 Contact Info". Screen readers will often read out a description like "telephone symbol" for that emoji, which might be fine or might be odd. Use such icons sparingly and be aware of what they announce. Alternatively, place them after the text so the important word is spoken first.

- Dynamic Content Changes: If something on the page changes without a full reload (like a section appearing after a user action), ideally the screen reader should be informed. Salesforce does handle some of this.

For example, if you use a Path and mark a stage complete, screen readers likely announce the stage change. If you collapse or expand an accordion, the state is announced ("expanded", "collapsed"). Just ensure to use native components for these so that's in place. If you ever use custom HTML in a Visualforce or LWC, you'd need to handle aria attributes – but we'll assume declarative mostly avoids that.

4. Forms and Error Messages: If users input data, are errors conveyed clearly? Salesforce will outline required fields in red and list an error message summary if a save fails validation. That summary is reachable by keyboard and is read by screen readers (it's an alert region). One tip: write validation rule error messages in plain language and indicate the field if possible. Instead of "Error: 527 bad", write "Error: The Close Date cannot be in the past." The user will then know what to fix. Also, grouping required fields logically and not hiding them unexpectedly helps ensure users don't miss them. For example, don't hide a required field behind a tab that the user might skip entirely – if it's required, keep it in the main flow or use conditional required logic carefully.

5. Inclusive Layout & Wording: Consider cognitive load. Simplicity (one of our design principles) benefits users with cognitive or learning disabilities. Clear headings, not too much info at once, and using simple language in instructions all help. Avoid acronyms or internal jargon in user-facing text unless you're sure the users know it – and if they do, they're likely not an accessibility issue, but clarity is good for everyone.

Accessibility Compliance Checklists: It can be useful to have a quick checklist when reviewing a Lightning page:

- Can I navigate all interactive elements with only a keyboard?
- Do all images/icons have alt text or labels (or are purely decorative)?
- Is the color contrast of text and important UI elements sufficient?
- Are my guidance and error messages understandable when heard (not just seen)?
- Have I avoided using color alone to convey meaning? (e.g., don't just say "fields marked in red are required" because a color-

blind person might not distinguish red; luckily Salesforce also uses asterisks for required).

- If using any video or audio in the UI (rare in declarative config, but maybe an embedded video in a Visualforce page?), are there captions or transcripts? (Probably not applicable in typical Lightning pages, but mention completeness.)
- Did I test with a screen reader (like NVDA or VoiceOver) at least once for a sample record? This is the ultimate test – hearing what a blind user would hear. It can be enlightening: you might find that if your page has many sections, the screen reader user might prefer a way to jump by headings. Consider adding meaningful headings via rich text if needed to chunk content.

One neat capability: If you have ChatGPT with vision (upload images) or an accessibility tool, you could even ask it to analyze a screenshot of your page for potential issues.

ChatGPT Prompt (Analysis of Screenshot):

User: uploads a screenshot of a Lightning page

"ChatGPT, do you see any accessibility concerns in this interface?"

ChatGPT: "Looking at the screenshot, most text seems high-contrast (dark grey on white). The layout looks clean. One concern: I notice a light green text on a white background in the KPI component – that might be low contrast, especially for users with color blindness. Also, the page uses a lot of acronyms like 'FY TDV'. If those aren't commonly understood, that could confuse some users (though not strictly an accessibility violation). The Save button and other controls seem present. Without interacting, I'd recommend ensuring that the collapsible sections (if any) are keyboard operable. Overall, it appears mostly accessible."

While AI might not catch everything perfectly, it can point out obvious color issues or jargon, supplementing your manual checks.

Compliance Standards: If your org must comply with formal standards (WCAG 2.1 AA is common target, or Section 508), Salesforce's core meets many of these, but you are responsible for customizations. Salesforce provides a VPAT (Voluntary Product Accessibility Template) that describes its compliance. If audited, you should be prepared to show that your customizations (like any additional code or significant UI changes) are also compliant. For declarative work:

- Keep any externally embedded content to a minimum or ensure its accessible.
- Use standard Lightning components (they've been vetted).
- If you build a community, use the Theme and components provided or check any custom Lightning component used there.

A11y (Accessibility) Tools: There are browser plugins (like Axe, Wave) you can run on a Lightning page. They might flag, say, some Salesforce code with warnings, but focus on issues related to your content. For example, Wave might flag if you have two headings with the same text (could be fine) or a low-contrast text. It's worth doing a scan on key pages. Often, you'll be in good shape thanks to Salesforce's baseline, but it's not automatic.

One example issue: In earlier Lightning, some combo box components weren't labeled properly for screen readers. Salesforce fixes such things over time, but if you encounter something like that, you can raise it to Salesforce support or find workarounds (maybe place an explanatory text nearby). As a declarative admin, you might not have much control over such components aside from being aware.

To conclude accessibility: It's like building a ramp to a building – everyone can use a ramp (someone with a wheelchair must, but someone with a bike or a stroller finds it helpful too). Similarly, an accessible Salesforce page is usually a better page for all users. High contrast, clear navigation, concise content – these benefit everybody.

Pro Tip: Involve users with disabilities in your testing if possible. If your company has employees who use assistive tech, get their feedback on your Salesforce configuration. They might point out things a sighted admin never noticed. For example, maybe a screen reader user says, "It takes me 10 tabs to get past the navigation to the content each time" – maybe you then show them the keyboard shortcut (there is a Lightning shortcut to skip to main content) or you consider reducing header elements. Little insights like that are gold.

We've now covered ensuring that our beautifully crafted UI is accessible and compliant. With this, we have closed the loop on user experience: from design and performance to guidance and inclusion. Let's now summarize what we've learned in this comprehensive chapter and set the stage for what comes next in the Salesforce Enterprise Thinking journey.

SUMMARY

In this chapter, we embarked on a deep dive into **User Experience and UI Configuration** using Salesforce's declarative toolkit, and we saw how thoughtful design can dramatically improve user adoption and satisfaction. We began with the **Lightning App Builder**, learning how to create modular record pages with components like tabs and accordions to organize information logically. By structuring pages with the user in mind – placing key data in the spotlight and deferring secondary details – we ensure that Lightning pages load faster and feel more intuitive. Real-world examples illustrated the difference: a well-designed sales dashboard page versus a cluttered one can mean the world in daily productivity.

We then mastered **Dynamic Forms**, Salesforce's answer to one-size-fits-all layouts. You saw how dynamic field visibility lets a single page serve multiple audiences and scenarios by showing each user only what they need. This reduces the reliance on dozens of page layouts and record types that used to clutter org configurations. For instance, instead of maintaining separate layouts for domestic vs. international accounts or standard vs. premium cases, one dynamic form can handle variations with smart filters. The result is a cleaner admin life and a more tailored user experience. Through our hands-on examples, you can now confidently set up dynamic forms to consolidate layouts and make your pages truly responsive to context.

Building on that, we explored **Dynamic Actions**, enabling us to streamline the buttons and quick actions users see. By configuring context-aware action visibility, we decluttered the record page header and ensured users have the right tools at the right time. No more "button overload" where users scratch their heads about which action to click – now the interface can guide them by only displaying relevant actions (like "Approve" for managers or "Request Renewal" only for expiring contracts, etc.). We saw this in scenarios from desktop to mobile and learned that we can even differentiate actions by device to optimize small-screen workflows.

Performance and responsiveness were a major focus. You're now equipped with best practices to **design Lightning pages for speed**: limiting initial load components, using tabs to lazy-load content, keeping related lists lean, and leveraging Salesforce Optimizer and the Lightning Analyzer to pinpoint slowness. We recounted how applying these techniques – such as moving heavy related lists off the default view or splitting a huge detail section into focused segments – can cut page load times significantly. In today's fast-paced world, these optimizations aren't mere technical tweaks; they translate to better user focus and less frustration. And with an eye on **mobile responsiveness**, you learned to tailor pages for phone users, ensuring that on-the-go access is just as efficient and friendly.

Crucially, we wove in **user-centric design principles** throughout our journey. We emphasized consistency (so users feel at home on every page), intuitive navigation (so processes flow naturally and with fewer clicks), and the importance of designing for the user's context and needs. We saw how features like Salesforce Path and Guidance for Success embody these principles by providing visual cues and helpful prompts in-app, reducing reliance on memory or external documentation. Our section on Path demonstrated how to guide a salesperson through an opportunity or a support agent through a case using stage-by-stage tips, which not only improves process adherence but also flattens the learning curve for new team members. We also touched on leveraging ChatGPT as a co-pilot in the design process – whether brainstorming layout ideas, drafting coaching tips, or even generating quick wireframes – showcasing a novel way to blend AI assistance with Salesforce expertise.

Finally, we turned our attention to **accessibility and compliance**, an area that underpins all the others. After all, a UI can only drive adoption if everyone can use it. You now have a checklist of accessibility considerations to apply to your configurations: from ensuring color contrast and keyboard navigability to providing alt text and screen reader-friendly labels. We reiterated that Salesforce's declarative tools give us a strong starting foundation (with SLDS components), but it's on us to use them wisely and fill any gaps (like writing clear error messages or avoiding color-only signals). By following these guidelines, you'll not only meet legal or policy requirements but truly live up to the ethos of inclusive design – making your Salesforce apps usable by the widest possible audience.

As we conclude Chapter 8, you've gained a 360-degree view of crafting exceptional user experiences declaratively. You've seen practical techniques and "Pro Tips" that you can apply immediately in your org, and you've been introduced to ways ChatGPT can assist you in this creative and technical endeavor. By implementing these best practices, your Salesforce UI will be more than just functional – it will be *engaging, efficient, and empowering* for users.

Looking ahead, our Salesforce Enterprise Thinking journey continues. In the next chapter, we will take these general UI/UX principles and apply them in a specific domain: **Sales Cloud Configuration Best Practices.** You'll learn how to optimize sales processes (like lead-to-opportunity) using declarative configuration, and you'll see how some of the tools and techniques from this chapter (like dynamic forms and paths) play a pivotal role in real-world sales scenarios. With a solid understanding of user-centric design from this chapter, you are well-prepared to dive into tailoring Salesforce for the Sales Cloud – configuring an environment that not only meets business requirements but also delights the sales teams who use it every day. Let's carry this focus on excellent user experience forward as we explore the nuances of Sales Cloud in the next installment.

9

SALES CLOUD CONFIGURATION BEST PRACTICES

Sales teams live and breathe in Salesforce, so configuring Sales Cloud effectively can turbocharge their productivity. This chapter explores how to declaratively customize Sales Cloud – **without tipping into over-customization** – to streamline the lead-to-opportunity lifecycle. We'll look at ways to capture and route leads faster, ensure data quality in opportunities, and enhance sales rep productivity through thoughtful configuration. The goal is to leverage Salesforce's standard tools (with a few smart tweaks) so that your system works *with* your sales process, not against it.

Imagine a scenario: new leads pour in from your website, and within seconds each is automatically assigned to the right rep or queue. No lead falls through the cracks or waits unattended. By the time a sales rep logs in, their **"My Leads" list is teed up perfectly** – every lead is where it needs to be. That's the power of effective lead management configuration. In this chapter's first sections, we'll cover how to achieve fast, accurate lead routing using declarative tools like assignment rules, queues, and flows.

Of course, not all leads are equal. Some are hot prospects ready to buy; others are tire-kickers. We'll dive into declarative lead scoring techniques to help your team **focus on the leads that matter most**. You'll learn how to set up scoring models using formula fields and Flows, all without writing a single line of code. These scores will guide reps to prioritize their follow-up, improving conversion rates.

Once leads convert, opportunities take center stage. Consistent opportunity management is critical – it keeps your pipeline healthy and your forecasts accurate. We'll discuss best practices for defining clear Opportunity stages, using Sales Processes and record types when appropriate, and enforcing data quality with validation rules. The theme here is standardization: when every salesperson follows the same well-designed process, you get reliable data and predictable outcomes.

Configuration isn't just about data; it's also about guiding people. We'll explore Salesforce's **Sales Path and Path Assistant**, which put **actionable prompts and key fields front-and-center** for sales reps at each stage of a deal. This declarative tool acts like a built-in sales coach – especially useful for onboarding new reps or reinforcing best practices. We'll show how to set up Path to drive consistent, productive sales behaviors, stage by stage.

Collaboration often makes the difference in closing big deals. Later in this chapter, we'll discuss enabling **Opportunity Teams and Account Teams** to support team selling. By configuring these features declaratively, you can clearly define each team member's role on a deal and extend sharing access without custom code. We'll cover default teams and how these tools foster transparency and accountability in the sales process.

Finally, we'll deep dive into two advanced areas: **Opportunity Splits and Collaborative Forecasting**, and tailoring the **Opportunity page layout with Dynamic Forms.** Splits allow multiple reps to share credit for a sale, aligning incentives in complex deals. Forecasting (when set up right) gives everyone from reps to execs a realistic view of future sales. And Dynamic Forms let you craft user-friendly, context-aware opportunity pages – showing the right information to the right people at the right time. This not only improves user experience but also reduces noise and confusion on the page.

By the end of this chapter, you'll have a toolkit of Sales Cloud best practices grounded in declarative configuration. Each section contains practical examples and even ChatGPT prompts to illustrate how an admin or consultant can get intelligent help from AI while building these solutions. Let's jump in, starting with the all-important task of lead management and how to handle new leads like a pro.

EFFECTIVE LEAD MANAGEMENT STRATEGIES

A healthy sales pipeline begins with effective lead management. In Salesforce, **lead management** involves capturing leads, routing them quickly to the right owners, and making sure no potential opportunity is lost or ignored. In this section, we focus on declarative strategies to improve lead handling: setting up Lead Assignment Rules and Lead Queues for automatic routing, using auto-response rules and other automation to engage leads immediately, and leveraging Flows for any custom logic needed beyond what standard rules offer. The goal is to configure Salesforce such that every new lead is processed promptly and accurately, boosting conversion rates and sales team efficiency.

When a new lead enters Salesforce (for example, from a web form or list import), the first decision is **"Who should handle it?"** Salesforce provides *Lead Assignment Rules* to answer that question. An assignment rule is essentially a list of criteria-action pairs, evaluated in order, that assigns the lead to a specific user or a queue. With a single active assignment rule, you can cover all your routing logic. For instance, you might route leads by region, product interest, or company size. Salesforce will evaluate each rule entry until it finds a match and then assign the lead accordingly. This happens automatically on lead creation – meaning reps don't have to spend time picking through new leads; the system does it for them.

Lead Queues play a vital role in this process. A queue is basically a holding area (with an owner that's not a person but a queue name) where leads can sit until a qualified rep claims them. Queues are great for scenarios like inside sales teams or regions – e.g., you could have a "West Coast Leads" queue that all west territory reps can access. Assignment rules can drop leads into a queue when criteria are met (like *State = California* might go to the "West Coast Leads" queue). Then, any rep who's a member of that queue can take ownership of those leads. This setup prevents a common problem: one rep getting overloaded while another has bandwidth. Instead, leads go to a shared queue and the team can distribute the work. **Don't forget to set a default lead owner** (user or queue) for leads that don't match any rule – this safety net ensures no lead slips through unassigned.

ChatGPT Prompt (Text Input): "How do I set up lead assignment rules to automatically assign new leads? Can you walk me through an example of routing leads by geography and product interest?"

Beyond assignment rules, Salesforce offers other declarative tools for lead management. **Auto-Response Rules** are one such tool – they send an automatic email to the lead (for example, "Thank you for your inquiry, we'll contact you shortly!") upon lead capture. While not directly affecting lead routing, auto-response messages improve the lead experience with immediate feedback. These can be configured declaratively with conditions (e.g., send different email templates based on lead source).

Now, what if your routing logic is too complex for a simple assignment rule? Or perhaps you want to do additional processing when a lead comes in – such as recording a timestamp, creating a follow-up task, or even auto-converting certain leads. This is where **Salesforce Flow** (specifically record-triggered flows on the Lead object) can augment your lead management. Flows allow for complex branching logic and actions. For example, a Flow could look at a lead's country and product interest and assign it to the correct queue or rep (mimicking assignment rules), but it could also create a task for the new owner to remind them to follow up within 24 hours, or update a custom "Lead Aging" field. Flows can even invoke the assignment rule: by setting the Lead Owner field to a special value and marking Assign using active assignment rule, the Flow can force Salesforce to run the standard assignment rule logic after any Flow logic.

Lead conversion is another critical step in the lifecycle. When a sales rep qualifies a lead, they convert it, which transforms the Lead record into a new Contact (and usually an Account and an Opportunity). Best practice is to ensure your *Lead Field Mapping* is configured in Setup – this maps key lead fields to account, contact, or opportunity fields upon conversion so that no information is lost during the switch. For instance, you might map a "Lead Industry" field to the Account's Industry field, or a custom "Lead Rating" field to a custom field on Opportunity, so the context carries over.

Pro Tip: If certain leads should be converted automatically (say, trial sign-ups that reached a threshold), consider using a Flow to trigger conversion. Salesforce's LeadConvert Flow action can convert a lead behind the scenes. Just be cautious – auto-converting without human review can create junk data if leads aren't truly qualified. Always include criteria (like lead score, which we'll discuss next) to decide if a lead warrants auto-conversion.

Let's tie these pieces together with a quick real-world narrative. Suppose **Cloud Kicks Inc.** gets 1,000+ leads per month from various sources: their website, events, and a purchased list. We configure an assignment rule such that: website leads go directly to a "Website Leads Queue" for speedy follow-up by an SDR team; event leads get assigned to the specific rep who worked at that event (using a custom field "Event Name" to match reps); and high-value leads (say, those from companies with >500 employees or "Hot" rating) go straight to a senior salesperson. Any lead that doesn't meet a rule condition lands in a default queue managed by the inside sales manager, who will review and redistribute them. We also set up an auto-response rule to send a polite acknowledgment email to every new lead, and a Flow that creates a task for the owner if the lead is from a top-tier account, prompting them to call within 1 business day. The result? Within moments of submission, each lead is in the right hands with next steps lined up, and no one wonders "what happened to that lead from last week" – it's all tracked.

ChatGPT Prompt (Voice Input): User (speaking into phone): "Hey ChatGPT, I'm getting swamped with leads. How can I use Salesforce to automatically distribute leads so my team can respond faster?"

ChatGPT (responding verbally): "You can start by defining Lead Assignment Rules with criteria like geography or product interest. For example, leads from APAC could go to an APAC queue. Let me walk you through creating a rule entry for that..."

By configuring efficient lead capture and routing, you set the stage for better conversion rates – after all, speed and accuracy in lead response often mean the difference between a deal and a missed opportunity. Now that we've ensured leads are getting to the right people promptly, the next step is figuring out *which* leads deserve the most attention. In the following section, we'll explore declarative lead scoring, a technique to prioritize leads based on their quality and fit, so your sales reps always focus on the hottest prospects.

Next we'll dive into lead scoring – how to set up point-based systems to rank and prioritize inbound leads.

CONFIGURING DECLARATIVE LEAD SCORING

Not all leads are created equal. Some leads will eagerly buy whatever you're selling, while others might never return your calls. **Lead scoring** is the practice of assigning points to leads based on certain attributes or behaviors, producing a score that estimates how qualified or sales-ready a lead is. In this section, we discuss how to implement a lead scoring system in Salesforce using declarative tools like formula fields, numeric fields with Workflow/Flow updates, and other no-code methods. We'll provide example criteria and scoring models, and show how automation can keep scores up-to-date. By the end, you'll have a blueprint for a simple but effective lead scoring solution that helps your sales team prioritize their outreach.

Imagine you're in a sales ops role and you're noticing that reps waste time on leads that are poor fits, while sometimes neglecting great leads. You decide to introduce a lead score that flags high-quality leads. Where to start? **First, define your scoring criteria.** Work with sales and marketing to identify what makes a lead more likely to convert. Common criteria include: job title (e.g. a CEO might be a higher value lead than an intern), company size (bigger companies might mean bigger deals), industry, engagement level (opened emails, attended a webinar, etc.), and demographic info (location, role, etc.). For example, you might decide: *"If the lead is a decision-maker (VP or above), add 10 points. If the company has over 500 employees, add 5 points. If the lead's* *industry is our target industry, add 5. If they clicked our pricing page, add 5."* Each factor contributes to the overall score.

Once you have criteria, you need to implement this in Salesforce **declaratively.** One straightforward way is to create a **formula field** on Lead called "Lead Score" (Number). This formula can be written to add up points for each criterion. For instance, a very basic formula might look like:

(IF(ISPICKVAL(Title, "CEO"), 10, 0) +

IF(EmployeeCount__c > 500, 5, 0) +

IF(ISPICKVAL(Industry, "Finance"), 5, 0) +

IF(Clicked_Pricing_Page__c = TRUE, 5, 0))

This adds 10 if Title = CEO, 5 if employee count > 500, etc. In practice, many scoring formulas also give points for non-blank fields (assuming if a lead provided info, they're somewhat engaged). For example, a simple score formula might award 10 points for any non-empty key fields. Salesforce even provides sample scoring formula snippets. The benefit of a formula field is that it's always up-to-date in real-time – whenever a field used in the formula changes, Salesforce recalculates the score automatically. It's also transparent to admins (you can open the formula and see all the criteria in one place).

However, formulas can get complex or hit limits if you have many criteria. Another approach is to use a standard Number field (say "Score__c") and maintain the score via automation. For example, you could use a **before-save Flow** on Lead that calculates the score in a formula resource and assigns it to Score__c whenever a lead is created or edited. Or you could use multiple **field updates**: e.g., a Flow that runs when a lead is created and increments the score field based on various conditions. For instance, one element in the Flow might say "If Title contains 'Chief' or 'VP', add 10 to Score__c; if Industry = Finance, add 5," etc. This is a bit more work to set up but can be easier to maintain if criteria change often (you could even store criteria in Custom Metadata and have the Flow loop through them – an advanced declarative trick for the ambitious admin).

What about *Einstein Lead Scoring*, you ask? Salesforce offers an AI-powered lead scoring feature (part of Sales Cloud Einstein) that automatically scores leads based on patterns it finds in your data. Einstein Lead Scoring is fantastic if you have it enabled: it's zero-effort (scores just appear) and it can uncover predictive factors you might not think of. But not every org has Einstein enabled, and our focus here is on **declarative (user-defined) scoring** that you control and understand. You can certainly mention Einstein to your stakeholders, but designing a custom scoring model can be done in any edition of Salesforce and tailored to your unique business nuances.

ChatGPT Prompt (Spoken): "I have an Excel sheet of leads with their conversion outcomes. Could you help analyze what factors correlate with conversions? I want to design a lead scoring model."

This user used voice input while showing ChatGPT a spreadsheet of historical lead data. ChatGPT analyzes it and suggests criteria: e.g., leads from certain industries and with certain titles converted at higher rates, so those should get higher points. The assistant also gently reminds to anonymize any personal data in the file before sharing, ensuring privacy.

Let's solidify this with a concrete example. Say your company, *Universal Containers*, sells B2B software. After some brainstorming, you settle on these scoring rules:

Job Role: If the lead's Title contains "Director" or higher (e.g., VP, CxO), +10 points.

Company Size: If Account Employee Count > 1000, +5; if 200-1000, +3.

Engagement: If the lead filled out a pricing request form, +10; if they attended a webinar, +5; if they just downloaded a whitepaper, +2.

Timing: If the lead was created in the last week, +3 (fresh leads are hotter).

Miscellaneous fit: Industry = target industries (say Healthcare or Finance), +5; Country = our selling countries, +5.

These points could sum up to, say, 40 or 50 for a very hot lead. We create a formula field Lead_Score__c that implements this. Some parts might use the CONTAINS() function for Title (to check for "Director", "VP", etc.), others a simple IF for numeric fields, and so on. The formula might be a bit long but it's all clicks not code. After testing on a sample of leads, we see scores roughly align with what salespeople intuitively felt – that's a good sign our model makes sense.

Now, how do we use the score? One way is to create a **list view or report** of "Hot Leads" where Lead Score > some threshold. Or use the score in assignment: e.g., if Score > 30, assign the lead to a senior sales queue for fast-track handling. You could even surface the score on the Lightning page and use **Dynamic Forms** or highlights panels to make it prominent (like a big colored number or an emoji indicator 🔥 for "hot"). Reps should be trained: a high score means this lead has high potential, prioritize it; a low score means lower priority or maybe needs nurturing before sales touch.

Pro Tip: Lead scores aren't static – they should evolve. As behaviors change (e.g., a lead suddenly starts clicking on your emails), update their score. Consider using a **Scheduled Flow** that re-evaluates lead scores periodically or triggers on key field updates (like on a campaign response). Additionally, involve Marketing: often, marketing automation tools (Pardot, Marketing Cloud, etc.) have their own scoring. Ensure your Salesforce score complements those or even incorporate marketing's score into Salesforce via integration or field sync.

One more declarative trick: use **grading or categorization** alongside scoring. For instance, you can create a formula field that grades leads as A, B, C based on score ranges (A = 30+ points, B = 15-30, etc.). This can be easier for sales to grok than a raw number. It's all doable via formulas (a CASE or nested IF formula).

ChatGPT Prompt (File Upload): The user uploads a document "Lead Scoring Criteria.docx" which contains a list of desired scoring rules. *"ChatGPT, here's our draft lead scoring criteria. Can you review it for any overlaps or missing factors, and suggest improvements? Also, ensure the criteria abide by privacy guidelines (we don't want to score based on sensitive personal data)."

To wrap up this section: declarative lead scoring in Salesforce is about defining a points system that captures what your business considers a hot lead, and using Salesforce's no-code tools to calculate and display that score. It's part art (deciding the points) and part science (using formulas/flows to implement). When done right, it's immensely helpful – sales reps can prioritize their day, marketing can qualify leads more objectively, and sales management can sleep better knowing the team is focused on leads that are more likely to turn into revenue. With our leads now routed and ranked, let's turn our attention to the next stage of the funnel: opportunities. In the next section, we will look at how to manage opportunities declaratively for consistency and high data quality, ensuring that once those top leads convert, they're handled in a disciplined way through to closing.

Next we'll move from leads to opportunities, exploring how to standardize opportunity stages, processes, and data requirements to keep the sales pipeline clean and predictable.

OPTIMIZING OPPORTUNITY MANAGEMENT

Converting a qualified lead is just the beginning of the sales journey. Once an opportunity is created, how you manage it in Salesforce can make a huge difference in forecasting and pipeline hygiene. **Opportunity management** best practices revolve around defining a clear sales process (with stages that everyone uses consistently), enforcing data quality (so that each deal record is complete and accurate), and using Salesforce features like record types or validation rules to accommodate different sales scenarios without chaos. In this section, we'll discuss setting up Sales Processes and Opportunity Stages, when to use multiple Record Types (and when not to), and how to use validation rules and automation to keep opportunity data clean. The outcome is a standardized approach to opportunities that gives sales leadership confidence in the data and helps reps understand what to do at each stage.

First, let's talk about **Opportunity Stages and Sales Processes.** Salesforce comes with a default set of stages (e.g. Prospecting, Qualification, Proposal, Negotiation, Closed Won/Lost) and each stage has an associated probability and a *forecast category*. Chances are, these defaults won't exactly match your business – and that's okay, because you can (and should) customize them. The key is to identify the distinct steps in *your* sales cycle. For example, a company might have stages like: **Discovery** (10%), **Demo** (25%), **Pilot** (50%), **Proposal** (75%), **Verbal Commit** (90%), **Closed Won** (100%), plus the lost stages. It's crucial that everyone understands what each stage means – define entry and exit criteria for each stage (e.g., "Proposal means a proposal has been delivered to the customer"). In Salesforce Setup, you can create a *Sales Process* that includes just the stages you want (in order) and then assign that to one or more Opportunity Record Types. Mapping stages to appropriate forecast categories is part of this configuration – for example, early stages map to "Pipeline" (unlikely to close yet), later ones to "Best Case" or "Commit" (depending on your forecasting approach), and of course Closed Won to "Closed" category.

Now, consider if you need **multiple Sales Processes or Record Types.** Not every org does – if all your sales deals follow the same basic stages, one process is fine. But let's say you have two distinct sales motions: one for new product sales, and another for renewals of subscriptions. New sales might have the full cycle of prospecting to close, whereas renewals might be simpler (perhaps just an "Renewal Offered" and "Renewal Closed" stage). In such a case, creating a second Sales Process and a corresponding Opportunity Record Type for renewals makes sense. Each record type can have its own stage picklist (via its sales process) and even its own page layout if different info is needed. **Best practice:** only introduce multiple record types if there is a **material difference in process or required fields** between them. Don't create record types for every minor variation – that leads to complexity and confusion. Keep it as simple as possible while still reflecting real differences.

Data quality on opportunities is paramount. An **opportunity record** should tell the story of a deal at a glance: who the customer is, what product or service they might buy, the amount, expected close date, current stage, and any other important details (like deal type, region, etc.). How do we ensure these fields are populated correctly? This is where **validation rules** come in. A validation rule can enforce that certain fields are filled out or have correct values *at certain stages*. For example, you might require that **Amount** and **Close Date** are set before an opp can move past the early stages. Or, when Stage = "Proposal", you might require that a field "Proposal Sent Date" is filled. Likewise, when

Stage = "Closed Won", perhaps a "Closed Won Reason" picklist must be selected (to capture why we won). Validation rules fire when a user tries to save a record that violates the rule, so they're great for preventing incomplete data. Be careful to align these requirements with the sales team's workflow – if you add too many mandatory fields too soon, reps might resort to putting in junk data just to save. A good approach is progressively adding requirements as the deal advances (often called "stage-gating" data requirements).

Pro Tip: Use **picklist values and dependency** to improve data quality. For instance, have an Opportunity "Type" (New Business, Renewal, Upsell) and make certain fields appear or be required only for certain types. You can do this with validation rules or with Dynamic Forms (more on that later). Also, implement **duplicate management** (Salesforce has duplicate rules and matching rules) to prevent the same opportunity or account from being entered twice. While duplicate rules are more often used for leads/contacts, they can help with opps in scenarios like channel sales where leads or deals might be registered multiple times.

Another powerful declarative tool for opportunity management is **automation via Flow.** For example, you can create a **record-triggered flow on Opportunity** that does things like: when an opp is moved to "Closed Won", automatically create a related Customer Success task or initiate a project (if you integrate with project management). Or if an opp hasn't been updated in 30 days and is not closed, have a time-based automation (either a scheduled path in flow or an old Workflow Rule with time trigger) that pings the owner or updates a "Stale" flag. These automations ensure that opportunities don't languish unattended and that the right follow-ups happen.

Consider an example of optimizing opportunity management in practice: **Global Industries Co.** identifies that a major problem is inconsistent stage usage – some reps jump straight from Prospecting to Closed Won, others linger in stages forever. We help them define a crisp set of stages and document what each means. We configure Salesforce so that when a rep tries to mark an opp as Closed Lost, they *must* fill the "Closed Lost Reason" (with a picklist of reasons) – this gives management insight into why deals are lost. We add a validation rule that **if**

Stage = "Negotiation" or later, the field "Legal Approved?" must be checked (ensuring legal review happened for late-stage deals). We also implement an "Opp Cleanliness" formula that gives a score or flag if key fields are missing (for dashboards). This way, at pipeline review meetings, sales managers can quickly see which deals are missing info.

ChatGPT Prompt (File Upload): *The user uploads "OpportunityStages.xlsx" containing a list of current stages and probabilities.* "ChatGPT, here is our current list of opportunity stages and how often deals actually close from each stage. Can you suggest improvements to our stage definitions and probabilities? Also, how should we map these to forecast categories in Salesforce?"

ChatGPT reviews the file. It notices, for example, that one stage has a 50% historical win rate but the probability is set to 75%, and advises to adjust it. It also suggests merging two redundant stages and mapping early stages to Pipeline category, late stage to Commit, etc., for consistency.

One often overlooked aspect of opportunity management is **training and adoption**. Even with perfect configuration, sales reps need to use the system correctly. This is where our next section, on Sales Path and Path Assistant, will help by providing in-context guidance. But even within standard opportunity page layouts, consider adding help text to fields or using the "description" property of stages (in setup) to clarify stage criteria. These little declarative additions can remind a rep what "Qualification" really entails, for example.

Opportunity management configuration sets the foundation for everything that follows: pipeline visibility, forecasting accuracy, and the effectiveness of sales coaching. By standardizing stages and enforcing key fields, we ensure that an opportunity logged in Salesforce isn't just **"there for show"** – it becomes a reliable representation of a deal. In the next section, we'll build on this by leveraging **Sales Path**, which takes those stages and key fields and turns them into a visual, interactive guide for salespeople on the opportunity page. This will further drive consistency and help reps follow best practices at each step of the sale.

Coming up: we'll enhance rep productivity and consistency by implementing Sales Path and the Path Assistant, turning our well-defined opportunity stages into a guided selling experience.

ENHANCING PRODUCTIVITY WITH SALES PATH AND PATH ASSISTANT

Even a well-defined sales process can falter if reps don't follow it. Salesforce **Sales Path** (with the Path Assistant's guidance) is a declarative feature designed to keep salespeople on track. It provides a visual indicator of an Opportunity's current stage and, more importantly, highlights key fields and advice ("Guidance for Success") for each stage. In this section, we'll discuss how to configure Path on opportunity records, what content to include as guidance, and tips for making the most of this feature. With Path, you essentially bake your playbook into Salesforce, so reps have the right coaching and focus at the right time – which boosts productivity and drives consistency across the team.

Sales Path on an Opportunity record, showing key fields and stage-specific Guidance for Success. Here, at the "Id. Decision Makers" stage, key fields are highlighted and tailored advice is provided to help the rep progress the deal.

In the image above, notice how the Opportunity stage is visually represented as a chevron at the top ("Prospecting → ... → Closed"), and below we see **Key Fields** and **Guidance for Success** for the current stage. This is exactly what Path offers: a quick, contextual snapshot of what needs to be done now. For example, at "Identify Decision Makers" stage, the key fields might be *Expected Revenue, Next Step, Main Competitor* – things the rep should definitely fill or update at this point. The Guidance might say: "At this stage, ensure you've identified all the decision makers and influencers. Log at least one Contact Role. Update Next Step with the date and topic of your next meeting." This kind of micro-instruction can be invaluable, especially for newer reps.

Configuring Sales Path is straightforward. In Setup, you find **Path Settings**, then create a new Path. You'll choose the object (Opportunity), the record type (if you have more than one), and the picklist field that represents the stages (usually "Stage"). Then you specify, for each Stage value, up to 5 Key Fields and a Guidance text (up to 1000 characters). A few best practices while doing this:

- **Choose at most 5 Key Fields per stage.** Salesforce limits it to five to enforce focus. Think about which fields absolutely must be reviewed or filled at that stage. Less is more – you're guiding attention, not overwhelming the user.
- **Write concise, actionable Guidance.** This isn't a place for a novel. Bullet points or a short paragraph work well. Use it to remind the rep of tasks ("Schedule a demo with all stakeholders") or quality checks ("Confirm the budget timeline with the prospect"). Some companies include a link to a Confluence page or internal playbook for complex stages (you can include hyperlinks in the guidance text).
- **Align with validation rules/required fields.** If you mark a field as key at a certain stage, it's wise to also ensure that stage can't be completed without that field (either through a validation rule or at least a report highlighting blanks). For example, if "PO Number" is a key field at Closing stage, perhaps require it when marking Closed Won.
- **Use consistent tone.** Since multiple people (sales ops, managers) might contribute to writing guidance, ensure it reads with one voice. Usually encouraging and direct: e.g., "Validate that the customer's security review is complete at this stage" rather than "Sales rep must have done security review".

ChatGPT Prompt (Operator Mode): "ChatGPT, use Operator to fetch some examples of great 'Guidance for Success' text for opportunity stages in SaaS sales."

ChatGPT's Operator agent browses Salesforce's documentation and community posts to pull a few examples. It finds guidance suggestions like focusing on discovery questions at early stages, MEDDICC criteria checks at middle stages, and proposal follow-up tips at later stages. The assistant compiles these and presents a refined list of ideas for the user's own guidance content.

Enabling Path on the Lightning page is the final step. Once you create and activate the Path, you just make sure the **Path component** is on your Opportunity Lightning Record Page (it usually is by default for standard objects – if not, you can edit the Lightning page and drag the Path component onto it, typically near the top). Voila – users will now see the Path when they view an opportunity.

The benefits of Sales Path can be **dramatic** for user adoption. Reps no longer have to remember what fields to update at each stage – it's right there in front of them. It's like having a checklist built into the page. New hires ramp up faster because the system itself coaches them. Experienced reps benefit from reminders that keep deals from slipping through the cracks (e.g., "Have you identified a champion? If not, do so before moving to next stage"). Managers love Path because it helps enforce process without constant micromanagement.

Consider **Northern Trail Outfitters** (NTO), a company in our Salesforce narrative. They implemented Path for their opportunities and defined guidance based on the MEDDIC sales methodology (Metrics, Economic buyer, Decision process, etc.). At each stage, the guidance reminds reps to gather certain info: for example, at "Qualification" stage, guidance says "Identify Pain: Document the customer's primary pain point. Identify Champion: Who at the customer is helping drive this deal internally?" Reps at NTO reported that this built-in coach saved them time – instead of flipping through training docs, the reminders were right where they worked. NTO also enabled **Celebrations** on Path – a fun feature: when a rep moves a stage to Closed Won, confetti can rain down on the screen (yes, Salesforce has confetti!). It's a small morale booster celebrating wins.

Pro Tip: Use Path on other objects too. Sales Path isn't just for Opportunities. You can enable Path on Leads (to guide through lead statuses), on Campaigns (maybe guiding through the campaign execution process), or on custom objects. Another interesting use: Path on Cases in Service Cloud to guide support reps through case statuses. The concept of guided path can improve process adherence in many areas.

ChatGPT Prompt (Talking to SORA): *Using SORA (OpenAI's video generation), a sales enablement manager says:* "Please generate a short animated video showing a sales rep using the new Sales Path on an opportunity, and highlight how the Key Fields and Guidance help them progress the deal."

The system produces a 30-second video: it shows a cartoon sales rep moving an opp from Qualification to Proposal, checking off Key Fields, and a voiceover explains the benefits of Path. This video is later shared in the company's internal newsletter to encourage reps to use the Path feature.

By implementing Sales Path and customizing the guidance to our sales process, we've taken our well-structured opportunity stages (from the previous section) and made them come alive for the end-users. It's declarative configuration directly influencing day-to-day sales behavior – a powerful win. Now, while Path helps individual reps navigate their deals, many sales efforts are team sports. Complex deals often involve multiple people: account execs, engineers, managers, partners, etc. How do we handle that in Salesforce declaratively? Enter **Opportunity Teams and Account Teams.** In the next section, we'll explore setting up these team structures to facilitate collaborative selling, ensuring that when multiple players are working on a deal or an account, Salesforce reflects it and everyone stays on the same page.

Next we'll configure Opportunity Teams and Account Teams for collaborative selling, defining roles and sharing access so that team selling is seamless in Salesforce.

FACILITATING COLLABORATIVE SELLING WITH OPPORTUNITY AND ACCOUNT TEAMS

Big deals often require a team approach – a salesperson might pull in a technical expert, a solutions consultant, an executive sponsor, etc. Salesforce's declarative answer to this is **Opportunity Teams** (for deal-specific collaboration) and **Account Teams** (for ongoing account management collaboration). In this section, we discuss how to set up and use these features effectively. We'll cover defining team roles, using default teams for efficiency, and how team selling configurations can extend sharing access without custom code. By leveraging Opportunity and Account Teams, you encourage collaboration and clarity: everyone knows who's involved in the deal and what their role is, and all team members have the access they need to contribute.

Opportunity Teams allow you to relate multiple users to an Opportunity, each with a specific **Role** (like "Account Executive", "Sales Engineer", "Executive Sponsor", etc.) and an access level (read/write access to that opportunity, for example). Once enabled in Salesforce Setup (it's an opt-in feature), you can customize the list of roles to fit your organization (e.g., add "Legal Advisor" if you often involve legal counsel in deals). Users can manually add colleagues to their opportunity's team, or better, use a **Default Opportunity Team**. A default team is a preset group of people that a user works with regularly; the user can define their default team in personal settings (or an admin can assist via API). For instance, a sales rep might always work with the same SE and the same account manager on deals – they can set those folks as their default team. Then with one click (Add Default Team) or automatically on opportunity creation, those team members get added to every new opp the rep owns. This saves time and ensures the team is looped in early.

Account Teams serve a similar concept but at the **Account (customer) level.** If your company does account-based selling or account management, Account Teams are useful to designate who covers what role for a given customer account – e.g., Account Owner (salesperson), Customer Success Manager, Technical Account Manager, etc. Much like Opp Teams, users can have a default Account Team that they apply to accounts they own. Account Teams can optionally grant record access to team members for not just the Account, but related Contacts, Opportunities, and Cases (configurable in Team settings).

One important thing to clarify: Opportunity Teams vs. Account Teams. Opportunity Teams are deal-specific and can change from opportunity to opportunity (even for the same account), whereas Account Teams are long-term and cover the entire relationship with the customer. Often, the Account Team might include a wider array of folks (maybe an executive sponsor or a support lead) who aren't involved in every single deal. Many organizations use both: Account Teams to manage territory and relationship assignments, and Opp Teams to manage who's on each deal's pursuit team. They are related but distinct. (In fact, enabling Opportunity Teams usually requires Account Teams to be enabled first, because the role definitions for opp teams piggyback on account team roles).

From a declarative setup perspective, here's what you do to **configure these teams**:

- Enable Account Teams in Setup and define the Team Roles picklist (common roles: Account Manager, Sales Ops, Customer Support, etc.).
- Enable Opportunity Teams (which will reuse the role list, though you can choose to have a slightly different subset for opps).
- Decide on **default access**: you can set it such that adding someone to an Opportunity Team grants them *Read* or *Read/Write* access to that opportunity (regardless of org-wide defaults). Similarly, for Account Teams, you can decide if team members get read or write on the account and related records.
- Encourage users (especially sales reps) to set up their default teams in personal settings. As an admin, you can help by providing instructions or even data-loading defaults for them if many reps have standard teams.
- Add the Opportunity Team related list to the Opportunity page layout (and Account Team to Account layout) so users can manage team members there.

Once configured, how does this **facilitate collaborative selling**? Let's illustrate: Suppose *DreamHouse Realty* (a customer) is in a big negotiation for a new product. The account's team includes an Account Executive (Alice), a Solutions Engineer (Bob), and a Customer Success Manager (Carol). On the specific Opportunity "DreamHouse – New Product Expansion Q3", Alice (the owner) adds Bob and Carol to the Opportunity Team, assigning Bob the "Solutions Engineer" role and Carol the "Customer Success" role. She also pulls in a pricing specialist, Dave, just for this deal (so Dave isn't on the account team generally, but he's needed here for a custom pricing model – she adds him with role "Pricing Analyst"). Now, on that opportunity record, anyone can see the team members listed and their roles. Bob and Carol automatically have access to the opportunity record to update it or log activities (because the act of adding to the team granted them access). They all collaborate via Chatter on the opp – tagging each other, updating deal information. There's no confusion about "who's doing what" because the roles are explicitly noted. If someone leaves or changes, updating the team list is simple and keeps everyone in the loop.

ChatGPT Prompt (SORA Video): "Create a short video illustrating how to add an Opportunity Team in Salesforce and explain each team member's role."

ChatGPT via SORA generates a quick how-to video: it visually shows a user editing an opportunity, adding colleagues with roles like 'Engineer' and 'Executive Sponsor'. It highlights how each role contributes. This video is used in training sessions for the sales org.

One often overlooked benefit of Opportunity Teams is how it ties into **Opportunity Splits and Forecasting** (which we'll cover next). When multiple people are on a deal, you may want to split credit or quota credit among them – splits feature works hand-in-hand with Opportunity Teams, since splits can be based on team members. So setting up teams is a prerequisite for using splits effectively.

Also, teams can feed into reports: you can report on opportunities where a certain person is on the team (useful for sales managers who act as executive sponsors on key deals – they can see all opps where they're on the team). In recent Salesforce releases, you can even use Opportunity Team members in list view filters and some automated processes.

Now, what about **Account Teams** specifically? Imagine a large account, *Universal Containers Corp.* They might have an Account Team consisting of: a Strategic Account Manager (primary owner), a Solutions Engineer assigned to the account, a Support Liaison, and an Executive Sponsor (VP level contact from your company). Each plays a role in the overall relationship. Account Teams in Salesforce allow those people to be associated with the account record, so anyone looking at the account can quickly see "Oh, here's the team working this account." It also means if a new opportunity is created under that account, Salesforce can prompt to add the Account Team to the Opportunity Team (handy feature – it'll ask "do you want to add these folks?"). This way, you maintain continuity from account planning to opportunity execution.

Pro Tip: Use automation (Flow) to add Opportunity Team members based on criteria. For example, if *Opportunity.Type = "New Product"* and it's over $100k, you might always need a Legal reviewer involved. You can create a record-triggered Flow that upon opp creation/update, if criteria met and if a certain user isn't already on the team, automatically insert an OpportunityTeamMember record for them. This ensures that big deals have the required support from all departments. It's declarative (Flow is clicks not code) and can significantly streamline team assembly for sales reps.

In summary, Opportunity and Account Teams empower **team selling** in a declarative way: clearly defined roles, easy visibility of "who's who" on deals, and built-in sharing rules so those team members have access to what they need. By configuring these features, we reduce the friction of collaboration – no more hitting permission roadblocks when someone tries to view an opportunity they're assisting on, and no ambiguity about responsibilities. Your sales org can sell as a coordinated unit, which often translates to smoother deals and happier customers (they're not getting mixed messages from uncoordinated internal teams).

Having set the stage for collaborative selling, we can now tackle how to share **credit** for those collaborations and forecast sales with accuracy. In the next section, we'll examine Opportunity Splits (to allocate revenue credit among team members) and declarative forecasting tools that help align the team on future revenue projections.

Moving ahead we'll set up Opportunity Splits and Forecasting, so that when teams sell together, everyone's contribution is tracked and management gets an accurate forecast of revenue.

LEVERAGING OPPORTUNITY SPLITS AND FORECASTING DECLARATIVELY

In many organizations, especially those with complex sales or team selling, it's not enough to just know "who is on the team" – you also need to track **how much credit each person gets** for the deal. That's where **Opportunity Splits** come in. And once deals are in flight, leadership cares about **forecasts** – predictions of revenue, usually by period and category. Salesforce provides Collaborative Forecasts to handle this, mapping opportunity data into a forecasting framework. In this section, we will discuss how to set up Opportunity Splits (including different types of splits like revenue and overlay), and how to configure Salesforce's forecasting feature declaratively. We'll explore forecast categories, quota setting, and aligning forecast data with your sales team structure. The end result: a system where shared deals are accounted for properly and future sales projections are visible and accurate.

Let's start with **Opportunity Splits.** By enabling splits, you can allocate percentages of an opportunity's amount to team members. Salesforce, out-of-the-box, offers two standard split types: **Revenue Split** (which typically must total 100% of the opportunity amount, dividing the pie among the sales team) and **Overlay Split** (which is often used for overlay roles like specialists or partners – this one can total more than 100% if you want, since it might represent additional credit that doesn't affect the core revenue figure). You can also define **custom split types** if needed (for example, maybe you want a split for "Referral credit" or others), each with its own rules. To set up splits, you must have Opportunity Teams enabled (which we did in the previous section). Then you go into Setup > Opportunity Splits, choose which split types to enable, and pick which fields they use (by default, the Amount field for revenue). You also decide if a split type should total 100% exactly or can be less/more.

In practice, here's how splits work: Suppose there's a $100,000 deal. Alice is the primary sales rep, Bob is the overlay specialist. They agree Alice gets 70% credit, Bob 30%. In the Opportunity Splits related list (or Lightning component), you'd add Alice and Bob, assigning 70 and 30 respectively to the Revenue Split (which would enforce that it sums to 100%). If you also have an Overlay Split type, maybe you give Bob 30% there as well or maybe 100% overlay to indicate he influenced the whole thing – it depends on your business logic. These percentages can later be used to credit commission or quota attainment appropriately.

Salesforce now supports **Product Splits** too (since Summer '23) , meaning you can split at the Opportunity Product level if needed. But that's an advanced scenario – most organizations are fine with opportunity-level splits.

Now on to **Collaborative Forecasting.** Forecasting in Salesforce allows you to project how much revenue (or quantity of product, or other metrics) is expected to close in a given period (month/quarter) and category (pipeline, commit, etc.). Declaratively, there are a few things to configure:

Forecast Hierarchy: By default, this mirrors the role hierarchy – it defines how forecasts roll up (i.e., a sales manager sees the sum of their team's forecasts). Ensure your role hierarchy is aligned with how you want to aggregate forecasts.

Forecast Types: This defines what you are forecasting. The common one is "Opportunities by Revenue" which sums the Amount of opps. But you might have one for "Opportunities by Quantity" or by Product family, etc. Also you can have forecast types that incorporate splits (e.g., "Opportunities by Overlay Split" – giving a specialist their own forecast based on split percentage of deals). When enabling splits, you can create forecast types that specifically use those split fields.

Forecast Categories & Stage Mappings: Salesforce has standard forecast categories (Pipeline, Best Case, Commit, Closed, and Omitted). Each opportunity stage is mapped to one of these categories. You should review and adjust these mappings to make sense. For example, maybe your first few stages map to Pipeline (meaning not likely enough to count in forecast), mid stages to Best Case, late stage to Commit. You can also add a "Most Likely" category if you use that (Salesforce lets you enable an optional category between Best Case and Commit).

Quotas: You can input quota amounts for each sales user (usually by month or quarter). This is done either manually in setup or via a data loader import to the Quota object. Quotas let you compare the forecast against targets in reports and the forecast view.

Enable Forecasting UI: Make sure the forecasting tab or feature is enabled and users have access to it. It provides a matrix-like view where each manager can see their commit, etc., and subordinates' commits.

Once set up, the typical use of forecasting is: each sales rep can review their opportunities, decide which ones they truly expect to close this period, and categorize them (often by just updating the Stage, which in turn updates forecast category, or explicitly adjusting a field if using override capabilities). Salesforce will show a roll-up: e.g., "Alice: Pipeline $200k, Best Case $150k, Commit $100k, Closed $50k against a quota of $120k for Q4." Her manager might have say 5 reps, and sees the rolled up numbers and can add their own judgment (managerial overrides can be enabled, allowing managers to adjust the commit up or down if they think reps are sandbagging or over-optimistic, for instance).

One important declarative feature in forecasts is filtering by product family or other dimensions. If your business wants separate forecasts for different product lines, you can set up a forecast type per product family, and the UI lets you toggle between them. If you want to forecast *new business vs renewals* separately (a common need in subscription businesses), you might use opportunity record types or a "Type" field and set up two forecast types (one that only includes new business opps, another for renewals). Salesforce allows multiple forecast types simultaneously.

Now, let's consider how **Opportunity Splits and Forecasts intersect**. If using splits, you may want each person's forecast to reflect their split, not the whole deal. Salesforce's answer: you create a forecast type for splits. For example, an "Overlay Forecast" where Bob (the overlay rep) sees $30k for that $100k deal we split earlier (since we gave him 30%). The system can sum Bob's split amounts for all his deals. Meanwhile, Alice would see $70k for that deal on her primary forecast. Essentially, splits let the pie get partitioned so each person's forecast is based on their portion. Internally, Salesforce achieves this by creating special split fields (like a Revenue Split Amount field behind the scenes) which get summed in forecasts. All declarative – you just tick the option to include splits in forecasts and set up the corresponding type.

Let's walk through a quick use case: **Express Logistics**, a company with a direct sales team and an overlay team of product specialists. They enable Opportunity Teams and Splits. On a big $1M deal, the direct salesperson (Raj) has 50%, and two specialists (Amy and Lee) have 25% each in overlay split (revenue split might still give Raj 100% if they treat specialists as extra credit, or they could do 50/25/25 in revenue split – depends how they manage quota). In the forecast, Raj (who has a $500k quarterly quota) will see $1M in his pipeline if revenue split is fully 100%

him *or* $500k if they used revenue split to split quota credit. Amy and Lee (overlay team with maybe their own overlay quotas) will see $250k each in their overlay forecasts. Management thus has a clear view: Direct forecast = $1M for Raj's area (which might roll to his manager's direct forecast), Overlay forecast = $500k total spread among specialists. This avoids double counting but also recognizes everyone's contribution.

An example of a simple forecasting chart: this bar chart illustrates how a sales team's forecast compares to their quota by quarter. Such visualizations can be generated from Salesforce data, helping teams track progress. In Q2 of the chart, for instance, the forecast (orange) exceeds the quota (gold), indicating the team is poised to hit their target, whereas in Q4 the forecast lags behind the quota.

The chart above is a generic visualization of forecast vs quota. In Salesforce, the Forecast tab provides a similar concept in table form for real data. The key for us is that by configuring forecast categories, stages, and quotas, Salesforce will crunch the numbers for us – no more manual spreadsheets that quickly go out of date. Reps update opps, and the forecast updates in real-time.

ChatGPT Prompt (Image Generation): "Create an infographic that explains Salesforce forecast categories (Pipeline, Best Case, Commit, Closed) in simple terms with icons."

ChatGPT uses DALL·E or a similar tool to generate a neat infographic: maybe a pipeline icon for Pipeline, a thumbs-up for Best Case, a handshake for Commit, a trophy for Closed Won. The infographic text defines each category's meaning. This is then shared in the internal wiki for training new sales reps on forecasting.

Quota management is also worth touching on. Quotas can be uploaded per user per period. A declarative tip: you can use a custom object or even the standard Quota object and build a Flow to mass-assign quotas if they follow patterns (e.g., every sales rep gets a default $100k quarterly quota except those in Enterprise segment get $200k – a Flow could auto-create those records each year). Alternatively, many companies just maintain quotas in a CSV and use Data Loader to upload annually.

Finally, consider the alignment needed for forecasting: You want your sales team to have a regular **cadence** (e.g., weekly forecast calls), a clear definition of what it means to put something in "Commit" (maybe the customer verbally said yes, or it's a late-stage with procurement), and discipline in updating close dates and stages. Our declarative setup can reinforce some of this (e.g., a validation rule that if Stage = Commit, Close Date must be in the current quarter, etc., to avoid sandbagging across quarters), but a lot is process and people management. Still, the beauty of Salesforce forecasting is that it gives a single source of truth where, with the right configuration, reps and managers can play from the same playbook.

Now, with opportunity splits and forecasts configured, our sales org can both **collaborate on deals and project future revenue confidently**. We know who gets credit for what, and we can see if we're on track to hit targets. The final piece of the puzzle is ensuring that the Salesforce user interface supports all these processes elegantly. Enter Dynamic Forms for Opportunities – in the next section, we'll explore how to declutter and tailor the opportunity page layout for different scenarios and user needs, making all this data and process easier to work with.

In our last section of this chapter, we'll tackle optimizing the Opportunity page itself using Dynamic Forms, to deliver a clean, role-specific experience for users and tie together the best practices we've configured.

TAILORING OPPORTUNITY PAGES WITH DYNAMIC FORMS

The more complex your sales process, the more fields and information your opportunities tend to accumulate. However, showing every field to every user all the time can overwhelm and confuse. Salesforce **Dynamic Forms** is a solution that allows you to design more intuitive, user-centric record pages by placing fields in flexible layouts and controlling their visibility based on criteria. In this section, we'll look at how to use Dynamic Forms to optimize Opportunity pages. We'll discuss splitting fields into sections (e.g., general info vs. renewal-specific info), using visibility rules so that only relevant fields show up for certain opportunity types or stages, and generally decluttering the UI. The aim is to support scenarios like new sales vs. upsells vs. renewals declaratively, giving each user just what they need to see to do their job, which improves user experience and data quality.

In the old days (and by old days, we mean Salesforce Classic or early Lightning), page layouts were pretty static: all fields in one "Details" panel, maybe arranged in sections, but everyone saw the same thing unless you made multiple page layouts and assigned by profile/record type. That was a blunt tool: for every slight variation you needed a new layout. **Dynamic Forms changed that paradigm.** Now you can take fields off the standard page layout and instead drop them as individual components anywhere on the page. You can have multiple Field Sections and arrange them alongside other components (like related lists, charts, etc.). Crucially, you can set **visibility rules** on both individual fields and entire sections. That means, for example, you could have a section called "Renewal Details" that only appears if Opportunity Type = Renewal AND Stage = Closed Won (maybe for capturing some post-win info). Or a simpler case: show the "Implementation Start Date" field only if the checkbox "Services Required" = TRUE. All without writing code – just point-and-click rules builder for each component.

How to set up Dynamic Forms on Opportunities? As of now (2025), Salesforce supports dynamic forms on standard objects including Opportunities. To migrate, you edit the Lightning page for Opportunity, and if you see an upgrade prompt (to convert the page to dynamic forms), you can do that. If not (perhaps you already did or built fresh), you simply ensure you have one or more "Field Section" components on the page. You drag fields from the palette into those sections. It's advisable to remove the old "Record Detail" component entirely once you've placed all needed fields via Field Sections , to avoid confusion (you don't want fields duplicated in two places). A Salesforce best practice is *not* to mix standard detail and dynamic fields on the same page.

Now, think about an Opportunity page. Perhaps break it down like this:

Primary Details: Always visible for everyone – e.g., Opportunity Name, Account, Close Date, Stage, Amount, Owner. These key fields could go at the top in a two-column section.

Qualification Info: This section might contain fields like Lead Source, Primary Use Case, etc., but you only show it if Stage is early (or always for internal users but maybe not needed for others).

Technical Info: If you have fields that only your sales engineers fill out (like "Solution Complexity" or "Integration Requirements"), you could set those to only show if the current user's profile = "Sales Engineer" or if a picklist "Needs SE Review" = Yes.

Renewal Info: Only for renewal opps – fields like "Original Contract End Date" or "Renewal Term (months)" should show when the opp type = Renewal (or if record type = Renewal).

Financial/Legal: Maybe a section that appears in later stages for everyone, containing "Legal Approval Date", "Discount %", etc., once an opp reaches Negotiation stage.

By crafting these sections and rules, the page adapts to the context of the deal. A brand new prospecting opp might only show a few basic fields to start with – nice and clean. As it progresses, more fields or sections can appear, guiding the rep to fill them when relevant (this complements Path, as earlier covered). If the opp is a renewal type from the outset, the renewal-specific fields are there from the start, but if it's not, those fields don't clutter the screen at all.

One of the greatest benefits here is reducing the need for multiple page layouts by profile or record type. For instance, you might previously have had a "Renewal Opportunity Layout" and a "New Business Opportunity Layout" because each needed some unique fields. With Dynamic Forms, you could potentially collapse that into one layout and use visibility rules to handle differences. This makes maintenance easier: one Lightning page to edit instead of two, and no need to juggle assignments.

ChatGPT Prompt (Wireframe): "ChatGPT, generate a text-based wireframe for an Opportunity page that includes sections for 'General Info', 'Technical Details', and 'Renewal Info'. Show how fields might be arranged, and indicate which sections are conditional."

ChatGPT produces a simple ASCII layout: it shows [General Info] with fields like Name, Close Date, Stage; side-by-side [Technical Details] with fields like Tech Requirements (visible if "Needs SE" checked); and [Renewal Info] in a separate section (visible for renewal opps). This wireframe helps the admin visualize the design before building it.

Let's consider an example: **ACME Corp Renewal Opportunity.** Say we have an opp for renewing ACME's contract. Because it's marked as a Renewal, our dynamic form automatically shows a section "Renewal Details" that includes fields like *Original Contract Value, Renewal % uplift, Customer Satisfaction Score* (perhaps pulled from a survey). If it were a brand new sale, that section wouldn't show – instead maybe a section "New Customer Profile" would show asking for *Customer Pain Points, Competitors in Deal*, etc. Meanwhile, the *Technical Details* section shows because the sales engineer on the team set a flag that custom integration is needed, revealing fields for *Integration Complexity* and *Estimated Professional Services*. ACME's opp is late-stage, so a *Legal/Approval* section is now visible prompting the rep to enter *Deal Desk Approval ID* after they got special pricing approved. The page only surfaces what's needed when it's needed, which keeps ACME's opportunity record easy to work with at each step.

Dynamic Forms also can improve **performance**. Loading a huge detail section with dozens of fields (some formula fields which query data, etc.) can slow a page down. By using visibility rules, especially if some formula fields are hidden until needed, you might reduce initial load. Also splitting into tabs (Lightning pages can have tabs) and putting sections in different tabs means they load on demand. For example, you might put seldom-used fields in a tab "Additional Info" which the rep can click only if needed.

Pro Tip: When using Dynamic Forms, regularly review field usage to avoid "visibility rule sprawl." Having too many complex rules can confuse admins later. Document your intentions (perhaps in field descriptions or an internal wiki). Also test with multiple scenarios to ensure fields appear/disappear as expected. One gotcha: if a field is required by validation but hidden by a rule, the user can't fill it – which leads to errors on save. Salesforce won't override validation even if field not visible. So be careful: if you hide a required field, make sure it's only hidden when it's not required by logic. For example, if you require "Reason Lost" when Stage = Closed Lost, don't hide that field when stage is Closed Lost or the user can't fill it and can't save. Usually you'd hide it when stage is not Closed Lost, and show it when needed.

Dynamic Forms are a purely declarative enhancement that can dramatically improve user adoption. Sales reps often resist filling in fields that seem irrelevant – and honestly, who can blame them? By showing them only what matters, you respect their time and attention, making it more likely they'll diligently fill in the important stuff. Many experienced admins and developers find that a well-designed dynamic layout **reduces training time** as well, since the interface itself cues the user what to do next.

We've now covered a wide range of Sales Cloud configuration best practices: from capturing and routing leads, to scoring them, to managing opportunities and guiding reps, to enabling team collaboration, splitting credits, forecasting revenue, and finally optimizing the user interface for it all. The common theme is using Salesforce's declarative features in smart, strategic ways to align the system to business processes. To conclude, let's recap what we learned and prepare to carry these insights into other parts of the Salesforce platform.

Up next: the chapter summary will review these key points and pave the way for our next chapter on Service Cloud configuration best practices.

SUMMARY

In this chapter, we explored how to get the most out of Sales Cloud using declarative configuration, all through the lens of Salesforce's "clicks, not code" philosophy. We began at the top of the funnel with effective lead management: configuring assignment rules and queues to ensure every new lead is captured and routed instantly to the right owner. By automating lead distribution and using flows for advanced scenarios, we set the stage for faster lead response and higher conversion rates. We then tackled lead quality through declarative lead scoring – designing a custom model with formula fields and flows so reps can focus on the leads most likely to turn into deals. The examples of point-based criteria illustrated how scoring can be implemented without a single line of Apex, and how ChatGPT can even assist in analyzing data to fine-tune those models.

Moving into opportunities, we emphasized standardizing the sales process. You learned the importance of clearly defined Opportunity stages and how to implement sales processes/record types only when necessary (to handle truly distinct scenarios). We applied validation rules to enforce data quality at key stages (like requiring certain fields before an opp can be Closed Won), thereby improving the reliability of pipeline data. Each opportunity now tells a consistent story, which feeds directly into better forecasting and analytics.

We also saw how to make Salesforce **guide the users** through these processes. By configuring Sales Path and Path Assistant on opportunity records , we put critical fields and tailored advice front-and-center for sales reps at each stage of a deal. This turns our static process documentation into an interactive, in-app coach – accelerating onboarding for new reps and keeping experienced reps on point. With ChatGPT prompt examples, we showed how an admin might generate or refine Guidance for Success messages, or even produce a short training video, ensuring the Path content is effective and engaging.

Collaboration was another key theme. We enabled Opportunity Teams and Account Teams to facilitate team selling, assigning clear roles and sharing access seamlessly. This declarative setup fosters transparency – everyone knows who is involved in the deal and what their responsibility is. Our "Pro Tip" on using Flows to auto-add team members demonstrated that even complex team assembly logic can be handled with clicks. By leveraging teams, and subsequently Opportunity Splits, we acknowledged the reality of group sales efforts and made sure credit is shared fairly. We configured splits so that multiple team members can get a piece of the pie on the revenue, and we saw how those splits tie into forecasting so each contributor's quota attainment is tracked.

On the forecasting front, we set up Collaborative Forecasts declaratively, mapping our opportunity stages to forecast categories (Pipeline, Best Case, Commit, etc.) and rolling up forecasts through the role hierarchy. We discussed setting quotas and how Salesforce can aggregate each rep's pipeline into a forecast that managers can use for decision-making. By doing so, we removed the need for the dreaded "spreadsheet forecast" living outside the system – now Sales Cloud itself becomes the source of truth for all revenue projections. The inclusion of a simple forecast vs.

quota chart in our content underscored how visualizing these numbers can drive home insights and motivate teams.

Finally, we turned to the user experience with Dynamic Forms. We learned how to declutter the Opportunity page by showing the right fields at the right time. Using visibility rules, we tailored the interface for different deal types (new business vs. renewal) and user roles, all declaratively. This not only makes life easier for the sales team – it also contributes to better data (since users are more likely to fill out fields when they're presented contextually and logically). The result is a Salesforce environment that feels customized for each scenario, yet maintainable for admins because it's driven by configurations rather than dozens of page layouts or custom code.

In essence, this chapter demonstrated a holistic approach to Sales Cloud configuration, covering the lifecycle from a raw lead to a closed deal and beyond (renewals), and weaving in productivity and AI tools like ChatGPT at each step to enhance outcomes. By applying these best practices, an organization can significantly boost sales productivity, data quality, and process consistency – all critical factors for scaling an enterprise sales operation.

As we conclude our focus on Sales Cloud, you're ready to apply similar declarative thinking to other areas. In the next chapter, we will shift gears to Service Cloud Configuration Best Practices, where we'll see how declarative tools can optimize customer support processes. The themes will feel familiar – routing work (cases instead of leads), guiding agents (with tools like Omni-Channel and Macros), and ensuring high-quality data – but the context and specifics will be new. With the solid foundation from Sales Cloud configuration in mind, you'll be well-prepared to tackle the challenges of configuring Salesforce for world-class customer service in Chapter 10. Let's continue our Salesforce Enterprise Thinking journey and bring the same level of thoughtful configuration to the service side of the house.

10

SERVICE CLOUD CONFIGURATION BEST PRACTICES

Monday morning at a busy tech support center: an inbox overflowing with customer emails, new cases logged through a web portal overnight, and a phone already ringing off the hook. The support agents grab their coffee and brace themselves. For a support manager, this scenario can either be a chaotic fire-fight or a well-choreographed routine – and the difference lies in how well Salesforce Service Cloud is configured. Service Cloud, when set up with best practices, acts like the air traffic control for customer issues, ensuring each case is logged, routed, worked, and resolved efficiently.

This chapter dives into the art of declaratively configuring Service Cloud for optimal customer support. We'll explore how to structure **Cases** (the core records of every issue) with a clear lifecycle, how to automatically assign incoming cases to the right team at the right time, and how to keep customers in the loop with timely communications. Along the way, we'll set up the *Service Console* to be an agent's best friend – a unified workspace that puts everything they need on one screen. We'll also configure productivity boosters like macros and quick text, so repetitive actions take seconds instead of minutes.

The focus here is on **clicks, not code** – using Salesforce's point-and-click tools to build a robust support system. But we're adding a twist: bringing in ChatGPT as a virtual assistant for the Salesforce admin. Throughout the chapter, you'll see how ChatGPT can help brainstorm process improvements, draft template text, generate training materials, and even act as a second set of eyes on your configurations. Our tone will be friendly and conversational (imagine getting advice from a seasoned colleague), but the guidance is grounded in real-world Service Cloud projects and the structured thinking of an expert business journal.

By the end of this chapter, experienced Salesforce professionals – whether you're a developer, admin, business analyst, or product owner – will have a toolkit of best practices for Service Cloud configuration. More importantly, you'll appreciate how leveraging ChatGPT in different modes (typing out questions, speaking them aloud, uploading files for analysis, or even using advanced features like Operator and SORA for automation and media) can supercharge your configuration process. We're going to cover everything from the fundamentals of case management to advanced topics like SLAs and team collaboration, all with an eye toward declarative solutions that scale.

Buckle up as we turn our attention to transforming customer service operations. First, we need to get the basics right – understanding the Case object and laying down a strong foundation for support processes. Like building a house, we start with a solid blueprint: let's begin with mastering case management fundamentals.

MASTERING CASE MANAGEMENT FUNDAMENTALS

At the heart of Service Cloud is the Case object. **Cases** represent customer issues or requests – each one a story that begins with a problem and ends (we hope) with a solution. Mastering case management fundamentals means setting up the Case object so that every support rep knows what's happening with a case at a glance, and no customer inquiry falls through the cracks. In this section, we'll define the core structure of a Case and establish clear processes that guide cases from creation to closure.

Key Case Fields: Salesforce provides standard fields on Cases that form the backbone of any support process. The most important is **Status** – a picklist that indicates where a case is in its lifecycle (for example, *New, Working, Escalated, Closed*). A well-defined set of status values is critical. Next, **Priority** tells how urgent or impactful the case is (common values are *Low, Medium, High*, or numeric rankings). Priority helps the team triage work – a "High" priority case might represent a major outage, whereas "Low" could be a minor question. **Case Type** is another standard field, often used to categorize the case (e.g. *Question, Incident, Feature Request*, or custom types relevant to your business). Additionally, **Case Origin** (Email, Phone, Web, etc.) tracks how the case was created, which can inform who should handle it or how responses are sent. Out-of-the-box you'll also see fields like **Subject**, **Description**, and **Contact/Account** (linking the case to the customer who reported it).

Configuring these fields means tailoring the picklist values and usage to match your support operations. For example, if you support multiple product lines, you might add a custom picklist field for "Product" or use *Record Types* to distinguish cases for different products or business processes. The golden rule of declarative configuration is to **use standard fields whenever possible** – they come with built-in functionality. So, if "Priority" means urgency to you, stick with Salesforce's Priority field instead of creating a duplicate custom field. The same goes for Status and Type.

Now, about that **Status** field: it's more than just open or closed. Best practice is to design a clear status flow that a case will follow. Typically, you start with **New** (case just logged), move to statuses like **Working** or **In Progress** (when an agent is actively investigating), maybe have an **On Hold** or **Waiting on Customer** (if you're blocked waiting for more info from the client), possibly an **Escalated** status (if it's been handed to a higher support tier or management), and finally **Closed** when the issue is resolved. Closed might not always be the final step – some organizations use a **Follow-Up** or **Resolved** status before truly closing, especially if they want to verify with the customer that the solution worked. The key is to keep this lifecycle logical and not overly complicated.

To manage variations of case lifecycles, Salesforce offers **Support Processes.** A support process is essentially a customized picklist set for the Case Status field, tied to specific case record types. This allows you to have different status options for different kinds of cases. For example, an "Issue" case might have statuses New > Troubleshooting > Fix Proposed > Verified > Closed, whereas a "Feature Request" case might have New > Under Review > Scheduled > Delivered > Closed. Using support processes ensures that each case type shows only the relevant statuses, keeping things clean for your agents. It's a best practice to define your status values and processes on paper first, then configure them – and make sure to document what each status means and when to use it.

ChatGPT Prompt (Voice): *Admin (speaking into phone):* "Hey ChatGPT, I'm setting up case statuses. What are some common status values I should use for a technical support case lifecycle?"

ChatGPT: "You might use statuses like **New** (just received), **In Progress** (agent working on it), **On Hold** (waiting for customer or resources), **Escalated** (needs higher-level attention), and **Closed** (issue resolved). Each status should have a clear meaning. For instance, *On Hold* could mean the support team is waiting for a customer reply or parts on order, whereas *Escalated* means it's been handed off to a specialist or supervisor. Defining these clearly will help agents and customers know exactly where the case stands."

Notice how ChatGPT's advice reinforces clarity. As an admin, you can even ask ChatGPT to **draft a one-page guide** for your support team explaining each status. It's a great way to ensure everyone uses statuses consistently. For instance, if you have a status called "Escalated," your team should know the criteria for escalation – perhaps the case is high priority or has been open more than 48 hours without resolution.

Let's talk about **Support Process configuration** step-by-step:

1. **List Out Status Values:** Write down all the statuses you think you need for your support process. Trim it to only those that represent distinct stages. (If you have more than, say, 5-7 statuses in one process, double-check if they're all truly necessary.)

2. **Create the Support Process in Salesforce:** In Setup, find *Support Processes* under the Case object management settings. Define a new

process (e.g. "Standard Tech Support Process") and select which status values belong to it (in the order they should generally be used).

3. **Assign to a Record Type:** Create a Case Record Type (e.g. "Tech Support Case") and associate it with the Support Process you created. You might also have another record type like "Feature Request Case" with its own process (fewer statuses, maybe no Escalated stage).

4. **Update Picklist Dependencies (if any):** If certain statuses should trigger other field requirements (like if a case is Closed, maybe a **Case Reason** or **Resolution Summary** field becomes required), set up validation rules or page layouts accordingly.

5. **Document & Train:** Publish an internal article or job aid listing each status and when to use it. Ensure agents and even customers (if they can see statuses on a portal) understand the lifecycle.

Setting up **Priority** and **Type** is usually simpler – decide on a few values that make sense and avoid overloading these fields with too many options. Priority often ties to service level commitments, so you might map "High" priority to urgent 2-hour first response, "Medium" to 8-hour, etc. Case Type can help you report on the mix of inquiries you get (problem vs. question vs. request), so define types that cover 80% of cases and use an "Other" for the rest rather than having 20 obscure categories.

One more fundamental: think about the **Support Lifecycle** beyond just statuses. Salesforce Cases have features like **Case Comments** and **Case Feed** for communication, and you can enable **Case Closure** reasons (a picklist to note why it was closed or the resolution type). These are optional, but consider using them if they add value to your process (for example, tracking common resolution reasons like "Answered by FAQ" vs "Bug Fix Provided").

Pro Tip: Don't create a dozen status values to cover every nuance. Keep statuses high-level (focused on stage, not detail). Use supporting fields for details. For instance, rather than having separate statuses for "Waiting on Customer" and "Waiting on Third-Party," you could have one *On Hold* status and use a checkbox or picklist to indicate who you're waiting on. This keeps the primary process linear and clear. Overly granular statuses can confuse agents and complicate reports – aim for simplicity first.

With the Case object structured properly – key fields in place and a clear status lifecycle defined – you set a strong foundation. Agents can log and update cases with confidence, and everyone speaks the same language about what stage a case is in. Now that we have our support processes mapped out for different case types, it's time to ensure those cases get into the right hands without delay. Next we'll tackle case assignment and routing, so that each new case automatically finds its way to the best suited support rep or team.

OPTIMIZING CASE ASSIGNMENT AND ROUTING

Once your cases are set up with the right fields and statuses, the next challenge is getting each case to the *right agent or team* as quickly as possible. In a small support team, you might get away with one big queue where everyone grabs what they can. But as volume and specialization grow, you need a smarter routing strategy. This section covers how to use Salesforce's declarative tools – **Queues**, **Assignment Rules**, and **Omni-Channel** – to automatically assign and route cases. The goal is to ensure no case sits untouched and that high-priority or specialized issues go straight to those best equipped to handle them.

Queues are one of the simplest routing tools. A Queue in Salesforce is basically a holding bin for records (like cases) that certain users have access to. For example, you might have a queue for "Tier 1 Support" that all level 1 agents can see, and another for "Billing Team" staffed by billing specialists. When a case is owned by a queue, it means any member of that queue can pick it up (take ownership) when they're ready. Queues are great for implementing a pull-based system: cases come in, land in a queue, and agents pull from the queue. The key is to create queues that correspond to logical buckets of work – often by product line, issue type, or urgency. Make sure each queue has an owner (could be a support manager or generic user) to receive notifications if needed, and consider using queue email alerts sparingly (nobody wants 100 emails saying "new case in queue" every hour).

Assignment Rules add automation on top of queues. An assignment rule lets you define criteria to decide *who* (or which queue) should get the case as soon as it's created (or updated). In Salesforce, you can have one active case assignment rule at a time, with multiple rule entries evaluated in order. For instance, you might set up:

- If *Case Origin* = "Web" and *Product* = "Networking" -> assign to **Networking Support Queue.**
- If *Case Type* = "Billing" -> assign to **Billing Team Queue.**
- If *Priority* = "High" -> assign to **Escalation Queue** (which pings a Tier 2 team).
- Otherwise (no criteria met) -> assign to **General Support Queue** (a default catch-all).

Each rule entry can assign to a User or a Queue. Best practice is to use queues for most assignments unless a specific person must always get certain cases. Queues give flexibility and visibility – if Joe is out sick, others can still see queue cases. When writing assignment rules, order matters: put specific criteria first and generic last. Also, be mindful that assignment rules typically run only on case creation by default. If agents manually change certain fields later, you might need other automation (like a Flow) to reassign ownership.

ChatGPT Prompt (Typing): *User:* "I have three support teams – Tier1, Tier2, and Billing. How should I configure assignment rules for new cases?"

ChatGPT: "Start by defining criteria for each team. For example: Tier1 gets all new cases by default. Tier2 gets cases that are escalations or perhaps cases from VIP customers. Billing gets cases where the case Type is Billing or the issue relates to invoices/payments. In Salesforce, create an assignment rule with entries such as: (1) If Type = Billing -> assign to Billing Queue, (2) If Account is VIP (a field/flag) -> assign to Tier2 Queue, (3) If Priority = High -> assign to Tier2 Queue (you might combine this with VIP if needed), and finally (4) Default entry -> assign to Tier1 Queue. This way, general cases go to Tier1, but important ones siphon off to Tier2 and billing issues to the Billing team. Remember to mark the last rule entry as 'stop' or make it catch-all so nothing slips through."

In addition to basic assignment rules, Salesforce's **Omni-Channel** feature is a game-changer for live routing. Omni-Channel pushes work to agents in real time based on their availability and skill set. Think of it like an automated dispatcher: agents indicate they are online/available, and Omni-Channel will route incoming cases (and also chats, messaging conversations, or other work items) directly to them. This ensures a first-come, first-served distribution and prevents situations where ten agents cherry-pick easy cases leaving the hard one sitting untouched.

Setting up Omni-Channel involves a few steps. First, you enable Omni-Channel in your org and define **Service Channels** for the objects you want to route – in our case, likely the Case object (and perhaps Chat Transcript or Messaging Session if you use live chat). Next, you create a **Routing Configuration** where you set priorities and capacities (e.g., an agent can handle at most 5 low-priority cases or 2 high-priority at once, and define push timeouts). You then associate the routing config and a queue to the Service Channel. For example, you might have a queue "Tier1 Omni Queue" and set that queue as an Omni-Channel queue. Now any case assigned to that queue will immediately be routed to an available agent who is a member of that queue. Optionally, enable **Skills-Based Routing** by defining Skills (like "Spanish" or "Product X Expert") and assigning them to agents and cases. Omni-Channel can consider skills so that, say, a case requiring Spanish language skill goes only to agents who have that skill.

A powerful aspect of Omni-Channel is that it can handle multiple channels concurrently. An agent could be getting cases, chats, and social media posts all through the same interface, prioritized by your rules. This is where mapping out your intake channels is important. If you offer phone, email, chat, etc., you likely want to set up Omni-Channel for each (cases from phone or email, live chat sessions, etc.) to ensure all come into one unified queue system.

One strategy: continue to use assignment rules for initial triage, but point them to Omni queues. For example, instead of assignment rule sending a case to "Billing Queue" which is just a holding bin, you could have "Billing Omni-Queue" that routes to agents immediately. Some orgs turn off assignment rules entirely and use **Omni-Channel Flow** (a special flow type) to route cases dynamically. But that's a more advanced configuration – a simple approach is: assignment rules decide *which team/queue*, Omni-Channel then auto-distributes within that team.

ChatGPT Prompt (Operator Mode): *User:* "ChatGPT, can you help test my assignment setup? Use Operator to create one case of each type (Technical, Billing, High Priority) so I can see if they go to the right queues."

ChatGPT (Operator): "Certainly! I will simulate creating three cases via the Salesforce UI/API: one technical issue case, one billing question case, and one high-priority case. *Creating cases...* The first case was assigned to **Tier1 Support Queue** (as expected for a general technical issue). The second went to **Billing Team Queue**. The third, high-priority case went to **Escalation Queue** for Tier2. All assignment rules fired correctly. Your routing is working!"

ChatGPT's autonomous Operator simulated test records so you didn't have to click through the interface three times. This confirms your configuration is correct.

In practice, you'd test using Salesforce's interface or by having a colleague submit some sample cases. But it's nice to imagine an AI helper clicking those buttons for you! The example above shows how ChatGPT's Operator mode could, in theory, execute repetitive test actions. While today you might still do some of this manually, the future is headed toward more AI-assisted admin tasks.

A few additional tips for case routing:

Round-Robin Assignment: Salesforce doesn't natively round-robin cases to users without code, but you can simulate it with multiple queues or an add-on. Omni-Channel can distribute based on least busy agent, which often achieves a similar outcome (workload balancing).

Assignment on Reopen: If a customer replies to a Closed case, consider using a case automation (like a Flow) to reopen it and maybe reassign to the previous owner or a queue. Assignment rules might not run on reopen automatically, so plan for that scenario to avoid reopened cases sitting idle.

Default Owner: In Support Settings, you can specify a default case owner (a user or queue) for cases that don't match any assignment rule. Always set this! It's your safety net if something slips past your rules – for example, assign it to a "Support Triage" queue monitored by team leads.

Pro Tip: For complex routing requirements, don't be afraid to mix approaches. You can use a simple assignment rule to bucket cases into a few broad queues (by product or region), and then use Omni-Channel within each queue to push cases to agents. This hybrid approach is often easier to maintain than writing a web of super-specific rules for every situation. Also, regularly review your assignment rules – as your business changes (say you add a new product line or a new support tier), update the rules so new cases continue to go to the right place.

With an optimized assignment and routing setup, your cases are now zipping to the right people without manual intervention. The support team can trust that when a case comes in, it pops into their view if it's meant for them, or lands in an appropriate queue. This reduces response times and ensures an even workload. Now that we've tackled the logistics of getting a case to an agent, let's make sure we're also keeping customers informed. Next we'll configure email templates and auto-responses, so every customer gets timely, professional updates on their case from start to finish.

STREAMLINING COMMUNICATION WITH EMAIL TEMPLATES AND AUTO-RESPONSES

Effective support isn't just about solving problems – it's also about communicating with the customer so they feel heard and informed. Salesforce Service Cloud provides **Email Templates** and **Auto-Response Rules** to help you send consistent, timely emails without needing to type each message from scratch. In this section, we'll configure automatic acknowledgment emails for new cases and craft template-based responses for common updates. The result: customers get a quick "we've got your request" confirmation and professional follow-ups, which boosts confidence in your support process.

Acknowledging New Cases: Have you ever submitted a help request and then heard nothing for hours? Auto-response rules prevent that silence. Similar to assignment rules, an auto-response rule evaluates each new case (or lead) and, if conditions match, sends out a pre-written email using a designated template. A typical setup is to send a "Thank you, we've received your case" email for every new case coming from a web form or email. This email can include the case number, the expected response time or support hours, and any next steps (like "here's how to track your case status online"). It's purely declarative to set up: you create an Email Template, then create an Auto-Response Rule with rule entries specifying criteria (for example, "if Case Origin = Web, use the Web Acknowledgment template").

Designing Email Templates: Salesforce supports text, HTML, and custom (Visualforce) templates, but the modern approach is to use **Lightning Email Templates** which come with a friendly editor and can pull in fields via merge fields. For instance, your template might say: "Hello {!Contact.Name}, we have received your case {!Case.CaseNumber}: {!Case.Subject}." Using merge fields ensures each email is personalized with the customer's name, case number, etc. Keep templates short and clear. It's often best to have a warm, professional tone – "Hello John, thanks for contacting our support team. Your case #12345 is being reviewed..." etc. You can also include helpful info like links to your help center or how to escalate if needed. If your support operates only certain hours, mention that ("Our team is available 9am-5pm Mon-Fri, we'll respond as soon as possible."). And if you've committed to an SLA for first response, you might explicitly say "You can expect an update within 4 hours."

ChatGPT Prompt (Typing): *User:* "Draft a friendly auto-response email template for new support cases. It should greet the customer by name, acknowledge their issue, give the case number, and promise a follow-up within one business day."

ChatGPT: "Hello {{Customer Name}},Thank you for reaching out! We've opened a case to assist you with your request. Your case number is **{{Case Number}}**. Our support team is on it – you can expect an update from us within one business day. In the meantime, feel free to reply to this email if you have additional details to share. We're here to help and will do our best to resolve your issue as quickly as possible. Sincerely, The Support Team"

ChatGPT provided a nicely formatted template. In Salesforce, you'd replace placeholders with actual merge fields like {!Contact.FirstName} and {!Case.CaseNumber}.

You can see how using ChatGPT to brainstorm or refine template wording saves time. In the above example, ChatGPT even added HTML formatting (line breaks, bold text for the case number). You might tweak that in Salesforce's template editor, but it's a great starting point.

Multi-Language and Tone Adjustments: If you serve customers in different languages, you'll need templates in each language. ChatGPT can help translate your English template to, say, Spanish or French, ensuring you maintain a friendly tone. Or maybe you want a more formal tone for enterprise clients versus a casual tone for consumers – you could ask ChatGPT to rewrite the template in a more formal style. This is far easier than writing each variant from scratch.

ChatGPT Prompt (Multi-turn):

User: "That acknowledgment email looks good. Now I need a shorter version for SMS notifications under 160 characters."

ChatGPT: "Sure. How about: 'We got your support request (Case # {{Case Number}}). Our team is on it and will contact you within 1 business day. – [Your Company Name]'"

User: "Great, and could you translate that to Spanish as well?"

ChatGPT: "Por supuesto. 'Hemos recibido tu solicitud de soporte (Caso #{{Case Number}}). Nuestro equipo lo está revisando y te contactará dentro de 1 día hábil. – [Tu Empresa]'"

In a few quick prompts, we obtained a concise SMS template and a Spanish version, thanks to ChatGPT.

Beyond the initial auto-reply, think about other emails in the case lifecycle. Common ones include:

Case Closed Notification: When a case is solved, send a closure email thanking the customer and perhaps asking for feedback or providing the resolution summary.

Status Update: If a case will take longer or goes through a major change (e.g., escalated to another team), you might send an update email. This can be automated via a Workflow Rule or Flow that triggers when Status changes to a certain value (like "Escalated") and uses an email template.

Customer Reply Received: If you get a reply from the customer on an ongoing case, some support teams send a "We got your update" note – though usually the agent would just personally respond, so use automation thoughtfully to avoid spamming.

For implementing these, you can use a combination of auto-response rules (which only fire on new case creation) and **Flows** or **Email Alerts** (part of Salesforce automation) for later-stage emails. All of these can use the templates you create. The beauty of templates is consistency: no matter which agent or automation sends it, the message stays on-brand and accurate.

Let's not forget configuration steps:

1. **Create Org-Wide Address:** If you haven't already, set up an org-wide email address like support@yourcompany.com. This ensures emails come "from" your support team name instead of an individual agent's email (which looks more professional and avoids confusion if that agent is out).

2. **Build the Templates:** In Salesforce Setup or the Email Template manager, create new Lightning Templates. Insert your text (maybe using ChatGPT drafts as baseline), add your company logo or footer if desired (keeping images small for email deliverability), and include key fields. Save and test the template by using the "Send Test" function to your own email to see how it looks.

3. **Set Up Auto-Response Rule:** Go to Support Settings -> Auto-response Rules for Cases. Create a new rule, add rule entries. A typical entry: if

Case Origin = Web (or if a specific web-to-case form, you can filter by case origin or other flags) then choose the template "Web Case Acknowledgment". You can have multiple entries if you want different templates for different scenarios (e.g., maybe a different reply for VIP customers vs. general).

4. Activate the Rule: Only one auto-response rule can be active, so combine all necessary entries into that one rule and activate it.

5. Test It: Simulate a case creation (e.g., submit your own web form or have someone send an email to your support address) and see if the auto-response arrives. Adjust any wording or formatting after seeing it live.

Pro Tip: When customizing email templates for Email-to-Case, always include the case reference number or thread ID somewhere in the email subject or body. Salesforce typically inserts a unique ref code in the email subject (e.g., a string like ref:00D... or the case number in brackets) by default. This is crucial: it allows customer replies to be automatically attached to the right case. If you remove or change those identifiers in your template's subject line, you risk breaking that email threading. So, if you use a custom subject like "Your case {{Case Number}} has been received", keep the case number part intact and Salesforce will handle the rest of the threading logic.

With auto-responses and templates in place, your customers will never feel ignored. They get immediate confirmation and consistent updates, without agents having to manually write each email. This declarative setup saves time for the support team and sets clear expectations for the customer. Now that communication flows smoothly, let's focus on empowering the agents on the other side of the equation. Next we'll look at the Service Console and productivity tools like macros and quick text, which help agents resolve those cases faster once they're engaged.

ENHANCING AGENT PRODUCTIVITY WITH SERVICE CONSOLE, MACROS, AND QUICK TEXT

Imagine being a support agent juggling multiple cases at once: you're answering an email on one case, the phone rings with another case, and you need to check a knowledge article for a third. Flipping between different browser tabs or screens slows you down. Enter the **Service Console** – a specialized Salesforce app designed as a one-stop workspace for support agents. When configured well, the Service Console can dramatically reduce clicks and page loads, helping agents resolve cases faster and with less frustration. Coupled with tools like **Macros** (to automate repetitive actions) and **Quick Text** (pre-canned snippets of text), it transforms the agent experience from a clunky multi-click process to a slick, streamlined workflow.

Service Console 101: The Service Console is essentially a Salesforce app with a tabbed interface. Instead of opening each record on a separate page, agents get a persistent navigation where they can have multiple records open as tabs or subtabs within the same browser window. For example, an agent can have Case #1001 open in one tab, and when they click the related Contact or a linked record, it opens as a subtab under that case – keeping context together. They might also have a second Case #1002 open in another tab, and easily switch between them. This is huge for productivity: no more losing your place or having 10 browser windows for 10 cases.

To set up a Service Console app declaratively, go to Setup -> App Manager and create a Lightning Console App. Choose the primary object (Cases) and any secondary objects you want available (like Contacts, Accounts, or custom objects your agents use). You can also configure a utility bar – a footer with handy tools like Notes, Macros, or Phone. In the console, enable features like **Split View**, which allows agents to see a list of cases on the left while viewing a specific case on the right. Also ensure the **Highlights Panel** on your case page layout shows key info (Status, Priority, Account, etc. at a glance at the top of the screen). The console supports keyboard shortcuts (for example, pressing "s" can save a record, or "e" to edit, etc.), which power users love.

ChatGPT Prompt (Design): *User:* "What's an ideal layout for my Service Console to maximize agent productivity?"

ChatGPT: "An effective Service Console layout is usually a three-panel design. On the far left, use **Split View** to show a collapsible list of cases or a queue – this helps agents navigate between cases quickly. The central panel is the main workspace: at the top, include a Highlights Panel with critical case details (like Status, Priority, Customer Name) so agents see the essentials without scrolling. Below that, have the Case Detail and related lists in tabs (one tab for Details, another for Case Feed/Activity, another for Related objects like Contacts or Assets). On the right side, include a sidebar for tools – for example, a **Knowledge** search component so agents can find help articles, and a **Macros** or **Quick Text** widget. Also consider an Omni-Channel widget if you're using Omni, so agents can accept new cases. This layout keeps everything in one screen: list on left, case in center, tools on right. Agents won't need to pop between screens, which boosts efficiency."

With your console structured, let's talk **Macros.** A macro in Salesforce is like a recorded script of actions that an agent can play back with one click. For instance, imagine a "Close Case" macro: when run on a case, it could update the Status to Closed, fill in a resolution text field with a standard note ("Issue resolved by following knowledge article XYZ."), and send a closure email to the customer – all in one go. Instead of the agent doing those 3-4 steps manually every single time, the macro does it in seconds. Macros especially shine for repetitive processes, like logging call details, sending a follow-up email, escalating a case, or categorizing a case with certain values.

Setting up macros declaratively is straightforward: In the console, there's a Macros utility (usually a widget at the bottom or side). Agents or admins with the right permissions can create a new macro, give it a name and description, then literally step through the actions: select the field to update and the value, or choose the email action and template to send, etc., and save the macro. Once saved, it can be run on any record of that object type. You can also make macros that are "bulk" (run on multiple records from a list view) for things like sending the same email to several cases at once (though use that carefully).

Quick Text is the simplest tool of the trio but a huge time-saver. Quick Text are predefined text snippets that agents can insert into emails, chats, or case feed notes with a few keystrokes. Instead of typing "Hello, thank you for contacting support, ..." every time, an agent can have a Quick Text named "Greeting" that contains that phrase. Or a Quick Text for common instructions ("Please restart the device and let us know if the issue persists."). In practice, when writing an email or chat, the agent can invoke quick text (e.g., by typing "/" or a keyboard shortcut, depending on setup) and select from a list. It instantly inserts the text inline.

Salesforce allows organizing quick texts by category and making them available in specific channels (some might be only for internal notes vs. customer emails, etc.). As an admin, you can preload the system with a library of quick texts for common scenarios, and you can empower agents to create their own (with approval if needed). Quick Text merges with record fields too – so you could create a snippet like "Hi {!Contact.-FirstName}," to personalize each greeting.

Let's see these productivity tools in action with ChatGPT's help:

ChatGPT Prompt (Quick Text): *User:* "Give me three examples of Quick Text snippets for a support agent: one for greeting a customer, one for asking for more information, and one for closing a case on a positive note."

ChatGPT: "Sure. Here are three suggestions:

1. **Greeting** – 'Hello {!Contact.FirstName}, thank you for contacting us. I'm sorry you're experiencing this issue, and I'm here to help.'

2. **Info Request** – 'Thank you for those details. Could you please provide [specific info needed]? Once we have that, we can move forward with a solution.'

3. **Closing** – 'I'm glad we could resolve your case. If you have any other questions or concerns, please don't hesitate to reach out. Thank you for choosing [Your Company]! Have a great day.'"

These can be created as Quick Text in Salesforce, with placeholders like {!Contact.FirstName} to personalize. Agents will then be able to insert them with a couple of clicks or a keyboard shortcut.

ChatGPT Prompt (Macro Design): *User:* "What steps should a 'Escalate Case' macro perform?"

ChatGPT: "An 'Escalate Case' macro might do the following: (1) Change the Case Status to 'Escalated'. (2) Update a custom field 'Escalation Level' to indicate it's at Tier 2. (3) Insert a Case Comment or Post to the case feed saying 'Escalated to Tier 2 support for further investigation.' (4) Send an email to the customer (using a template) to let them know their case has been escalated for specialized attention. (5) (If using case teams) Add the Tier 2 queue or a specific user as a Case Team member. Once these actions are recorded in the macro, a single click by the agent will execute all five steps in sequence."

This outline can guide how you actually build the macro in Salesforce's Macro builder. ChatGPT essentially pseudo-coded the process for you.

Now, think about training your support team to use these tools. Change can be hard for some users – they might be used to the old way. One approach is to create a quick tutorial or demo. This is where ChatGPT's **SORA** (text-to-video) capability could be a creative ally. You could literally ask ChatGPT to generate a short video that demonstrates using a macro in the console.

ChatGPT Prompt (SORA Video): *User:* "Create a 30-second video script showing a support agent using a macro in the Service Console to close a case."

ChatGPT: "*[Scene opens with an agent at their computer]* **Narrator:** This is Amy, a support agent, about to close a case using a macro. *[On screen: Amy clicks 'Macros' and selects 'Close Case'.]* With one click, the macro changes the case status, inserts a resolution, and sends a thank-you email to the customer. *[On screen: The case status field updates to Closed, an email preview pops up]* **Narrator:** Amy saved time and ensured nothing was missed. *[Scene ends with Amy smiling]* **Narrator:** Empower your team with macros in the Service Console, and make every support interaction lightning-fast!"

ChatGPT's SORA would generate this as a video with voice-over and on-screen actions. You could use such a video to train or inspire your agents.

Even if you don't use SORA or a fancy video, the point is clear: showing agents the benefit of these tools encourages adoption. Many times, experienced agents will come back and say, "Wow, can we also have a macro for X situation?" or "I wrote a quick text for that tricky answer we give repeatedly." That's when you know the productivity culture is taking root.

Pro Tip: Organize your macros and quick texts so they're easy to find. Use naming conventions like "Macro – Close Case" or folders/categories like "Quick Text – Greetings" vs "Quick Text – Troubleshooting". As your library grows, good organization ensures agents aren't overwhelmed searching for the right snippet. Also, involve your support reps in creating these assets – they know their day-to-day tasks best, and their input will lead to more useful macros and texts.

By configuring a tailored Service Console and equipping your team with macros and quick text, you dramatically reduce the "time to click" for common tasks. Agents can focus on the customer and the problem, not on navigating Salesforce or typing the same sentences again and again. The result: faster case resolution and happier support reps (who aren't exhausted by mundane actions). Next, we'll build on this efficiency by ensuring agents have the *knowledge* they need at their fingertips. Next we turn to Knowledge base management – setting up a repository of answers and making it seamlessly accessible during case work.

EFFECTIVE KNOWLEDGE BASE MANAGEMENT

A well-maintained knowledge base is the secret weapon of a high-performing support team. It allows agents to quickly find solutions that have already been discovered, and even better, it enables customers to self-serve answers to common questions (reducing the number of cases filed in the first place). Salesforce Knowledge, the module for creating and managing support articles, can be configured declaratively to fit your organization's needs. In this section, we'll cover setting up Knowledge article types, organizing content with data categories, establishing a publication workflow, and making sure agents can easily find and attach articles during case work.

Setting Up Salesforce Knowledge: If you haven't enabled Knowledge yet, you'll need to do that in Setup (ensure you have Knowledge User licenses for your support agents and check the "Knowledge User" permission on their user records). In Lightning Knowledge, there is a unified Article object where you can use **Record Types** to represent different kinds of articles. For example, you might have a "FAQ" record type with fields for Question and Answer, and a "Troubleshooting Guide" record type with fields for Symptoms, Cause, Solution, etc. Plan out what types of content you'll have. Many teams start simple: one article type for all support articles, using a rich text field for the content. More mature implementations might have multiple types for better structure and search (e.g., differentiate how-to guides vs. reference documents).

Data Categories: Think of data categories as tags or folders that help classify articles. They are organized in hierarchies (category groups and subcategories). For instance, you might have a category group "Product Line" with categories for each product (Product A, Product B, Product C), and another group "Issue Type" with categories like Installation, Configuration, Troubleshooting, FAQ. An article can be tagged in zero, one, or many categories in each group. Categories serve two purposes: (1) **Filtering/Search** – agents or customers can filter knowledge search results by category (e.g., show me only articles about Product A), and (2) **Access Control** – you can restrict visibility of articles by category to certain audiences (for example, partner users see only articles in the "Partners" category, or customers see only the "Customer" category articles while internal articles are tagged "Internal" and hidden from customers). Setting up categories requires a bit of upfront thought about how to slice your knowledge base logically. Keep it simple at first – you can always refine the taxonomy later.

Article Lifecycle and Approval: How do articles get created and published? Best practice is to treat knowledge articles like living documents that go through review. A common workflow:

- Support agent (or technical writer) creates a new article as a **Draft** (not visible to others yet).
- A subject matter expert or manager reviews the draft. Salesforce Knowledge can use the standard Approval Process for this: e.g., when the article is ready, the agent submits for approval, the manager approves it, and only then it becomes Published.

- Once **Published**, the article is available to the chosen channels (internal app, customer portal, etc., depending on its settings).
- Over time, articles might need updates – you can create a new version of an article, which goes through the same process, while the old version can remain published until the new one is ready (then the new one replaces it).
- If an article becomes obsolete, you can archive it (so it's no longer shown in searches for users).

This process ensures quality control – you don't want half-baked or incorrect solutions being used. It also creates a sense of ownership and curation in the knowledge base.

Making Knowledge Easily Accessible: In the Service Console, you should add the Knowledge component so agents can search the knowledge base without leaving the case. Typically, when viewing a case, an agent sees a sidebar or a widget to search for articles, possibly pre-filtered by context (Salesforce can automatically search using the case's subject or description keywords – this is called Suggested Articles). Agents can then read an article and click "Attach to Case" if it's relevant, which links the article to the case (for tracking and for easy reference). They can also send article content to the customer – either by inserting a Knowledge article into an email (which can copy the text or link), or by using a built-in "Email Article as PDF" feature. Training agents to leverage this is important: rather than writing a long explanation, they can say "I've attached a knowledge article with detailed steps." It saves time and ensures consistent info.

On the flip side, consider exposing knowledge to customers directly. If you have a customer portal or help center (perhaps built on Experience Cloud), you can publish certain articles to the public or logged-in users. For example, create a self-service FAQ page where customers can search your knowledge base. For this, each article can be marked with visible channels – e.g., an article could be internal-only, or also visible to customers. Data categories again help segment what different audiences see.

Managing a knowledge base also means measuring it. Salesforce Knowledge has reports for article usage: which articles are attached to cases most, which are viewed the most in the self-service portal, etc. Keep an eye on those – they tell you where knowledge is working and where there might be gaps. If an issue is coming up often but no one's written an article for it, that's a signal to create one.

Now, let's leverage ChatGPT for knowledge management tasks:

ChatGPT Prompt (File Upload): *User uploads a PDF titled "Product XYZ Troubleshooting Guide".* **User:** "ChatGPT, I've uploaded a technical guide. Can you draft a Knowledge Base article from this? Focus on the main issue types and solutions, in a concise format."

ChatGPT: "Sure. I've scanned the guide. Here's a draft article:

Title: Troubleshooting Guide for Product XYZ – Common Issues and Solutions

Issue 1: Device not powering on

Symptoms: Device shows no lights or response when pressing the power button.

Cause: Often caused by a depleted battery or faulty power cable.

Solution: 1) Ensure the power cable is firmly connected... 2) Try a different outlet... (additional steps) ... If the device still doesn't power on, contact support for potential hardware replacement.

Issue 2: Cannot connect to Wi-Fi

Symptoms: ... (and so on)

I've included two common issues as an example. You can continue similarly for other issues in the PDF. This draft focuses on concise symptoms, causes, and solutions for each problem."

ChatGPT pulled out the key problems and resolutions from the technical PDF and structured them into a typical "troubleshooting" article format. An admin or content writer would then refine and publish this in Salesforce Knowledge.

Note: When uploading internal documents to ChatGPT, ensure no sensitive personal data is included, or use ChatGPT's privacy controls/enterprise version so your data isn't retained. Always conform to your company's privacy policies.

As we see, ChatGPT can act like a first-pass content creator. It distilled a lengthy guide into a draft article with a clear structure. This accelerates knowledge creation significantly – instead of writing from scratch, your team can review and edit ChatGPT's draft.

Another way to use ChatGPT is for analyzing knowledge gaps. For example, you could feed it case data or transcripts and ask what issues lack good knowledge articles. ChatGPT might notice, say, many cases about "password reset policies" and suggest creating an article "How to Reset Your Password Securely." It's like having a data analyst comb through support logs for you.

Pro Tip: Maintain your knowledge base like a product. Assign someone (or a small committee) to own knowledge quality – regular reviews, updates, and archiving of old content. Encourage agents to flag articles that need improvement or to suggest new articles for novel issues. A simple practice is a monthly audit of the top 10 articles and the bottom 10 (by usage or rating). Update the top ones to keep them fresh, and either improve or remove the bottom ones if they're not helpful. By keeping the knowledge base curated, you ensure it remains a go-to resource rather than a dusty library.

With a robust knowledge base in place, agents can solve problems faster (often by simply sharing an article) and customers can even help themselves (deflecting cases). This dramatically improves efficiency and satisfaction. Now we've covered proactive knowledge sharing, our next focus is making sure we honor our promises to customers about support timing. Next we'll configure entitlements and milestones to manage Service Level Agreements (SLAs), ensuring no case silently breaches its deadline for response or resolution.

MANAGING SLAS WITH ENTITLEMENTS AND MILESTONES

If you've promised your customers certain service levels – like "we'll respond within 4 hours" or "we'll resolve critical issues within 2 days" –

you need a way to track and enforce those commitments. Salesforce's **Entitlement Management** feature is built for this. It allows you to define **Entitlements** (the services or support contracts a customer is entitled to) and **Milestones** (key steps or time-bound targets on cases, such as First Response or Resolution times). With declarative configuration, you can have Salesforce automatically tick down the clock on each case, warn you when a deadline is approaching, and even take action if an SLA is violated. Let's break down how to set up and use entitlements and milestones to stay on top of your SLAs (Service Level Agreements).

Entitlements: In Salesforce terms, an entitlement is typically an item that lives on a customer's account or contract record indicating what kind of support they get. For example, Account ABC might have an "Gold Support" entitlement (24/7 support with 1-hour response) and Account XYZ has "Standard Support" (business hours support with 1-day response). Each case can be linked to an entitlement (often Salesforce will auto-link if the case's account has one active). Entitlements carry with them the rules (entitlement process) that define the SLAs. You might also use **Entitlement Templates** to automatically apply a certain entitlement to new cases based on criteria (so agents don't have to manually choose the entitlement on each case).

To use entitlements, first enable them in Support Settings. Then decide how you'll organize them: by account, by product, or by support level. Many choose to have one active entitlement per account that represents their support level. Others might have multiple (if they sell support per product or per purchased package). Keep it as simple as possible while reflecting reality.

Milestones: These are the specific targets attached to entitlements. Common milestones include "First Response Due" (the time by which you must send the initial response to the customer) and "Resolution Due" (the time by which the case should be resolved or at least next steps established). You define each milestone with a duration (say 4 business hours for first response on standard cases, or 1 hour for urgent cases). You also can specify criteria for milestones, so maybe you have a milestone that only applies if Priority = High, etc.

Milestones are grouped into an **Entitlement Process.** Think of an Entitlement Process as the SLA policy. For example, "Gold Support Process" could have:

- First Response in 1 hour
- Resolution in 8 hours
- Escalation in 30 minutes (maybe an internal milestone to escalate to managers if not assigned quickly) – just as an example.
- Whereas "Standard Support Process" might have:
- First Response in 8 hours
- Resolution in 48 hours (2 business days)
- Perhaps no automatic escalation milestone.

You can create these processes declaratively. In Setup, find Entitlement Processes, and define one for each support tier or scenario. For each process, you add milestone entries:

- Pick the Milestone (like "First Response") and set its target time (e.g. 8 hours) and whether to use business hours or 24/7 clock.
- Add any criteria if needed (for example, if you want a different first response time depending on case priority, you might actually create two milestone entries "First Response – High Priority" and "First Response – Standard" with criteria Priority = High on one, and Priority = Medium/Low on the other, each with different hours).
- Attach actions to the milestone if desired: Salesforce allows you to define automated actions when a milestone is approaching or breached. For instance, when 50% of time is gone you could have it send an alert email to the case owner ("Reminder: you have 4 hours left to respond to case..."), and at 100% (violation) maybe automatically escalate the case or notify a manager. These actions are set up via the Entitlement Process UI (which behind the scenes uses timers and workflow logic).

On the case page, you'll want to add the **Milestones** component (in Lightning, there's a standard component that shows active milestones, countdown clocks, and whether they're completed or violated). This gives agents a visual cue: e.g., "First Response due in 3:25 hours" counting down, which turns red if overdue.

Now, configuring entitlements can be tricky to plan. Let's use ChatGPT to help outline a scenario:

ChatGPT Prompt: *User:* "Our company offers Silver and Gold support plans. Silver: 8x5 support, first response 4 business hours, resolution 3 business days. Gold: 24x7 support, first response 1 hour, resolution 1 day. How should I set this up in Salesforce?"

ChatGPT: "You should create two Entitlement Processes – one for Silver, one for Gold.

Silver Entitlement Process: use your default business hours (8x5). Add a 'First Response' milestone with 4 hours (business hours) and a 'Resolution' milestone with 3 days (business hours). Add any actions (e.g., email reminders) as needed.

Gold Entitlement Process: create or use a 24/7 business hours calendar to accommodate round-the-clock. Add 'First Response' milestone with 1 hour (24/7) and 'Resolution' with 24 hours (1 day, 24/7). Possibly also add an 'Initial Response Escalation' milestone at maybe 1 hour as well, to immediately flag if not responded in that time.

Next, create Entitlement records or templates: one called 'Silver Support Entitlement' linked to the Silver process, another for Gold linked to the Gold process. Assign these to the respective Accounts (or use Entitlement Templates to automatically apply Gold to Gold-tier Accounts' cases).

Train the team that cases for Gold clients will show a 1-hour timer, etc. Also, set up reports or dashboards to monitor milestone compliance for each tier (e.g., % of milestones met)."

ChatGPT has outlined a clear implementation plan: two processes, matching business hours to support hours, linking to accounts, etc. This blueprint can guide an admin step-by-step.

As ChatGPT notes, Business Hours are important. In Salesforce, you define business hours (e.g., weekdays 9-5 except holidays). Milestones can be configured to pause outside business hours – so an 8-hour response target doesn't tick during the night or weekend if that's outside support hours. Gold being 24x7 requires either setting business hours to 24x7 or telling the milestone to ignore business hours (run continuously).

You might wonder: what if a case doesn't have an entitlement? Salesforce offers a setting to enforce that every case must have one (or you can default one via automation). It's wise to make sure every case falls under some SLA process, even if it's a generic one for free support. That way, every case has measurable timelines.

Another concept is **Escalation Rules** (a separate older feature from entitlements). Escalation rules can also trigger when a case is open too long based on criteria. If you use entitlements, you might not need escalation rules, or use them for things not covered by entitlements (like maybe escalate cases that are High priority and haven't been touched in 2 hours regardless of formal SLA). But be careful not to double-escalate via both mechanisms.

Monitoring SLAs is key: use Salesforce reports on milestones (e.g., "Open cases with upcoming or breached milestones") to see how often you meet vs. violate deadlines. There's nothing like data to show improvement or highlight issues. For example, if you see 90% of first response milestones are met but only 60% of resolution ones are, that could drive process changes or more staffing for complex issues.

Pro Tip: Even if you don't have paying customers with formal SLAs, consider using entitlements and milestones internally. You can set targets like "respond to all internal IT helpdesk tickets within 4 hours" and track them. It instills discipline and provides metrics. Additionally, always review the assumptions in your SLA settings – if you find that a 4-hour response time is frequently missed, investigate why. Maybe the target is unrealistic or you need more coverage. Use the data from milestones to continuously improve your support process or adjust expectations.

With entitlements and milestones in place, your team gets an automated guardian for time commitments. No more forgotten cases languishing for days – Salesforce will make it obvious which cases need attention and when. This leads to more trust from customers (they see you consistently meet your promises) and less fire-fighting for managers (because issues are escalated before they become crises). Now, we've put in place almost every piece of our Service Cloud puzzle. But complex cases often involve multiple people working together, which is where our final topic comes in. Next we'll explore Case Teams and how to use them for collaborative case resolution.

COLLABORATIVE RESOLUTION THROUGH CASE TEAMS

No support agent is an island – tough cases often require input from multiple people with different expertise. Maybe a support rep needs a developer's help to analyze a bug, or perhaps a sales rep should be looped in to manage customer expectations for a big account. Salesforce **Case Teams** is the feature that enables this kind of collaboration right on the case record. By setting up case teams, you can formally designate who is involved in resolving a case and what role each person plays, all while controlling their access to the case.

Enabling and Configuring Case Teams: First, an admin needs to turn on Case Teams (in Support Settings) and define **Case Team Roles**. Roles are labels like "Case Owner," "Support Engineer," "Escalation Manager," "QA Analyst," etc., which describe each member's function on the case. You might mirror your escalation path: for example, a role for Tier 1 Agent (who might usually be the case owner), a role for Tier 2 Specialist, and maybe one for an Executive Sponsor for major clients. You also decide for each role what access they get (read-only or read/write on the case). Typically, anyone actively working on the case gets read/write, whereas someone who's just an observer might be read-only.

Once roles are set, you can manually add Case Team members on a case-by-case basis. On the Case page, you'll see a Case Team related list (after enabling) where an agent can click "Add Member," choose a user and a role for them. For example, add Alice as "Support Engineer" and Bob as "Product Specialist" to the case. This will give Alice and Bob access to the case (if your org-wide sharing on cases is private, case team membership opens up the record to them) and they can now view and update the case just like the owner can. There's an option to notify new team members via email when they're added – useful to ping them that their help is needed.

Case Team Templates: If you frequently need the same group of people on certain types of cases, you can set up a template. Say every Sev-1 (Severe outage) case should have an Escalation Manager and a Support VP involved. You can create a template "Sev-1 Response Team" with those roles filled by specific people or placeholders. Then on a new Sev-1 case, an agent can apply that template and boom – all the designated folks are added in one go. This saves time and ensures consistency in who gets involved for critical issues.

Using case teams doesn't by itself automate who to add – it provides the structure. You might use other automation (like a Case Trigger or Flow) to automatically add a team template or specific users when certain conditions occur (e.g., if Priority = Critical, auto-add the "Sev-1 Response Team" template). That moves into a bit of custom automation, but it's doable declaratively with a Record-Triggered Flow that uses the "Add to Case Team" action or by invoking an Apex-defined action if needed. For this chapter, just know it's possible.

Collaboration in Action: Once multiple people are on the case team, encourage them to use the case's Chatter feed or Case Comments to communicate, so all updates stay with the record. Each team member can also get case update notifications if they follow the case. Case teams basically create a defined group around a case, which is much cleaner than a flurry of separate emails or side conversations. Everyone on the team can see the latest status, files, and communications right on the case.

What about reporting? You might want to track, for example, how many cases required escalation to Tier 2 (which you could infer by a Tier 2 role being populated on the case team). You can create a custom report type for Cases with Case Team Members to analyze this. Or simply run a report filtered by Case Team Role = "Escalation Manager" to list all cases that had an escalation manager involved. This can be an indirect measure of case complexity or process adherence.

ChatGPT Prompt: *User:* "What Case Team roles would you recommend for a software company's support cases?"

ChatGPT: "Consider these roles:

Case Owner: The primary support agent responsible for the case.

Technical Support Engineer: A Tier 2 or specialist who provides deep technical insight.

Escalation Manager: A supervisor who oversees high-priority cases and coordinates resources.

Quality Assurance: Someone from QA if the issue might relate to a bug that needs verifying.

Customer Success Representative: If the case is for a big customer, a rep from the customer success/account management team might be looped in to manage customer communication.

You can adjust based on your needs, but ensure each role has a clear purpose. For instance, you might not use QA on every case, only on those involving potential product defects. Keep the roles limited to the key players to avoid confusion."

With ChatGPT's suggestions, you can refine your Case Team Role setup. Perhaps you'll implement 3-5 roles initially that cover most scenarios.

One thing to clarify: Case Teams are mostly an internal coordination tool. Customers don't see case team members (unless you choose to expose that via a portal, but usually not). So you still want one person, usually the Case Owner, to be the main voice to the customer. The case team's purpose is to let multiple people work on the backend. It's a good practice for the case owner to update the case with any insights from other team members, or even better, have team members post their updates on the case (so the history is logged and visible).

Pro Tip: If you have a set of people who often swarm on urgent issues, use a Case Team Template to add them quickly. For example, an "Incident Response Team" template could include a Lead Support Engineer, a Communications Manager, and a Dev Lead. The moment a P1 outage case comes in, apply this template to the case – instantly those three people are added and notified. It's faster than trying to remember who to email or ping in the moment of crisis.

Case Teams, when used well, provide transparency and accountability. Each member knows they are part of the resolution, and you can even define responsibilities (perhaps in your internal process docs, e.g., "Escalation Manager ensures weekly updates to the customer on long-running cases"). By having the structure in Salesforce, nothing is ad hoc – it's clear who is involved and in what capacity. This approach to collaborative support is sometimes called "swarming" (multiple experts working concurrently rather than a linear tiered escalation). Salesforce can support swarming via case teams combined with Chatter for communication, which many find more effective than traditional tiered support in complex cases.

With that, we've covered a wide range of Service Cloud declarative best practices: from the basics of case management to the fine details of SLAs and team collaboration. Let's wrap up what we've learned and prepare to move forward.

SUMMARY

In this chapter, we navigated the landscape of Service Cloud declarative configuration and learned how each piece contributes to a world-class support operation. We began with the fundamentals of Case management – establishing a clear case lifecycle with well-defined Status values, priorities, and support processes. This foundation ensures that every case is tracked consistently from New to Closed. We then implemented intelligent case assignment, using queues and assignment rules (and Omni-Channel for real-time routing) to guarantee that each new issue finds the right owner or team immediately. No more cases falling into a black hole; the system routes them where they need to go.

We also streamlined communications by deploying email templates and auto-responses, so customers are kept informed at each step. A simple acknowledgment email or timely update can vastly improve customer satisfaction, and we've made those communications automatic and uniform. To boost agent efficiency, we configured the Service Console as an optimized workspace, and introduced productivity tools like macros and quick text – reducing multi-click tasks to single-click actions and giving agents instant access to canned responses. Our support team now has a cockpit (the console) and time-saving gadgets (macros/quick text) to speed through cases without sacrificing quality.

Furthermore, we set up a robust knowledge base, organizing articles with sensible categories and approval workflows. By integrating knowledge into the case process, agents (and customers via self-service) can quickly find answers, leading to faster resolutions and case deflection. We tackled SLA management through entitlements and milestones, making sure we track critical timelines like first response and resolution due dates. This adds a layer of accountability and helps us meet our support commitments reliably – Salesforce will flag and escalate any case that's at risk of running late. Finally, we fostered collaboration using case teams, enabling complex cases to be swarmed by multiple experts at once. With case teams, everyone necessary can access and contribute to the case, ensuring that tough problems get a 360-degree effort without chaos.

Throughout these sections, we also illustrated how ChatGPT can be a helpful co-pilot for Salesforce admins and consultants. We saw it brainstorm status values, draft polite customer emails, outline macro steps, generate knowledge article drafts from a document, and even simulate configurations. This shows the potential of blending AI assistance with your declarative toolkit – it can enhance creativity, catch details, and accelerate documentation or planning. The combination of solid Salesforce best practices with AI support can elevate your solution design process to new heights.

By applying the best practices in this chapter, your Service Cloud implementation will be more organized, efficient, and responsive. Customers will notice faster responses and resolutions, agents will enjoy a more streamlined workspace, and managers will have clearer visibility and control over service performance. We've turned Service Cloud's powerful features into a coordinated system that embodies Salesforce Enterprise Thinking – aligning technology, people, and process for superior support outcomes.

As we move to the next chapter, keep in mind that even the most well-designed configurations need validation. In **Chapter 11: Testing and Quality Assurance**, we'll learn how to rigorously test our Salesforce declarative solutions (including all the Service Cloud setups we've done) to ensure they work as intended and continue to deliver high-quality results. After all, a change in one area can have ripple effects – proper testing will catch issues before they impact your users or customers. Let's head into the realm of testing strategies and tools to safeguard the excellence of our Salesforce configurations.

11

TESTING AND QUALITY ASSURANCE

Quality assurance is the cornerstone of any reliable system, especially in a platform as dynamic as Salesforce. In a world where businesses expect their apps to run without a hitch, even a small misconfiguration can have ripple effects. Case in point: Salesforce's own 2024 outage was estimated to have cost the company around $20 million. That's the price tag when a system goes down unexpectedly.

It's the era of "clicks, not code." Salesforce admins have powerful tools that make deploying features easier and faster than ever. But all that speed can be a double-edged sword. Rapid changes mean more moving parts, and each new validation rule or automated flow could break another business process. One expert put it plainly: this low-code boom has "placed the burden on Salesforce Admins to keep their apps running smoothly" – a task that grows "increasingly difficult" without disciplined QA.

So how do we catch problems before they hit production? Testing is the answer. In Salesforce, testing is absolutely critical; it's the only way to ensure your automations, validation rules, and page configurations work together to serve real users. Manual testing can feel overwhelming – one admin even described deployments as a "tangled ball of yarn". Fortunately, structured testing cycles, the right tools, and even AI assistants (like ChatGPT) can turn that chaos into confidence.

In this chapter, we'll take a practical tour through **testing and quality assurance for declarative Salesforce projects.** We'll start by planning an effective QA strategy, then walk through each stage of the testing lifecycle. You'll learn how to unit-test your flows, rules, and formulas; how to craft integration tests that ensure multiple components play well together; and how to engage real business users in formal UAT. We'll also explore Salesforce-native QA tools (Optimizer, DevOps Center, etc.), set up comprehensive regression tests for ongoing releases, and even stress-test your system's performance and security.

Think of this chapter as your QA co-pilot. Each section builds on the last: by the end you'll be designing test plans, guiding stakeholders through trials, and using tools (like wireframe sketches and metadata analyzers) to boost confidence. With clear signposts and practical examples, our goal is to make testing not just a checkbox but a confidence-builder. Let's dive in and make sure the Salesforce automations we build behave exactly as intended.

DEVELOPING AN EFFECTIVE TESTING STRATEGY

In any Salesforce project, first and foremost is a solid testing strategy. This isn't about checking a box — it's about building confidence in your release. As one Salesforce expert bluntly noted, *"Testing is absolutely critical in Salesforce"* because it helps ensure the platform functions flawlessly for users. With the right strategy in place, you catch mistakes early, minimize rework, and avoid costly downtime (remember the $20M outage example).

Your strategy should outline testing cycles for different types of tests. Typically this includes:

Unit Testing: Focus on individual declarative components in isolation. For example, if you have a validation rule on an Account object, write tests to confirm that rule behaves correctly for a variety of inputs (both valid and invalid data). Use the Flow Debugger, debug logs, or simple record creation in a sandbox to verify that outcomes match expectations. You might ask ChatGPT for help by saying, *"Generate edge-case data to test this phone number validation rule."* The AI's suggestions can help you craft thorough test inputs.

Integration Testing: Verify that multiple components and systems work together as expected. For instance, imagine a lead-to-cash process: creating a Lead, converting it to an Opportunity, applying an Order, and having an automated email notification. Integration testing means running through such end-to-end scenarios to ensure no step breaks the chain. Use a sandbox with realistic data or mock external systems. Write test scripts that describe each user path. You could even prompt ChatGPT: *"List possible end-to-end testing scenarios for a Salesforce lead-to-order process in a retail company."* This might generate useful scenario ideas to include.

Regression Testing: Whenever you modify an existing feature or add new functionality, re-run tests for all previous scenarios to confirm nothing broke. In other words, ensure that the new update "does not adversely affect existing functionality". Over time, you'll build a regression test suite — a collection of smoke tests or automated scripts covering critical workflows. Before each release, run this suite (ideally automated, or at least as a documented checklist) to catch unintended side effects.

Performance and Security Testing: Finally, validate how the system behaves at scale and under different user profiles. For performance, you might simulate bulk data loads or many simultaneous users in a full or partial sandbox. Because Salesforce is multi-tenant, you need to respect governor limits and design tests that mimic real load patterns. For security, systematically review profiles, permission sets, and sharing rules. Include tests where users in different roles try to perform actions (or view data) they shouldn't, to confirm your sharing model is airtight.

These testing cycles should be woven into your release process. A simple approach is to define your test plan up front, then run unit tests as you build each feature, followed by an integration pass, and then a regression pass before final deployment. In Agile teams, you might have a quick test-review meeting every sprint or a final UAT sprint before production.

Always tailor the strategy to your project's complexity. For a small change (like a single flow or page layout tweak), your plan might just be a checklist and a few sandbox runs. For a big rollout spanning departments, create a formal test plan document. That plan should list environments (which sandboxes, what data), responsible owners (who executes which tests), and timelines (when to refresh sandbox, when to freeze changes for testing). Prioritize tests based on risk: focus first on the parts of the system that handle critical data or complex processes. For example, if you're in healthcare, patient data validations and privacy rules would take top priority. In finance, compliance and permission changes might be the riskiest. A clear priority matrix in your strategy helps the team allocate effort effectively.

Modern teams often use tools or even AI to assist strategy planning. For instance, provide your requirements document (or a concise feature list) to ChatGPT and ask it to outline test cases or risk areas. For example:

ChatGPT Prompt (Text): "We are planning a Salesforce release for an insurance application. It includes new flows, validation rules, and some formula fields. Outline an effective testing strategy that includes unit, integration, regression, and security tests."

The response can highlight test focus areas, suggest a cycle plan, or warn about common pitfalls. Of course, always review and refine any AI output against your real requirements. **Pro Tip:** Save these ChatGPT interactions in a secure note with your release plan. You can then refine the prompts next time.

For large teams, consider a "shift-left" approach: involve testers and admins from the very start. Use whiteboard sessions or user-story workshops to identify testing needs. If you have user stories documented (in Word, PDF, or Jira), another tactic is to upload that document to ChatGPT's file upload mode and ask: *"Identify the key business scenarios from this doc to prioritize for testing."* The AI might pick up scenarios you hadn't explicitly listed. Always scrub any sensitive data before uploading; for example, replace real customer names with fictitious ones or omit personal information.

Ultimately, an effective strategy is a living document. As your declarative configuration evolves, your strategy will evolve too. Review it before each release and after major changes. With a structured QA plan in hand, you're ready to start testing individual components. In the next section, we'll dive into **unit testing declarative components**, putting this plan into action one component at a time.

UNIT TESTING DECLARATIVE COMPONENTS

Once your strategy is set, it's time to write unit tests for each declarative piece. This means testing validation rules, flows (including record-triggered and screen flows), formula fields, and any behind-the-scenes automation. The goal of unit testing is to ensure each element behaves correctly in isolation, so you can trust it as a building block in larger processes.

Validation Rules: Review each rule and write tests for both triggering and non-triggering scenarios. For example, if you have a rule enforcing that an Opportunity's Close Date cannot be in the past, create an Opportunity with yesterday's date (should fail) and another with today's or future date (should pass). Test picklist rules by trying each picklist option. For each test, note the expected error message or behavior. You can use the Salesforce UI or data import tools (like Data Loader) to insert records and watch for errors.

Need help designing comprehensive test inputs? ChatGPT can assist. For example, prompt it:

ChatGPT Prompt (Text): "List valid and invalid test data for this validation rule: Account.Phone must be 10 digits."

The AI might return several example phone numbers (with and without country codes, letters, etc.). Copy realistic examples and run them through your org's rule. Always review its output – for instance, remove any suggestion that looks like real personal data. **Pro Tip:** If your validation rule involves an email or name field, double-check cases like leading/trailing spaces or special characters – these often cause unexpected behavior.

Flows and Process Actions: Next, test any flows, process builders, or workflow rules individually. For record-triggered flows, use the Flow Debugger: open the flow in a sandbox and click **Debug** to walk through it. Enter sample values as prompted and check each path's outcome. If the flow has a screen, simulate user input for each field. Cover every decision branch and include negative paths – Salesforce advises that "if your flow uses Decisions, be sure to create tests that apply to each branch and also include negative tests". For example, if a flow has a decision on Priority = High vs. not high, create test cases for both. Also test related automations: if you have a Process Builder or Workflow Rule, make the triggering change (create/update a record) and ensure the action fires (e.g., email sent or field updated).

You can even enlist ChatGPT for flows. Try a voice prompt:

ChatGPT Prompt (Voice): "I have a flow that creates a follow-up Task when a Case priority is High. Talk me through how to test this flow.

ChatGPT might respond as if it were guiding you: "Create a Case with priority High and confirm a Task is created; then create a Case with priority Low and confirm no Task." This interactive style can highlight steps you might forget, like verifying Task details.

Formula Fields: Test formula fields by manually calculating expected results. For each formula, consider boundary values. For example, if a formula multiplies two fields, try a few numeric values (including zero and negative) to confirm the result is correct. If a formula uses date arithmetic, test around year boundaries (e.g., a date on Dec 31) to see how it behaves. Use the Developer Console's Query Editor to run SOQL and fetch formula outputs for your test records. You can also ask ChatGPT for edge-case ideas:

ChatGPT Prompt (Text): "For the formula field TotalPrice_c = Quantity_c * Price_c * (1 - Discount_c/100), what edge cases should I test?"

The AI may remind you to test zero quantity (TotalPrice should be 0), 100% discount (TotalPrice should be 0), negative prices (if allowed), and large values. Incorporate these into your tests.

If you have many flows or formulas, another trick is to compile a list of them and let ChatGPT analyze. For example, create a CSV of flow names and triggers and:

ChatGPT Prompt (Upload): "I uploaded a CSV listing our flows and their trigger conditions. Identify which flows might never be executed based on these conditions."

ChatGPT (in Operator/SORA mode) will scan the conditions and might highlight flows with no matching criteria, pointing out unused automation.

You can also generate quick diagrams to clarify logic. For example, ask ChatGPT's image mode:

ChatGPT Prompt (Generate Image): "Draw a flowchart: 'When a new Case with Priority = High is created, assign it to the High-Priority queue.'"

It will produce a simple flowchart diagram (via DALL·E) with decision boxes. Visualizing the logic can help you spot missing steps.

Finally, always do unit tests in a sandbox or developer org, never in production. After running tests, Salesforce usually rolls back the data (no need to cleanup). If something goes wrong, check the debug logs for clues. You can even add temporary debug actions in a flow (like creating a Debug__c record with key variable values) to record how data moves. **Pro Tip:** Leverage the Flow Debugger's logs: set your log levels to DEBUG for flows and Apex, and inspect them carefully when a test fails.

By the end of unit testing, each component should be rock-solid. You'll have validation rules that fire exactly as intended, flows that execute all their branches correctly, and formulas that produce accurate values across edge cases. In the next section, we'll shift to **integration testing**: combining these pieces into real-world user journeys to see if the whole system works together.

CONDUCTING INTEGRATION TESTING

With individual components verified, the next step is integration testing: validating entire business processes end-to-end. Imagine a sales process: a Lead is captured, converted to an Opportunity, products are added, and an order is sent to billing. Integration testing means running through such scenarios and confirming every step worked correctly. This is typically done in a Partial or Full Copy Sandbox with integrations connected. In that environment, you should mimic production as closely as possible – same objects, fields, automation, and connected systems (ERP, payment gateways, email services, etc.).

Start by outlining the key user journeys. For each journey, write a test script: detailed steps that a user would take, along with expected outcomes. For example, in a customer onboarding process:

1. Create a new Account with test data. Ensure the system automatically provisions an initial Order record.

2. Verify that the correct email notifications are sent (check email logs or your inbox).

3. Update the Order status to "Approved"; confirm that an invoice record is generated and a Chatter post appears.

As you define these scripts, tap ChatGPT to help:

ChatGPT Prompt (Text): "Write an integration test plan for the following scenario: 'A new Deal registration in Salesforce should create a Customer record, generate an Order, and notify the Sales Manager via email.' Include steps and expected results."

ChatGPT might respond with a numbered list of test steps and checks (e.g. "1. Create Deal record → Verify Customer created," etc.). Use this as a draft to refine your actual test cases.

For systems integrated with Salesforce via APIs, you'll want to test the external side too. If an Order syncs to an ERP system, for instance, run a test order and then query the ERP (or its sandbox) to confirm the data arrived. If no test system is available, mock the response. ChatGPT can even help generate mock data:

ChatGPT Prompt (Text): "Generate a sample JSON payload for a new customer creation API with fields: Name, Email, Phone, Address."

You can use this JSON to simulate the external callout and ensure your Salesforce side handles it as expected.

Be sure to include negative paths. What happens if a required field is missing in an integrated record, or the external API returns an error? For example, deliberately cause a 500 error on the ERP endpoint and confirm that Salesforce logs it or retries as designed. Document each outcome. Again, ChatGPT can outline these tests:

ChatGPT Prompt (Text): "Describe test cases for a REST API integration where Salesforce must retry on failure and notify an admin if it still fails."

It might suggest steps like "Induce an API 500 error, verify Salesforce retries after X minutes, then check that an error email was sent to the admin." Add these tests to your plan.

You can also have ChatGPT talk through a scenario to unearth gaps. For example:

ChatGPT Prompt (Voice): "Act as the QA team and talk me through testing the entire Quote-to-Cash process in our Salesforce instance."

ChatGPT might narrate steps ("First, create a Quote and add products; next, convert it to an Order..."), and it could ask about edge cases (like multi-currency or discounts) that you should include. This conversational approach often reveals missing steps in your own testing checklist.

If you have documentation (like a requirements PDF or spreadsheet of integrations), use ChatGPT's upload feature for analysis. For instance:

ChatGPT Prompt (Upload): "I've uploaded our integration requirements document. List all data mappings and trigger conditions we should test."

ChatGPT will extract items such as "Account_ID → Customer_ID mapping" or "Opportunity Stage = 'Closed Won' triggers invoice creation." This automates converting specs into test points.

Throughout integration testing, use realistic data. If possible, seed your sandbox with a subset of production data (scrubbed to remove PII) so you test with real volumes and combinations. If that's not feasible, at least create sample data for each scenario (e.g. one customer in each region, one order of each type, one contract for each business unit). Tag test records clearly (like adding a "Test__c" checkbox or prefix to names) so you can identify and clean them up later.

Pro Tip: Maintain a running log of defects found during integration testing. If a test fails, record it immediately (with steps to reproduce) and add it to your project board or issue tracker. Once the team fixes it, verify the fix and mark the test as passed. This loop prevents new changes from reintroducing old bugs.

After integration testing, ideally you have one more dry run with business users – the formal UAT phase (next section). But even before that, ensure any major fixes from integration tests have been re-verified. When all the paths work correctly, you're ready for user acceptance testing, confident that your declarative components are solid together.

ENGAGING BUSINESS USERS WITH USER ACCEPTANCE TESTING (UAT)

User Acceptance Testing (UAT) is where business users verify the solution meets their needs in plain language. UAT generally happens late in the project, using a sandbox or pre-prod environment with realistic data and scenarios. In essence, it lets end users "give the thumbs up" before go-live. Salesforce guidance emphasizes that UAT should always be part of your testing strategy, so stakeholders can confirm the configuration satisfies the requirements.

Begin UAT planning by gathering your user stories and acceptance criteria. These are often written as "As a , I want so that." Turn each into a scripted test. For example, a story might be: "As a Sales Rep, I want to record a client's interest level so I can prioritize follow-ups." A corresponding UAT script could be:

1. Log in as a Sales Rep.

2. Create a new Lead and set the "Interest Level" field to "High."

3. Verify that the Lead saves successfully and the Interest Level displays as "High" on the Lead detail page.

Write each step in plain business terms – avoid Salesforce jargon. Provide these scripts (as a document or spreadsheet) to testers so they know exactly what to try. Include the expected result in user-friendly language (e.g., "System should allow save and show 'High'").

ChatGPT can help translate technical requirements into business-focused test steps. For instance:

ChatGPT Prompt (Text): "Convert this requirement into a UAT test case: 'Automatically assign an account manager by territory when creating a new customer account.'"

You might get: "Create a new customer with a Texas address. Check that the account manager is set to the Texas region rep." This gives you a ready-to-use script to hand to testers.

Next, design a simple test results form. A classic format has columns for Test ID, Description, Steps, Expected Result, and Pass/Fail. You can even prompt for a layout sketch:

ChatGPT Prompt (Generate Wireframe): "Draw a sample UAT test case template with fields: Test Case ID, Description, Steps, Expected Results, Actual Results, and Pass/Fail."

ChatGPT will produce a conceptual form (boxes and labels) which you can reproduce in a doc. The goal is to make it easy for non-technical users to follow each case and mark outcomes.

Then, run walkthrough sessions. You might sit with a user and click through each test, or have them do it independently while you observe. Provide context in business terms: for example, instead of "verify validation rule X," say "try saving a contact without an email – the system should show an error prompting for email." Encourage testers to verbalize what they expect. Sometimes have them "think aloud" as they test; this can reveal missing help text or unclear instructions.

During testing, capture feedback systematically. Use your form and ask testers to note any failures or questions. Afterward, gather their comments. You can even use AI to organize feedback. For example:

ChatGPT Prompt (Text): "Summarize these tester comments into action items: 1) I saved an order but it didn't appear on the report. 2) I couldn't find the email notification on the dashboard."

ChatGPT will turn this into a clear list of bugs to fix (e.g. "Fix issue where closed orders aren't showing in the sales report; ensure the system sends notifications and they appear on user dashboards"). Review and refine its output.

It's often helpful to communicate with users during UAT. For example, set up a Slack or Chatter group for UAT feedback. You could even ask ChatGPT to craft an announcement:

ChatGPT Prompt (Text): "Draft an email update summarizing UAT results: '95% of test cases passed, 2 critical issues found, ready for go-live plans.'"

The AI can produce a professional status email you adapt for stakeholders. For visual reporting, ask ChatGPT for chart ideas:

ChatGPT Prompt (Generate Image): "Create a pie chart graphic showing 90% passed tests and 10% failed."

The actual DALL·E image may be low-res, but it will inspire a chart you can make in your presentation.

Pro Tip: Make the experience user-friendly. Provide data so testers aren't stuck; e.g., pre-create a few records they will need. If possible, enable "Login As" so they can experience the system as different users. Celebrate their progress — a quick chart of tests completed can be motivating.

After UAT, compile all outcomes. Any "Fail" results become defect tickets or work requests. Fix them, then get quick sign-off from the business sponsor (usually via a simple email or shared checklist). This formal acceptance of UAT means the system is ready for final regression checks and deployment.

In summary, UAT is about collaboration. We write test scripts in plain language, guide business users through them, and listen to their feedback. This final check ensures the system delivers the value you promised. Next, we'll explore **Salesforce tools for ongoing quality assurance** – ways to catch issues even before users notice.

LEVERAGING SALESFORCE TOOLS FOR QUALITY ASSURANCE

Salesforce provides built-in tools to help you monitor and improve the quality of your declarative configuration. One of the most powerful is **Salesforce Optimizer**. This app analyzes your org's metadata and produces a "health check" report. Optimizer flags issues like unused fields, overly long page layouts, complex formulas, and inactive automations. For example, it may note that dozens of custom fields on the Account object have never been used, suggesting you could remove them to declutter your schema. It might highlight a validation rule or flow that's never fired, or a field that's no longer on any page layout.

Once Optimizer gives you its findings, act on them. Use the exported.xlsx file to sort by the most critical issues. You might find a page layout with 50 fields and a 10-second load time – a strong signal to simplify it. Remove obsolete fields, merge duplicates, and deactivate stale automations. Not only does this make the UI cleaner, but it also reduces your testing surface. With fewer fields and rules, regressions are less likely. In short, Optimizer guides cleanup that pays dividends in quality and performance.

Another key tool is the **Lightning Usage App**. It shows metrics like page load times and component usage. If a particular Lightning page is slow to load, you can drill in to see which components or fields are causing the delay. Use that insight to optimize. Similarly, the **Setup Audit Trail** tracks the last 180 days of configuration changes, so you can review recent edits that might introduce risk. For example, if a last-minute field was added on the weekend, you can quickly see if that might have caused a new error.

For deployments, Salesforce offers **Change Sets** and the new **DevOps Center** (in pilot). Change Sets are the classic drag-and-drop method to move metadata between related orgs. However, they are manual and lack version control. DevOps Center is Salesforce's modern answer: it provides a guided pipeline in Setup, connects to a Git repository, and enforces checks and approvals. If you have DevOps Center enabled, try creating a project and linking it to your source repo. It will automatically track changes, make deployments repeatable, and maintain a history of what was moved. Even if you stick with change sets, consider scripting: use Salesforce CLI (SFDX) or Ant to retrieve and compare org metadata, ensuring nothing is missed.

You can also use external tools or scripts to analyze metadata. For instance, run an SFDX command to list all custom fields on an object and feed that list to ChatGPT:

ChatGPT Prompt (Text): "Here is a list of custom field names on the Account object: [AccountID, AnnualRevenue, Revenue, Total_Annual_Revenue__c, Sales_Region, Region__c]. Identify any fields that appear redundant."

ChatGPT might highlight "AnnualRevenue" and "Revenue" as overlapping, suggesting a cleanup. This kind of analysis can catch inconsistencies in naming or duplication across objects.

Don't forget Salesforce's **Health Check** tool (Setup > Security Health Check). It scores your org's security settings against Salesforce's baseline. Run Health Check periodically and address any Critical or High issues. Security misconfigurations are a quality risk too – they could allow data exposure or login issues.

Finally, be proactive about change monitoring. You could schedule Optimizer to run monthly (its app can auto-run on a schedule) and have it email you a report. Store the history of Optimizer runs (it even saves an XLS in Files) and create a simple dashboard in Salesforce or an external tool to trend your improvements. For instance, track the number of inactive fields over time, or how many Validation Rules fired. If you see a spike in unused fields, it may indicate a growing technical debt that needs attention.

ChatGPT Prompt (Upload): "I uploaded the latest Optimizer report. What are the top five issues I should address first?"

In file-upload mode, ChatGPT can scan the report and advise priorities (e.g. "You have 30 unused custom fields on Lead and 20 on Contact – consider reviewing those"). Always verify its summary with the raw report before acting.

Leveraging these tools means catching many problems automatically. By combining Optimizer, Usage metrics, audit trails, and deployment pipelines, you create a culture of continuous improvement. And of course, you can use ChatGPT to help interpret tool outputs or automate mundane checks. The net effect is a cleaner org and easier testing.

In the next section, we'll dive into **Implementing Effective Regression Testing:** techniques to ensure new changes don't break existing functionality.

IMPLEMENTING EFFECTIVE REGRESSION TESTING

Regression testing is the safety net that catches unintended side effects whenever you change your Salesforce configuration. It involves re-running a suite of tests on existing functionality to ensure that new updates haven't broken anything that used to work. In a Salesforce org that evolves with frequent enhancements and seasonal releases, regression testing is critical to maintaining stability and user trust. Every new field added, flow updated, or validation rule changed could potentially impact a business process elsewhere. By systematically retesting core functionality, you ensure that new updates do not negatively impact the current system, maintaining stability and performance. In other words, regression testing helps identify unintended consequences early, before they impact users in production.

Building a Reusable Regression Test Suite

An effective regression suite is a reusable set of test cases for your most critical processes, run consistently on every release instead of reinvented each time. Key tactics include:

Identify Core Business Processes: List the critical user journeys in your Salesforce org – the sales pipeline from lead to opportunity, customer support case lifecycle, invoicing and payments, etc. Prioritize processes whose failure would be highly disruptive to the business.

Document Test Cases Clearly: For each process, write step-by-step test cases with expected outcomes. Include variations and edge cases (for example, converting a lead with and without an associated campaign). Many teams keep this documentation in spreadsheets or testing tools. *ChatGPT can assist here* – you might ask it to draft test case outlines given a user story or requirement description.

Leverage Cross-Industry Insights: Don't limit scenarios to your org's past issues. Tap into broader industry experience. If you're in healthcare, consider high-volume transaction scenarios from retail, or complex approval flows from finance. ChatGPT can highlight common failure points in similar processes across sectors.

Keep the Suite Up to Date: Treat your regression scripts as living assets. When new features launch, add or update test cases. When old features retire, remove those tests. Regular maintenance keeps the suite relevant.

Automate Where Feasible: Running a full regression manually can be time-consuming. Use tools or scripts to automate repetitive steps. For example, Salesforce testing tools (Provar, Tosca) or Selenium can handle routine click-throughs and verifications. **Pro Tip:** Save human effort for tests that need judgment (like evaluating if a screen is user-friendly), and automate repetitive data entry and validation.

ChatGPT Prompt (Typing): "Draft a regression test plan for our Lead Conversion process. Include 5 key scenarios (e.g. converting with related records, missing data, validation rule triggers) and expected results."

This shows how describing your process to ChatGPT can yield a full test plan. ChatGPT often suggests scenarios you might overlook, and you can always ask it to refine or add cases (for example, "Now include some negative tests for bad data inputs").

ChatGPT Prompt (Voice via Operator): User speaks: "Hey ChatGPT, we changed our discount approval flow for large deals. What existing processes should we test to make sure nothing else broke?"

Using ChatGPT's voice interface, an admin can get quick guidance without typing. ChatGPT might respond with a checklist of related processes to verify: standard quote approvals, order processing with discounts, renewal quotes, and any reports or dashboards tied to discount fields, so nothing is overlooked.

Sandbox Strategies and Data Seeding for Realistic Tests

A regression test is only as good as the environment it runs in. Executing tests on a sandbox that closely mirrors production helps catch issues real users might face. Salesforce provides several sandbox types:

Full Copy Sandbox: A replica of production, including all data. It's ideal for regression testing because it has real data volumes and configurations. If possible, refresh a Full sandbox before major testing – you'll uncover issues that smaller sandboxes might miss. **Pro Tip:** Use a Full sandbox for final regression tests whenever you can, especially if your org has complex data relationships or millions of records. It's the closest to production behavior.

Partial Copy Sandbox: Contains a sample of production data based on a template. It's useful for testing core functionality without the overhead of a full dataset. You might include a few weeks of recent records or a representative subset of accounts and cases. Partial sandboxes can be refreshed more frequently.

Developer Sandbox (Dev/Dev Pro): These include configuration and code but little to no data. They're great for unit tests and builds, but for regression testing you'll need to load data into them (via scripts or anonymized extracts) to create realistic records.

Seeding test data means populating a sandbox with records so your test cases have meaningful scenarios to run through. When seeding, mirror real conditions: if one test involves converting a lead, ensure you have leads with various attributes (with tasks, without, duplicates, etc.).

ChatGPT Prompt (File Upload): The user uploads a CSV of sample Accounts plus a data dictionary. "ChatGPT, look at this sample data. What additional records or variations should I load to cover edge cases in testing?"

ChatGPT might notice missing scenarios and suggest adding those records, ensuring your test data covers all critical cases. This guidance helps you avoid leaving out an important use case when preparing data.

Always practice good data hygiene. Anonymize customer names, emails, and personal details in sandbox data. If you share logs or files with

ChatGPT, scrub confidential info first, and use ChatGPT's privacy mode or Enterprise tier to ensure sensitive data isn't retained.

Validating Changes to Prevent Breakages

With your regression suite ready and a proper test environment in place, finally run the tests and verify nothing has broken:

Run the Full Suite for Major Changes: Whenever you deploy significant updates (a new app, major process change, or even a Salesforce seasonal release), execute your entire regression suite. A small tweak might seem harmless, but in Salesforce's interconnected world, a new Opportunity validation rule could unexpectedly impact an existing Case flow or report.

Compare Results to Baselines: Document test outcomes and compare them to expected results. If you have pre-change screenshots or reports, confirm that post-change outputs match (e.g. a report still shows the same totals, an approval still goes to the correct manager). For automated tests, review logs for failures; for manual tests, note any anomalies.

Use ChatGPT for Quick Diagnosis: If a test fails and the cause isn't obvious, bring in a fresh perspective. Paste the error message or debug log snippet (sans sensitive info) into ChatGPT and ask what might be wrong. ChatGPT can often interpret the error and suggest likely causes. For example, it might highlight that a "Discount Code" validation rule prevented a record save – revealing what needs adjustment.

Continuously Improve the Suite: Treat each regression cycle as a learning opportunity. If a bug slipped through to production, add a test case for that scenario going forward. Over time, your suite becomes more comprehensive, and your confidence in release quality grows.

ChatGPT Prompt (Sora Video Generation): "ChatGPT, use Sora to create a short video summary of our latest regression test results. Show how many tests passed, and highlight any critical failures."

With ChatGPT's Sora feature (AI-generated video), you can turn a routine test report into an engaging visual update. A short video can visually summarize test results with a voice-over – an appealing way to keep stakeholders informed.

Cross-Industry Regression Testing in Action

Regression testing best practices apply across all industries. For example:

Finance (Banking): A bank adds a new loan product in Salesforce. After deployment, they rerun tests on existing loan approval processes to ensure calculations, workflows, and document generation still work correctly.

Healthcare: A hospital introduces a new patient intake flow in Salesforce Health Cloud. Before go-live, they retest scheduling, prescription refills, and insurance billing to confirm the new intake form doesn't disrupt existing processes.

Retail: An e-commerce retailer enables a new promo code feature for call center reps. They re-run tests on the entire order workflow – from adding items to cart through payment processing and inventory updates – to ensure the discount code doesn't break the fulfillment process.

Pro Tip: Maintain a small "smoke test" suite of core happy-path tests that you can run quickly in a sandbox for every minor change. It's like a chef tasting the soup before serving – a quick check to catch glaring issues. For major releases, follow up with the full regression banquet by testing every feature thoroughly.

With rigorous regression testing guarding against unexpected breakages, you can deploy changes with confidence. Your team and users get new features while trusting that existing functionality remains solid. Now that the foundation is secure, we can push the envelope. Next up is advanced performance and security testing – making sure your Salesforce configuration stays speedy, scalable, and safe as your organization grows.

ADVANCED PERFORMANCE AND SECURITY TESTING

Having solid fundamentals in place, it's time to push your Salesforce org to its limits and ensure it holds up. This section tackles two critical areas: performance testing for large-scale usage and security testing to protect data. Using advanced strategies – and a bit of AI help – you can verify that your org stays fast and secure as it grows.

Performance Testing Under Load

Salesforce's cloud platform is built to scale, but you need to test that your declarative configurations handle heavy data and user loads. In Salesforce terms, **Large Data Volumes** (LDV) often mean millions of records or thousands of users. Even if you're not there yet, it's wise to simulate future growth.

Identify likely performance bottlenecks early: long-running automations (flows or processes looping through many records), complex report filters on big datasets, pages with many components, or bulk data imports. Next, **simulate realistic load** in a Full sandbox (the only type that supports true performance testing). Use tools or scripts to mimic multiple users and large record batches. For example, you might use Apache JMeter or the Bulk API to create thousands of records in a short time. Watch the system behavior – check debug logs or Event Monitoring for signs of strain like governor limit warnings or slow SOQL queries. If you detect a flow hitting limits at high volume, consider optimizations (e.g. removing unnecessary loops or using scheduled paths to spread work out).

Make sure your automations are **bulk-safe.** Salesforce processes records in batches, so test by, say, updating 200 records at once and see if all related flows and processes execute properly. If something fails or slows down with a batch, you may need to refactor that automation (for instance, have a single flow update many records at once, rather than one flow instance per record). Also measure the user experience: track page load times and response times under load. If a Lightning page with many components takes too long to display, try simplifying it (fewer components or filtered data) and test again.

ChatGPT Prompt (Load Test Plan): "Help me design a load testing plan. We expect 500 concurrent users at peak, mostly creating and updating orders. Suggest key test scenarios and metrics to monitor."

ChatGPT can outline a plan covering various angles of stress testing. It might suggest simulating peak read/write transactions (like a flash sale for a retail site or a large batch of case creations in a call center), as well as monitoring metrics like transaction response time, error rate, and Salesforce server limits. This AI-generated plan ensures you don't overlook critical scenarios unique to your business model.

Declarative Security and Access Testing

A fast org means little if data isn't properly protected. Declarative security in Salesforce – profiles, permission sets, roles, sharing rules, login policies, etc. – must be tested to confirm users only see and do what they should. The principle of **least privilege** should guide your checks.

Use dedicated test user accounts for each profile or role and verify their access in a sandbox. For example, log in as a test Sales Rep: confirm they cannot access admin-only features or see records that should be private. Spot-check that sensitive fields (like Salary or SSN) are hidden or read-only for profiles that shouldn't see them. **Pro Tip:** Create a "permission matrix" of critical actions vs. profiles, and systematically test each combination (e.g. can a Marketing User export reports? Can a Support User delete Cases?). This structured approach catches unintended access rights.

Next, validate your **sharing rules** and any **restriction rules**. If a sharing rule should allow a certain group to see specific records, set up those conditions and verify with a test user in that group. Conversely, ensure that users outside the group cannot see those records. For restriction rules (which hide records even if sharing otherwise grants access), confirm that a user who meets the restriction criteria truly cannot find or report on the affected records. For instance, if volunteers should only see Cases from their own region, test with a volunteer user to make sure cases from other regions are invisible.

If you use **login flows** or special authentication steps, test those too. Make sure the flow (like a terms-of-service prompt after login) appears only for the intended users and doesn't block others. Confirm that any mandatory multi-factor authentication (MFA) or IP restrictions are working by trying to log in under various conditions (e.g. outside allowed IP ranges) and verifying that Salesforce appropriately allows or denies access.

Think like a malicious actor for a moment: try URL manipulation (guessing an ID in the URL to access a record you shouldn't) or use a guest user link to see if any data is exposed without login. Salesforce's security model should prevent these, but your configurations (like a mis-set guest user permission) could open a hole.

ChatGPT Prompt (Security Audit Plan): "Generate a checklist for a Salesforce security audit focusing on profiles, permissions, data visibility, and org settings."

ChatGPT will produce a comprehensive audit checklist. It typically covers reviewing profile permissions, analyzing permission sets for unnecessary access, checking field-level security on sensitive data, evaluating sharing rules for appropriateness, confirming MFA and network access settings, and running tools like Health Check. This AI-made list can serve as a second set of eyes, ensuring you don't miss any aspect of a thorough security review.

Tools and AI for Ongoing QA

Salesforce's own tools combined with ChatGPT create a powerful arsenal for quality assurance:

Salesforce Optimizer: This tool scans your org and identifies performance and complexity issues. It might flag things like very large page layouts, unused fields, or profiles with broad permissions. Use these insights to clean up configurations. You can even ask ChatGPT to interpret Optimizer results – for example, *"Optimizer says 3 pages have over 100 fields and one profile has too many admin permissions; what should I do?"* The

AI can suggest specific actions (streamline the pages, tighten the profile's access) based on best practices.

Security Health Check: The Health Check in Setup gives you a security score and lists which settings differ from Salesforce's recommended baselines. Focus on high-risk items (like password expiration, session security, clickjacking protection) and bring them in line with best practice. If you're unsure why a setting matters, ask ChatGPT. For instance, "Why is 'minimum password length = 8' a risk?" and it will explain the potential vulnerability and how a longer password mitigates it. Health Check lets you fix many issues with a click, but understanding the "why" helps you prioritize.

Metadata Exports & AI Analysis: For complex orgs, consider exporting security metadata for analysis. You might pull all profile and permission set data into a spreadsheet. ChatGPT (especially with a code interpreter plugin or by feeding it summaries) can highlight anomalies or outliers. It could find, for example, that one custom profile has "View All Data" on an object whereas none of the similar profiles do. This might prompt you to investigate and correct an over-permission. Similarly, an export of field-level security could be scanned to find any fields that are visible to too many people.

Cross-Industry Examples in Action

Retail (Performance): An e-commerce company simulates a Black Friday sale in a sandbox. The load test reveals a flow in the checkout process that times out when hundreds of orders hit simultaneously. The team examines debug logs and uses ChatGPT to get tips on optimizing the flow (like moving some logic to a scheduled job). After adjustments, a re-test shows the flow handles the peak load successfully.

Finance (Security): A bank conducts a quarterly security audit of its Salesforce org. Using Security Health Check, admins discover the session timeout was set longer than policy allows and correct it. They also export profile permissions and let ChatGPT scan for oddities – it flags that a "Loan Processor" profile had unnecessary Delete rights on customer records. The team swiftly revokes that permission. These steps, combined with role-based testing (verifying that retail bankers cannot see wealth management data, etc.), keep the org compliant.

Pro Tip: Make performance and security testing a regular part of your release cycle. Even small changes can have big effects. For example, if you deploy a new automation, do a quick load test on related processes to ensure response times stay acceptable. Likewise, whenever you add or modify profiles and sharing rules, re-run a mini security audit. Consistent vigilance, aided by tools and ChatGPT's analyses, ensures that your Salesforce org remains both **efficient** and **trusted**.

With this comprehensive approach to testing and quality assurance, you've built confidence that your declarative Salesforce solutions work correctly, perform under pressure, and keep data safe. Now that we've mastered testing and QA, Chapter 12 will focus on building performance and scalability into your Salesforce solutions – ensuring your implementation not only passes tests today but continues to thrive as data grows and user demand increases.

SUMMARY

In this chapter we walked through a comprehensive approach to testing and quality assurance in declarative Salesforce projects. We started by designing an overall **testing strategy**, identifying the need for unit, integration, regression, and performance/security tests. We emphasized structured cycles (for example, per sprint or release) and risk-based planning, even using ChatGPT to brainstorm scenarios and edge cases.

We then drilled down into **unit testing** each component: validation rules, flows, formula fields, and other automations. We discussed using sandboxes, debug logs, and new Flow Test tools to verify each piece. We sprinkled in many *ChatGPT prompts* – from generating test data and scenarios to reviewing formula logic – to speed up the process and help both admins and business users think through cases.

Next came **integration testing**, where we stitched components together in realistic end-to-end scenarios. Here again, we used chat prompts to draft integration test scripts, simulate external API calls, and even analyze requirement documents for missing test points. We stressed running these tests in a partial/full sandbox with sample data and connected systems, and tracking any issues we found.

Involving actual end users through **User Acceptance Testing (UAT)** was our fourth step. We converted technical requirements into plain-language test cases, guided users through scripts, and collected their feedback. We described formats for test result capture, how to walk through sessions, and how AI could help summarize tester comments. This ensured the business itself could approve the solution before go-live.

We also leveraged Salesforce's built-in QA tools (Optimizer, Usage metrics, Health Check) and modern DevOps practices to keep our org healthy. Optimizer reports highlighted cleanup opportunities, and DevOps Center or source control helped us package changes safely. We showed how to feed tool output into ChatGPT to prioritize fixes or plan actions.

Throughout, we maintained a friendly, real-world style: think of us as a guided co-pilot helping the team, not a lecture on theoretical testing. By the end, you should feel equipped to *plan* your test efforts, *execute* detailed unit and integration tests, *engage* users in UAT, and *monitor* your org's health continuously.

In the next chapter, we'll build on these foundations by tackling performance and scalability in detail – ensuring that every new change truly adds value without breaking what already works. The journey to a rock-solid Salesforce deployment continues!

12

PERFORMANCE AND SCALABILITY

Performance and scalability are the unsung heroes of successful Salesforce implementations. It's easy to build a solution that works well with 100 records and a handful of users; the true test is whether it stays fast and reliable with millions of records and hundreds (or thousands) of users. In this chapter, we'll explore how to keep your Salesforce org running like a well-oiled machine as it grows. We will see why even "clicks, not code" solutions must be designed with the platform's limits and architecture in mind, and how smart declarative configuration choices can prevent slowdowns and failures down the road.

In plain terms, **performance** is about how fast and efficiently your Salesforce apps respond (page loads, reports running, automations executing), and **scalability** is about sustaining that performance even as usage increases. A retail org's order management app might work fine with 500 orders a day, but what about Black Friday when that spikes to 50,000? A university's recruiting app could handle 1,000 student records in year one, but can it handle 100,000 after a few years of admissions cycles? Planning for scalability ensures your org can handle growth without degrading the user experience.

This introduction sets the stage for a deep dive into Salesforce performance best practices. We'll start by grounding ourselves in platform fundamentals – understanding why Salesforce imposes certain limits and how multi-tenancy works. Then, we'll move through specific areas: optimizing data retrieval for large volumes, avoiding common configuration anti-patterns that hurt speed, implementing data archiving strategies to trim bloat, using asynchronous processes (like scheduled flows and Platform Events) to handle heavy lifting in the background, managing concurrency to avoid record locking, and finally monitoring and measuring performance proactively. Throughout, we'll include cross-industry examples – from healthcare to finance to nonprofit – to illustrate these concepts in real-world scenarios.

Crucially, we'll also demonstrate how **ChatGPT can be a co-pilot** in this journey. Using well-crafted prompts, you can have ChatGPT help identify issues (like unselective queries or inefficient formulas), design strategies (like an archival policy or an async processing plan), and even create diagrams or explanations for stakeholders. We'll use a variety of ChatGPT interaction modes – text, voice, file uploads, video (SORA), and image generation – to showcase how you might leverage AI to ensure your Salesforce org not only works, but works *well* at scale.

By the end of this chapter, you'll have a toolbox of performance and scalability best practices for declarative Salesforce development. These principles will help you keep your org running smoothly as data volumes grow and user counts climb. Let's dive in, starting with the fundamentals of Salesforce's architecture and why understanding it is key to building high-performance declarative solutions.

UNDERSTANDING SALESFORCE PERFORMANCE FUNDAMENTALS

Every Salesforce practitioner should start with a solid grasp of *why* the platform imposes limits and how it manages resources behind the scenes. Salesforce is a **multi-tenant** cloud platform – meaning many customers (tenants) share the same infrastructure. This is like an apartment building where all residents share water, electricity, and internet. If one apartment starts hogging resources (e.g. blasting the AC and downloading huge files 24/7), everyone suffers. Salesforce prevents such "noisy neighbors" through **governor limits** that cap resource usage for each org. These limits apply to everything from how many database queries you can run in one transaction to how much memory your process can use. They ensure no single org (or automation) can monopolize the shared CPU, database, or network to the detriment of others in the same "pod."

Understanding governor limits isn't just an academic exercise – it's crucial for designing efficient declarative solutions. Even though you're using clicks not code, the flows, formulas, and other automations you configure execute on the Salesforce platform under the hood. They must play by the same rules as Apex code. For example, a **record-triggered Flow** that queries thousands of records will hit the same query row limits as an Apex trigger would. Salesforce will throw a runtime error if you try to retrieve more than 50,000 records in a single transaction – the infamous "Too many query rows: 50001" error. Similarly, a complex screen flow that performs dozens of queries or updates might hit the limit of 100 SOQL queries or 150 DML operations per transaction. If any limit is exceeded, Salesforce doesn't just slow down – it *fails* the operation and **rolls back the entire transaction** to protect data integrity (so no partial data gets saved if a limit triggers an error).

Beyond limits, Salesforce's architecture influences performance through **transaction management and resource allocation.** A *transaction* in Salesforce is a single unit of work that can involve multiple operations – for instance, a user saving an Opportunity could fire a flow that updates related records, sends emails, and creates tasks, all in one transaction. Salesforce will allocate a slice of CPU time and memory to that transaction. If the sum of all those actions takes too long or uses too much CPU, the platform will terminate it (governor limit on CPU time, typically 10 seconds for synchronous transactions). This is why seemingly "light" declarative tasks can slow down if chained together excessively. A best practice is to keep transactions lean – only do what's necessary in each save or each automation context – and offload heavier processing to asynchronous processes.

Another foundational concept is **selectivity of operations**. Salesforce employs a sophisticated query optimizer under the hood. If you design a report, list view, or flow filter that isn't *selective* (meaning it can't use an index and would have to scan a huge chunk of data), the platform may refuse to run it or take a very long time. Understanding which fields are indexed (e.g. record IDs, foreign keys, some formula fields if marked as External ID or unique, etc.) helps you craft filters that Salesforce can execute efficiently. We'll dive deeper into selective queries in the next section, but keep in mind that the platform is optimized for retrieving small slices of data quickly – it's *not* meant to dump your entire 5 million row table in one go.

Finally, let's touch on why knowing these fundamentals is key to performance: **if you know the rules of the platform, you can design within them instead of stumbling into them by accident.** An admin who knows about the 50k query limit will instinctively add a filter or limit to a flow's "Get Records" element when dealing with Large Data Volumes, avoiding an error. A business analyst who understands multi-tenancy will plan batch data loads during off-peak hours to be a "good neighbor" in the cloud. And a product owner who is aware of transaction limits might break a complex process into a few smaller steps, rather than one monolithic transaction that risks timing out. Knowledge of the platform's workings lets you **anticipate and prevent performance issues** instead of reacting to them after users complain.

Pro Tip: Salesforce's official limits can change with releases, so always refer to the latest documentation or use the **Limits Quick Reference** in Setup. For a declarative developer, it's also useful to know that many Apex governor limits apply to Flows and Process Builder too. If you suspect a Flow is hitting limits, consider using **Debug logs** with debug level set to "workflow." The log will show how many queries or DML operations the flow executed and any limits it approached. This is a great place to have ChatGPT help interpret logs – you can copy the relevant log lines and ask ChatGPT to summarize where the bottlenecks might be.

User (typed): "What are the most important Salesforce governor limits I should watch out for in my flows and automations?"

ChatGPT: "Some of the key governor limits to keep in mind are: (1) **SOQL query rows** – a maximum of 50,000 records can be retrieved in a single transaction. Exceeding this will cause a runtime error. (2) **SOQL queries** – up to 100 queries per synchronous transaction (this includes queries done by Flows, formulas needing to fetch data, etc.). Hitting this limit usually means you need to simplify or combine queries. (3) **DML operations** – up to 150 inserts/updates/deletes per transaction. If a declarative process tries to update records one by one in a loop, it can hit this limit; bulkify those operations. (4) **CPU time** – each transaction gets a maximum of 10,000 milliseconds of CPU time. Complex formulas or multiple flows firing together can consume CPU time and hit this limit. (5) **Concurrent Apex limit** – if you have too many simultaneous long-running transactions (like batch jobs or synchronous saves taking over 5 seconds), Salesforce may throttle them. This can happen declaratively if, say, a huge Flow batch runs while users are editing records. By being mindful of these limits and designing within them, you'll keep your org's performance solid."

To wrap up this fundamentals section: think of Salesforce as a shared environment where efficient, **governor-abiding** design is rewarded with fast performance and happy users. Neglecting these fundamentals can mean designs that work in small-scale testing break spectacularly in production. In the next section, we'll apply these principles specifically to data queries and retrieval, learning how to optimize queries when dealing with Large Data Volumes – a common scenario as orgs grow.

Next, we'll dig into optimizing queries for large data volumes – ensuring your reports, list views, and automation filters remain lightning-fast as your data scales.

OPTIMIZING QUERIES FOR LARGE DATA VOLUMES (LDV)

One of the most frequent performance pain points in Salesforce is retrieving data efficiently when you have **Large Data Volumes (LDVs)**. An org with 50 million transaction records or a nonprofit with 5 million donation records can easily run into query performance issues if searches and filters aren't designed carefully. This section will show you how to keep queries selective and fast. We'll cover how Salesforce indexing works, how to structure filters in reports, list views, and flows, and we'll illustrate best practices with examples from finance, healthcare, and retail domains.

When Salesforce has to fetch data – whether for a report, list view, or automation – it uses a **query optimizer** to decide how to get results. The optimizer's goal is to use an index to narrow down results whenever possible (just like how an index in a book helps you jump to the right page instead of reading every page). A **selective query** is one that uses filters that hit an index and reduce the result set significantly (Salesforce's rule of thumb: the filter should typically narrow down to <10% of the data, otherwise it might be considered non-selective). An **unselective (non-selective) query** is like asking for "everyone in the state of California" in a phone book – it's too broad, and Salesforce might have to scan too many records. In fact, Salesforce will outright **block** a query that is non-selective on an object with over 100k records, returning an error if you try to run it. This often manifests as a list view or report that just won't load, or a flow/Apex error "Non-selective query against large object type... consider an indexed filter" when a filter isn't tight enough.

So, how do we ensure queries are selective? Here are some key tactics:

Filter on Indexed Fields: Salesforce automatically indexes certain fields (Record IDs, OwnerId, lookup and master-detail fields, audit dates like CreatedDate, some formula fields marked as External ID, etc.). You can also ask Salesforce Support to add **custom indexes** on fields, or mark fields as External ID/Unique yourself to index them (with some limitations). Design your list view/report filters to use these indexed fields. For example, filtering Opportunities by CreatedDate THIS YEAR will use an index on CreatedDate, making it selective, whereas filtering by a non-indexed checkbox might not. *In a finance scenario:* say you have a Loan object with 1 million records. A report that filters loans by Loan_Status = 'Active' (assuming Loan_Status is a picklist not indexed) might be slow if 500k loans are active. But if you also filter by Region = 'West' and Region is indexed (perhaps a text field marked External ID), that combined filter could be selective (maybe only 50k loans in West are active, 5% of total). Always ask: "Is there a way to filter this data by something indexed?" If not, consider adding an index or redesigning the query.

Avoid Extremely Broad or Null Filters: Queries that look for "all records" or use NOT conditions (e.g. Status != 'Closed' or Field != NULL) tend to be non-selective. For instance, a healthcare org might have an Appointment object with 2 million records. A list view filter like "Show all appointments where Cancelled__c != TRUE" is basically fetching almost all records except a few – not selective at all. It's better to positively filter what you want (e.g. Status = 'Scheduled' OR Status = 'Checked In') which targets specific subsets. Also, avoid filters that allow null, like "Phone != null" when you have millions of contacts – that essentially returns "all contacts that have any phone," which in a large org is probably most of them. That was exactly the issue in one real case: a flow filter included "Phone is not null" and caused a non-selective query error. When data volumes are large, even a filter on an indexed field can fail selectivity if the value you're filtering on is extremely common (e.g. filtering on a picklist value that 80% of records have). In such cases, you might need to add another filter to narrow it down further.

Leverage Selective SOSL and Search where appropriate: If users need to search across text fields, using a **SOSL** search (via global search or API) might be faster because Salesforce's search engine can handle large volumes differently. For example, retail customer service reps looking up orders by customer name might get faster results using global search (which is indexed and optimized for name fields) rather than a list view filtered by a name field. However, keep in mind SOSL has its own limits and will only return the first 2,000 results by design – which is usually fine for user-facing searches.

Optimize Filters in Automation (Flows etc.): When building Flows that loop through records, always include **filters in Get Records** elements to fetch only what's needed. It's easy to accidentally grab "all records" of an object and then filter in the flow logic – a big no-no in LDV contexts. For example, an education nonprofit might have a nightly scheduled Flow to update student data. Instead of getting **all** student records and then checking which ones need updates, build the criteria into the Get Records (e.g. "Status = 'Graduating' AND Last_Modified_Date > Last_n_Days_1" to only get recent grads). Also, use the new Spring '25 feature to limit the number of records retrieved by Get Records if you only need a sample or top X records. And whenever possible, use **subfilters** (like in a Flow, you can filter related records via a subquery) to avoid nested loops and extra queries.

Use Query Plan and Index Tools: For Salesforce developers, the Query Plan tool in Developer Console can show if a SOQL query is using indexes and give cost estimates. Admins can achieve similar insight by using Salesforce Optimizer or the **"Analyze" button on list views/reports** (if available) to check if a filter is selective. If you suspect a report is slow, try copying its filters into a SOQL query (using Workbench or Developer Console) and prepend EXPLAIN to see how it's being executed. This might be more on the advanced side for an admin, but it's a good trick to work with a developer on. Alternatively, you can describe the scenario to ChatGPT and ask for advice:

User (voice): "I manage a hospital's Salesforce org. Our Appointment records (about 3 million of them) are slow to load in list views. One list view filters by Department (a text field) and Date (a formula field that equals the appointment date). It often times out. How can I optimize this?"

ChatGPT (responding via voice): "It sounds like your list view filters might not be using selective fields. First, check if the **Department** field is indexed – if it's a picklist or an external ID field, it could be, otherwise consider indexing it if you frequently filter on it. The **Date formula field** is likely not indexed (formula fields generally aren't indexable unless you request a special index and it's a deterministic formula). If possible, filter on a standard ActivityDate or a date field instead of a formula. For example, use a filter like Appointment Date >= THIS WEEK or a specific range, which might hit an index if Appointment Date is a normal date field. Also, ensure you're not showing too wide a range – filtering to the last 30 days of appointments, for instance, will be far more selective than showing all past appointments. Another tactic: create a **scheduled job** that tags or moves old appointments to an archive (more on that in the Archival section later) so that your primary object has fewer records to search through. In summary, use indexed fields for Department (or get it indexed), avoid filtering on a formula, and narrow the date range to make the query selective."

In that example, the advice touched on something we'll explore later – archival – but importantly highlighted indexing and filter scope. Let's solidify best practices: **Always aim to reduce the number of rows the query has to scan.** Use **AND conditions** with selective fields to progressively narrow down results. Avoid using too many OR conditions on different fields (which can confuse the optimizer and often prevent using indexes). If you have a complex filter logic, sometimes breaking it into two queries might be better (e.g. first query narrow set A, then filter A for subset B in memory), though this is more relevant in Apex than declarative.

Different industries often have different large data challenges:

- In **finance**, consider a bank with an object for financial transactions (maybe millions of records for payments). A report for compliance might need to pull transactions for a specific high-risk customer or a date range. The admin should ensure that report has a filter like "Customer = John Doe AND TransactionDate = LAST 90 DAYS" rather than just "Customer = John Doe" (which could span many years of data). Also, enabling **divisions** or **partitioning** data by year could help (Salesforce has a feature called Divisions to logically partition data – not widely used, but worth mentioning if truly massive data sets).
- In **retail**, a common case is an Orders object with every order placed online. A customer service agent might have a list view "Recent Orders" which should filter status (e.g. only "Open" orders) and date (e.g. last 6 months) to be selective. If they had a list view "All Orders Ever" it would rightfully never load if there are millions of orders – and that's good! Instead, maybe provide a filtered list by region or use dashboards that aggregate data rather than raw lists.
- In **healthcare**, perhaps you have a Cases object for patient support cases. If you accumulate hundreds of thousands of cases, make sure your automated processes (like an assignment Flow that routes cases) include selective criteria. For example, if a routing flow looks for cases that are unassigned, it should also

maybe only look at new cases (Status = 'New' and CreatedDate = TODAY) rather than scanning all unassigned cases in history.

One more tool in our toolbox: **Skinny Tables and Index Tables** – these are more behind-the-scenes and usually involve Salesforce support. But for extremely large objects, Salesforce can create a skinny table (a slim replica of the main object with fewer fields, optimized for certain queries) or you might use external big data systems. This is advanced and often not in the declarative realm, but as an admin or architect you should know these options exist if standard indexing isn't enough.

Pro Tip: Use the **"Report on Reports"** trick to identify slow reports. Salesforce has a standard Report called "Report Performance" (or you can create one on the Report Usage object) that shows how long reports take to run and how often they're used. If you see a report that consistently takes >30 seconds to run, that's a candidate to optimize filters or add indexing. Similarly, in Setup, the **"Query Monitor"** (if you have Event Monitoring, or via Workbench) can show you which queries are taking the longest or hitting limit warnings. These tools, combined with a bit of ChatGPT assistance to interpret them, can guide you to the hotspots in your org.

Optimizing queries is often the quickest win for improving Salesforce performance. Users notice when a list view loads in 2 seconds versus 20. By applying these LDV query optimizations – selective filters, indexing strategy, and savvy data partitioning – you can keep those wait times low even as data grows. In the next section, we'll look at the flip side: not just how to write good queries, but how to **avoid common configuration patterns that inadvertently cause bad performance.** These are the "anti-patterns" that can sneak into orgs and how to recognize and prevent them.

Next, we'll identify common performance anti-patterns in declarative design – and how to avoid or fix them to keep your org running efficiently.

RECOGNIZING AND AVOIDING PERFORMANCE ANTI-PATTERNS

Salesforce's declarative toolbox is powerful, but with great power comes great responsibility – it's easy to configure something that *works* functionally but quietly drags down performance. This section is all about those **anti-patterns** – design choices that seem convenient initially but can hurt your org's speed and scalability. We'll cover a variety of them: overusing formula fields, monstrous nested IF formulas, flows that update too many records at once (or repeatedly), too many roll-up summaries, and even user interface issues like overloaded page layouts and Lightning pages. Importantly, we'll discuss how to *detect* these issues (using tools like Salesforce Optimizer, or even ChatGPT's assistance) and how to refactor them into healthier patterns.

1. **Formula Field Frenzy (Too Many or Too Complex Formula Fields):** Formula fields recalculate on the fly every time a record is accessed or a report runs. They are incredibly useful for real-time calculations, but they come at a cost. Each formula field is effectively a little piece of code that runs when needed. If you have dozens of formula fields on an object, viewing a single record means the system might be crunching all those formulas at once. A particularly gnarly formula (say, a nested IF or a REGEX on a long text) can significantly slow down record retrieval. As evidence: formula fields are **not stored in the database**; they compute at read-time, which "can make things slow down when reading the records in some cases". Imagine an insurance company's Policy object with 40 formula fields computing various premium adjustments and risk scores – opening a Policy record could take a noticeably long time as the system evaluates all 40 formulas. The anti-pattern here is using formula fields for everything because it's "convenient," without regard to performance. The fix is to **use formulas judiciously**. For frequently needed, seldom-changing values, consider **pre-calculating and storing the value** (e.g. use a before-save flow or nightly batch to calculate a value and store it in a number field). That way, reading a record doesn't have to constantly recalc it. Also, break up very complex formulas. Instead of one formula with 10 nested IFs, maybe use a couple of intermediate formula fields or helper fields. However, note that referencing one

formula field inside another can also slow things (and can't be indexed, so don't use a formula field as a report filter if you can avoid it).

2. The IF/CASE Mountain (Nested Logic in Formulas or Validation Rules): This is related to the above. A deeply nested IF or CASE statement in a formula field or even a validation rule can chew up CPU time. Sometimes business logic gets very complex (e.g. a university admissions formula calculating eligibility could branch into 15 conditions). If you find yourself with a formula hundreds of characters long with nested conditions, consider alternatives. Options include splitting logic into multiple fields, or using a **Flow** or **Apex trigger** to calculate and set a value instead (especially if the logic doesn't need real-time recalculation and only changes on certain events). This moves the cost from "every record read" to "only when something changes," which is often more efficient.

3. Overactive Flows (Too Many Record Updates or Unnecessary Loops): Flows (and previously Process Builder and Workflow Rules) can inadvertently cause performance issues if not designed for bulk and efficiency. A classic anti-pattern is the "update loop": a record-triggered flow that updates related records one by one inside a loop. For example, a non-profit has a flow: "When a Donation is marked as refunded, loop through all related Payment records and update their status." If there are 100 Payment records, and the admin updates each with an Update Records element inside the loop, that's 100 DML operations – likely hitting the 150 DML limit if combined with other things, and certainly slower than necessary. The better approach is to **collect records in a collection variable and do a single Update** outside the loop (bulk operation). Salesforce has improved flow performance, but the admin must still design with bulk in mind. Always assume your flow could handle 200 records at once (since record-triggered flows process batches from bulk loads) and make sure it doesn't e.g. SOQL query or update inside a loop for each record. The anti-pattern is "N+1" queries or updates in flows. The pattern to use is "bulkify": queries before loops, and single updates after loops.

Another flow anti-pattern is kicking off too many automations in sequence. E.g. you have a record-triggered flow on Contact that updates the parent Account, which triggers an Account flow, which maybe triggers something on Opportunities... Suddenly one user edit fans out into a cascade of operations. This can consume a lot of CPU time and increase risk of locks. If you see chain reactions, consider combining logic or using a Platform Event to decouple (we'll get to that in async section). Use **Salesforce Optimizer** or even simply the Setup "Flows & Processes" list to audit how many flows/triggers are on one object – too many can be a smell. ChatGPT can assist by analyzing an exported Flow (via XML or screenshot) to identify inefficiencies:

User (upload): Uploads a Flow summary or XML file. *"ChatGPT, here's a description of a Flow: it loops through 500 Opportunity Line Items and inside the loop, it does a 'Get Records' to find the related Product record, then updates a field on that Product. Is this a performance risk?"

ChatGPT: "Yes – this flow is performing a SOQL query (Get Records on Product) inside a loop for each of the 500 line items, which means 500 queries. That will almost certainly exceed the governor limit of 100 SOQL queries per transaction. It's a performance anti-pattern. A better design would be to query all needed Products in one go before the loop (perhaps gather the unique product IDs from the line items, then do one Get Records filtering by those IDs). Then you can update the Products in bulk after. This way, you avoid repetitive queries and updates. The current design is likely to hit limits or at least slow things down significantly."

4. **Roll-Up Summary Overload:** Roll-up Summary fields are handy for aggregating child data (total Opportunities on an Account, etc.), but they have limits. Salesforce allows up to 25 roll-up summaries per object (and more by request in some cases), but just because you *can* have 50 doesn't mean you *should*. Each roll-up summary on a parent means that whenever a child record changes, Salesforce recalculates that roll-up. If an Account has 10 roll-up fields (summing different subsets of Opportunities, for example), then every time an Opportunity is inserted or updated, Salesforce does 10 calculations. That's overhead on the transaction. If you insert 200 opportunities in bulk for one account, the system will churn through all those recalcs. In a high-volume scenario (think a retail site where an Account might get hundreds of Order records a day), too many roll-ups can become a bottleneck. The anti-pattern is having lots of roll-ups where perhaps reports or on-demand calculations could suffice.

Best practice: keep roll-ups to the truly necessary aggregates that users need on records for operational purposes. For more analytical aggregates (like total orders last year), consider using nightly batch calculations or just running reports/dashboards. Another idea: use **Declarative Lookup Rollup Summaries (DLRS)** or Apex triggers which can sometimes be more flexible (e.g. roll up on criteria that standard roll-ups can't). But note, if not designed carefully, those can also impact performance.

5. Page Layout Bloat and Too Many Lightning Components: The user interface itself can be a source of slowness. Salesforce Optimizer often flags page layouts with an excessive number of fields. If your record pages have 8 sections with 100 fields total, it's going to load slower – especially if many are formulas or have inline calculations. In Lightning, the default recommendation is to keep the number of fields in the *Details* component to a reasonable number – Salesforce's own guideline is *about 60 or fewer fields on a page* for optimal load times. If you have more, consider using **Dynamic Forms** to show sections only when needed, or move some fields to a secondary tab. Likewise, each **Related List** on a page is another query behind the scenes. A page with 10 related lists (cases, orders, contacts, opportunities, etc.) will load all those lists (or at least the first batch of them) when the page opens. Best practice: limit to maybe 3 related lists on the main tab, and put others under secondary tabs.

Lightning pages also allow many custom components. We've seen "home pages" or record pages where overzealous builders toss in 15 dynamic components, charts, embedded reports, etc. It's rich, but it can be slow. A known guideline is to target no more than ~25 components on a page (and honestly fewer is better). Use **tabs and accordion** components to lazy-load content that isn't immediately needed. For example, on a Contact page for a university, you might have one tab for "Overview" (basic info, key related lists) and another tab for "Background" (which has a component showing all past interactions). If the user never clicks "Background," those components don't load, saving time. Salesforce Ben's tips on Lightning performance echo this: break up pages into tabs, keep the initial load light, and only load heavy components on demand.

6. Too Much JavaScript or Third-Party on Pages: This might be less in declarative scope, but if you've added a lot of third-party Lightning components or Visualforce with heavy scripts, they can slow things. Identify if any one component is a culprit by using the "Analyze" button in Lightning App Builder. The Page Analyzer will highlight components that take significant time or are outside best practices. If something is consistently slow, consider removing or replacing it.

How do we *discover* these anti-patterns in an existing org? A few tips:

Salesforce Optimizer: Running Optimizer (an in-app tool now) gives an interactive report. It flags things like "Too many fields on layout" or "Too many triggers on object" etc. For instance, it will warn if a Lightning page has a very high estimated load time, or if you have many unused fields (which indirectly helps performance if you clean them up). The Optimizer is a great starting point for a health check.

Event Monitoring and Lightning Usage App: These can highlight slow page times. The Lightning Usage App's "Slowest Desktop Record Pages" section shows which pages have the highest load times. Often a pattern emerges (e.g. perhaps your Case record page is consistently slow – time to trim it down).

User Feedback and Observation: Don't underestimate just talking to users or observing. If support agents joke that they "click Case and go get a coffee," it's a red flag that page or process is slow. Identify what on that page might cause it (lots of related lists? many formulas?).

Debug Logs for flows/processes: As mentioned, turning on debug logs for a sample transaction can surface "SOQL 101 errors" or show if multiple flows are firing. Look for repeated queries or long execution times in logs.

ChatGPT assistance: If you can export metadata (like a list of all formula fields on an object with their formulas, or an XML of a flow), you can have ChatGPT analyze it for complexity. For example, you could feed it a particularly hairy formula and ask, "Can this be simplified or is it too complex?" or "What is the computational complexity of this formula?" You might be surprised – ChatGPT could suggest a more concise formula or point out that it references 5 other fields (which themselves might be formulas).

Let's illustrate fixing an anti-pattern with a cross-industry example: A **nonprofit** org found that their Donations page was very slow for major donors. They had put 12 roll-up summary fields on the Account to show various donation stats (total this year, total last year, largest single donation, etc.). Additionally, the Account page had 7 related lists (Contacts, Donations, Payments, Activities, etc.) all on the main tab. This caused the page to take 8-10 seconds to load. The solution was multi-fold: remove a few non-critical roll-ups (they moved "largest donation" and some others to a report instead of a field), and redesign the Lightning page by moving most related lists to a secondary "Details" tab (only keeping a key "Recent Donations" related list on the main tab). They also enabled "Dynamic Related List" to only show certain donations (like only the last 5) instead of a long list. The result: the page load dropped to ~3 seconds, and user satisfaction improved. This shows how avoiding the anti-patterns (too many rollups, overloaded pages) and following best practices can yield tangible performance gains.

Pro Tip: When dealing with lots of automation, consider using the **"One Process per Object"** approach. Instead of multiple record-triggered flows on the same object, consolidate into one flow where possible. Fewer triggers/flows means less overhead per transaction. This also avoids order-of-execution confusion. If you must have multiple, document their purpose and ensure they don't do redundant checks. And always turn on **"Filter Conditions"** in flows to limit when they run. A flow that runs on every update but only sometimes does something is still consuming time each run. It's better to put a criteria so it only runs when needed.

By staying vigilant for these anti-patterns and proactively refactoring them, you keep your org healthy and performant. In summary: use formula fields wisely, simplify complex logic, bulkify flows, be cautious with too many roll-ups, and slim down the UI. Your users will thank you with higher productivity (and fewer complaints).

Now that we've talked about trimming and optimizing what you have, the next question is: what about data you don't need at your fingertips anymore? Old data can be a silent performance killer – it bloats reports, slows searches, and eats up storage. In the next section, we tackle **data archival strategies** – when and how to remove or offload data from Salesforce to keep things lean and fast, without losing what the business needs.

Next, we'll explore effective data archival strategies to manage data growth – ensuring old records don't weigh down your org's performance and scalability.

IMPLEMENTING EFFECTIVE DATA ARCHIVAL STRATEGIES

As Salesforce orgs mature, they accumulate vast amounts of data: years of cases, closed opportunities from a decade ago, outdated campaign members, etc. While Salesforce can technically store all this data, not all of it needs to live in the fast lane of your production org. **Archiving** is the practice of moving or removing older, less-used data to improve system performance, reduce storage costs, and meet compliance requirements without deleting information outright. In this section, we'll learn when to archive, what tools Salesforce provides (such as Big Objects), and how to implement archival declaratively. We'll also see examples from education and nonprofit sectors where archiving can be especially useful, and provide ChatGPT prompts to help design an archival policy.

Why archive? Over time, excessive data can lead to *sluggish reports, slow list views, and even longer backup and deployment times.* According to a Salesforce data management article, too much data can result in "prolonged report load times, sluggish processes, and potential system failures". Archiving improves performance by trimming the working set of data the platform has to sift through on a daily basis. It's like cleaning out a closet – the stuff you rarely use can be packed away, making it faster to find the stuff you do use regularly. Moreover, many industries have **compliance mandates** to retain data for X years, but not necessarily in the live system. Archiving allows you to meet retention policies by storing data in a cheaper or slower repository.

When to Archive: There's no one-size-fits-all answer, but common triggers include data that's *historical* (e.g. closed cases older than 3 years), *infrequently accessed* (e.g. old high-volume logs), or *out of scope for current operations* (e.g. a project's data after the project ends). In an **education** context, consider student records: a university might decide that once a student has graduated and 5 years have passed, their detailed records (course registrations, housing, etc.) can be archived. Current students and recent alumni remain in the main system for advising and services, but older alumni data is moved elsewhere to keep the system responsive. In a **nonprofit** scenario, perhaps donations need to be retained for 7 years for auditing, but anything older can be archived to an external storage. The active org then only contains the last 7 years of donations, which is plenty for daily fundraising operations.

What are the options for archiving in Salesforce declaratively? Here are a few:

Big Objects: Salesforce Big Objects are meant for storing massive data volumes within Salesforce in a somewhat "append-only" manner. You can create a custom Big Object, e.g. Archived_Case__b (Big object names end with __b). Big Objects have the advantage of being part of Salesforce (so you can use platform features like API to query them), but they have limitations: they must have a defined index, and you can only query them by that index (no fancy SOQL unless the fields are in the index). They're great for archival because you can throw billions of records in there and it won't count against normal data storage or slow down your regular objects. For example, you might archive Case records older than 5 years into a CaseArchive big object. You'd define an index maybe on Case Id or Year, and you can still retrieve those records via queries or custom Lightning components if needed (though not via standard UI easily). The process could be a **scheduled Flow** (or an Apex batch for heavy lifting) that once a month moves eligible cases to the big object and deletes them from core.

External Archives (Data Warehouse or External Objects): Many enterprises use external databases or data warehouses for archiving Salesforce data. This might involve extracting old data and storing it in, say, AWS S3 or an Azure database. Salesforce Connect (External Objects) could even let you *link* that data back in if needed via OData. Declaratively, you could use **Outbound Message** or **Platform Event** to send data to an external system as it ages out, or use an ETL tool (not purely declarative, but admin-friendly tools like MuleSoft Composer or Informatica Cloud can do it with low-code). For example, a healthcare org could push old patient visit records to a HIPAA-compliant archive outside Salesforce after 10 years, keeping Salesforce snappy for current patient info.

Thin the Data In-Place: Sometimes archiving is as simple as *deleting* data that is truly no longer needed. Salesforce now offers a feature called **Data Archive** (for Field History beyond 18 months, as an example) or you might just do an Export and Delete. If deletion is allowed (no need to keep the data), that's the simplest way – but many orgs prefer to retain data somewhere.

Scheduled Archival Flows: On the declarative front, you can set up a **Schedule-Triggered Flow** that runs say monthly. It can find records older than X (use a Get Records with filters like CreatedDate or ClosedDate < X) and then do one of two things: either hard delete them (caution: this is irreversible, but frees storage), or copy them to an archive location then delete. Copying might require an Apex action if writing to an external DB, or if writing to a Big Object (currently, I think big objects can be inserted via Async Apex, not sure if Flow supports directly – likely need an Apex invocable to insert to big object). Another purely declarative trick is to **export data to CSV and then use an external system** – for instance, periodically use the Data Export service, and then use an automation (maybe outside Salesforce) to pick up the files and store them.

A key part of archiving is planning: which objects, what criteria (age, status), and how to retrieve archived data if needed. Let's say we have an **education example:** A university wants to archive "Events" data (campus events, attendees) older than 2 years. They decide to use a Big Object called Archived_Event__b with fields for Event Name, Date, and a JSON blob of other details (one strategy to store flexibly). A scheduled flow runs monthly to find events with EventDate < "2 years ago". For each event found, the flow creates an Archived_Event__b record (maybe via an Apex action since big object insertion might require it), then deletes the original Event record. The archived data isn't visible in Salesforce UI by default, but if someone truly needs it, an admin could run a query or build a Lightning component or use an integration to fetch it. The result: the Event object in Salesforce stays a manageable size, keeping reports and filters fast, while older data lives in the archive.

For a **nonprofit example**, consider a charity that has a custom object "Donation__c" with millions of records over many years. They only report on the last 5 years for active fundraising. They might use an external archiving service (like Odaseva or Salesforce's own archiving tools) to move anything older than 5 years out of Salesforce. They keep a summary (like each Account's total historical giving) so they don't lose the insight, but the detailed old donation records are off the platform. This can drastically cut down report run times and storage usage.

It's worth noting that archiving isn't just about performance, it also helps in org **maintenance** – fewer records means faster deployments (e.g., filters and sharing recalculations have less to crunch through), and even things like weekly exports or sandbox refreshes become quicker. However, archiving should be done with caution: always back up data before deleting, and ensure stakeholders agree on what's considered "archive-worthy."

Let's engage ChatGPT to sketch an archival policy:

User (SORA): Uses ChatGPT's SORA video generation to create a short video for stakeholders explaining the archival plan. *"ChatGPT, generate a brief outline for a 20-second video explaining our Salesforce data archival strategy for old cases."

ChatGPT (video narration): "[Scene: Filing cabinets closing] In our Salesforce org, old support cases are getting a new home. Cases older than 5 years will be securely archived – moved out of the main system but kept for reference. [Scene: Speedometer rising] This means a faster Salesforce for everyday use – quicker search and reports. [Scene: Shield icon] Don't worry, archived data stays safe and can be retrieved if needed for audits. [Closing scene: Team celebration] By cleaning up old data, we keep Salesforce agile and our team efficient!"

The above is a conceptual voiceover; the actual SORA-generated video would depict the metaphor of archiving files and speeding up.

That example shows how you might communicate the concept. Now, practically, how do we implement? Suppose you are not comfortable writing Apex for moving data. You can use **third-party apps:** there are Appexchange tools for archiving which give point-and-click interfaces to select objects and criteria and then handle the data move (often storing in big objects or external store). If you prefer in-house, you might coordinate with IT to set up a database and use an integration tool (like MuleSoft or even a Python script periodically run) to transfer data out of Salesforce. Declarative-minded folks can still own the design (deciding criteria, scheduling, monitoring success) even if a bit of code is used to execute.

It's also important to define **how to retrieve archived data** if needed. Perhaps you create a Lightning page or tab that searches the archive (via an API call or Connect). Or simply have a policy: "Contact IT for retrieval of archived records within 30 days" or something. For many cases, retrieval is rare, but make sure legal/compliance is satisfied that you can get the data if required.

Pro Tip: Utilize **Field History Archiving** (if you have Field Audit Trail add-on, Salesforce automatically can move field history older than 18 months into a special archive big object). This is one less thing for you to worry about performance-wise. Also, if certain objects are just growing too fast, consider if **Chatter Feed** or **Email Messages** are piling up; those can be archived or purged with specific tools if they become a performance concern.

Before moving on, one more cross-industry anecdote: A large **financial services** firm had a custom trade transactions object with tens of millions of rows. Regular users only needed the last 2 years of data for trading analytics; older trades were for compliance storage. The firm used a combination of Big Objects and an external data lake. They archived anything older than 2 years to the data lake (accessible via a separate analytics tool) and also pushed a summary per account to a Big Object for quick reference in Salesforce (like "Lifetime Trades Count"). By doing so, their average report time in Salesforce dropped from 15 seconds to 3 seconds, and they avoided having to constantly increase Salesforce data storage (saving cost).

Effective archiving keeps your org lean without losing valuable information. It's a key part of scalability: plan not only how to grow, but how to **shed load** gracefully. With our data volume under control through archiving, we next turn to making sure heavy processing doesn't bog down users. That means leveraging asynchronous processes – letting the system handle big tasks in the background. In the next section, we'll examine how to use Scheduled Flows, Platform Events, and other async mechanisms to maintain snappy performance for end users even when doing a lot of work.

Next, we'll look at asynchronous processing and platform events – techniques to handle large-scale operations without impacting real-time user experience.

LEVERAGING ASYNCHRONOUS PROCESSING AND PLATFORM EVENTS

In a Salesforce org of any size, there will be tasks that are too heavy or slow to do while a user is waiting. Imagine a retail system where saving an Order triggers updates to inventory, sends a confirmation email, calculates loyalty points, and posts a message to an ERP system – doing all that *while* the user waits for the "Save" to complete would make for a terrible experience. Instead, Salesforce offers ways to do work **asynchronously**, meaning in the background, on separate threads, or after the main event is done. In declarative configuration, we can leverage things like **Scheduled Flows**, **Autolaunched flows kicked off by Platform Events**, and integrations with external systems that don't block the user interface. This section will explain asynchronous strategies and how to use them declaratively, with examples from retail and manufacturing where these patterns shine. We'll also show how you might have ChatGPT create an orchestration diagram or identify processes that are good candidates for async.

First, let's clarify the value of async (asynchronous processing). When you process something asynchronously, you essentially "fire and forget" – you initiate a process and allow Salesforce to handle it in its own time, separate from the original transaction. The immediate benefit is improved UI responsiveness: the user isn't stuck waiting. A secondary benefit is that asynchronous processes often have higher resource limits. For example, a **Scheduled Apex** or **Queueable** can run longer and use more CPU time than a synchronous save. Even in declarative land, a **Scheduled Flow** runs in a system context and can process batches of records with higher limits (like more SOQL rows) than a single trigger context might allow. Essentially, async lets you scale by doing more work behind the scenes without bogging down the front-end.

Declarative Asynchronous Options:

Scheduled-Triggered Flows: These are flows that you set to run at a certain time (daily, weekly, etc.). They're great for batch jobs – e.g. nightly data quality updates, recalculating scores, or archiving data as we discussed. From a performance perspective, scheduled flows help you *remove heavy lifting from interactive processes.* For instance, instead of calculating a complex "customer lifetime value" every time a record changes (which could slow down each save), you might have a scheduled flow recalc all customers' lifetime values at midnight. Users see updated values next day, and their save forms are fast because they didn't do that calc on the fly. Recent updates to Salesforce even increased the limits for scheduled flows (e.g. as of Spring '25, scheduled flows can handle more records per run with some higher limits on CPU/DML in async context). So they are quite powerful.

Platform Event–Triggered Flows: Platform Events are a publish/subscribe model within Salesforce (and even between Salesforce and external systems). You declaratively define a **Platform Event** (like an object, e.g. Order_Placed__e). You can then have a flow that subscribes to that event (flows have a trigger type "Platform Event" where you choose the event). When something publishes the event, the flow runs, processing the event message. The beauty performance-wise: the publishing and the processing are decoupled. Using our retail example, when an Order is placed (say an Order record is saved), instead of that record-triggered flow doing everything, we publish an Order_Placed__e event carrying the Order ID and maybe some details. That publish happens *after the order save commits* (Salesforce events are published after commit to ensure integrity). The user's Order save completes quickly. Meanwhile, the platform event is delivered to the subscribing flow, which runs in a separate transaction to handle, for example, updating inventory and sending confirmation. If that process is heavy or even fails, it doesn't directly affect the user's Order save (the user is not locked waiting on it). This pattern drastically improves perceived performance and decouples processes to reduce lock contention (since inventory updates happen separately, they won't hold locks that delay the order save).

Retail Example: A large retailer uses Platform Events to orchestrate order fulfillment. When an order is placed in Salesforce, a platform event "Order_Placed" is fired. One subscriber flow updates the warehouse stock levels (maybe by creating a "Inventory Reservation" record or calling an API to the warehouse system). Another subscriber (could even be an external system subscribed through CometD) sends the confirmation email and text message. These all happen asynchronously. The user who clicked "Place Order" in Salesforce gets confirmation almost instantly because Salesforce isn't doing all those things in the same transaction.

Manufacturing Example: Think of a factory IoT integration. Machines send readings that create "Sensor Event" records in Salesforce (or directly publish platform events to Salesforce). You don't want each of those triggers to immediately update dozens of related records (could be thousands of events per hour). Instead, publish a Platform Event for "SensorReadingReceived" that a flow or an Apex handler processes in batches. That way, the intake of events is lightweight (just record or event creation), and the heavy processing (like recalculating machine health or triggering alerts) is handled async. This pattern is much more scalable.

External Service Callouts Asynchronously: Salesforce has a feature for external callouts in flows (using "External Services" or recently an HTTP callout action in Flow Builder). However, callouts can be slow or unpredictable due to network. Ideally, integrate external calls in an async way. If using platform events, an external system can subscribe and respond. Or within Salesforce, you might use a **Future method or Queueable Apex** invoked from a flow for a callout. Declaratively, one might use an Action that enqueues an Apex job for callout, which immediately returns control to flow, and the Apex does the slow call in the background. The user isn't waiting for a response. (We mention

Apex here because pure declarative might not handle all scenarios, but an admin could install a ready-made Apex action for this.)

A key concept when designing with async in mind is **identifying what doesn't need to happen right away.** Not every process can be async; if the user must see the result immediately (e.g. calculating a quote price when they click a button), that's interactive by nature. But many things can be deferred. A rule of thumb: if it's a notification, an audit entry, a long calculation, or a heavy data sync – consider doing it asynchronously. Another consideration: use async to avoid hitting synchronous limits. For example, if you need to update 10,000 records based on a single trigger, you simply can't in one go (you'd hit DML limits). But you could fire off a batch or platform event that processes those in chunks async.

Prompting ChatGPT for Async Opportunities:

User (typed): "I have a process where when a Case is closed, we update 5 different related objects (custom objects for surveys, follow-ups, billing adjustments, etc.), and some of those have their own automations. Users sometimes see saving a Case take 5+ seconds. How can I improve this?"

ChatGPT: "This is a classic scenario for using asynchronous processing. Instead of updating all 5 related objects in the same transaction as the Case closure, consider publishing a **Case_Closed** platform event. Have separate **flows or handlers** for each of the 5 follow-up actions subscribe to this event. When a case is closed, commit that change and immediately publish the event. The user's Case close action will finish quickly (since we're no longer doing all those updates on the spot). Then, in the background, subscribers will handle updating the survey record, creating the follow-up task, making the billing adjustment, etc. Each of those will be a separate transaction (or grouped as needed), which also means if one fails, it doesn't rollback the others or the original case. This decoupling through events will greatly improve the user experience. Alternatively, for actions that can wait a bit longer, you could use a **Scheduled Path** in a record-triggered flow: for example, 15 minutes after case closure, execute some actions. That's another declarative way to defer processing slightly. By leveraging these async tactics, you reduce the immediate load and avoid long wait times on the UI."

ChatGPT's mention of **Scheduled Paths** is a good call – within a Record-Triggered Flow, you can actually add a scheduled action (like "X hours after trigger date/time"). It's not exactly the same as a full scheduled flow, but it's another declarative async tool. For example, in manufacturing, upon a "Production Order" status change, you might set a scheduled path to, say, 2 hours later, check if certain conditions are met and then do heavy updates (perhaps hoping that off-peak hours or just giving a buffer helps).

Creating Orchestration Diagrams: Sometimes it's hard to explain async processes to stakeholders or team members. A diagram can help. You could ask ChatGPT to generate a description or even attempt an ASCII/mermaid diagram of an event-driven process:

User (image generation): "Can you generate a simple diagram showing an Order save triggering a Platform Event, and two subscribers (Inventory Update flow and Confirmation Email) processing that event?"

ChatGPT: "Certainly! Imagine a flowchart:

User action: Save Order –> **Salesforce:** Order Record Saved –> **(Publish)** Order_Placed Event. From that event, two arrows go out: one to **Inventory Update Flow** (updates stock levels), another to **Notification** (sends confirmation email). Both of these happen asynchronously after the Order save. The user sees Order saved immediately."

If using an advanced markdown, ChatGPT might even output a mermaid code for the flow. In a real setting, you might use a tool like Lucidchart (which was even referenced in the Architects diagrams) to draw it, but the key is, ChatGPT can help conceptualize it.

Cross-Industry Asynchronous Use Cases: We touched on retail and manufacturing. Here are a couple more:

Finance: A bank's lending app – when a loan application is submitted, it might kick off credit checks, fraud screening, and document generation. Those can happen asynchronously. The user gets a "Application submitted" quickly, and then background processes do the heavy work. If something crucial comes up, they can update the record later (or send an alert).

Nonprofit: After a major donation is recorded, perhaps a sequence of actions like sending a thank-you letter, updating an external finance system, and alerting the donor's account manager need to occur. Those can be done via a platform event "Major_Donation__e" so that the main donation save isn't delayed.

Education: When a student applies online (creating an Application record), you might need to generate a student profile, send a welcome email, and queue them for an interview. Rather than one massive flow, you can do some immediate (e.g. create basic student record) and then spin off the rest asynchronously.

Cautions with Async: It's not a silver bullet. Asynchronous processes have their own limits (just higher ones). Also, because they run later, you need to design for eventual consistency – there's a slight delay, and multiple async processes might run in parallel. For example, two platform event subscribers could run at the same time on different servers, so if they both update the same record, you could get a collision (though less likely than in one transaction). There's also no guarantee exactly when a scheduled or queued job runs (Salesforce aims for scheduled flows to start close to their time, but it could vary). Salesforce Architects warn that async has "no SLA" – no guaranteed time. Usually it's quick, but not always. So don't use async where real-time is a must.

In summary, adding asynchronous techniques to your declarative toolkit massively improves scalability. You free up the user interface from heavy lifting and use Salesforce's ability to process in the background. This leads to more **scalable concurrency** (multiple operations can proceed without stepping on each other) and better user satisfaction. We have one more major consideration on that note: even with async, sometimes many users or processes at once cause contention – specifically, *record locking and concurrency issues*. In the next section, we'll focus on managing high user concurrency and minimizing contention, ensuring that as your org scales in number of users and processes, they don't lock each other out.

Next, we'll explore strategies for managing high user concurrency and avoiding record locking bottlenecks in a large Salesforce org.

MANAGING HIGH USER CONCURRENCY AND CONTENTION

Scalability isn't just about handling more data or tasks – it's also about handling more users simultaneously. In large Salesforce implementations (think enterprises with thousands of users or integrations running in parallel), you can run into issues where processes conflict with each other. The common culprits are **record locking**, **shared resource contention**, and something known as **data skew** that can exacerbate both. In this section, we'll look at how to minimize these pitfalls. We'll explain record locking and how to avoid it through design (like partitioning data ownership and careful automation ordering), and we'll illustrate with scenarios from nonprofits and education where multiple users or systems collide on the same data. We'll also include prompts for spotting concurrency bottlenecks with ChatGPT's help.

Record Locking Basics: Salesforce, like any database system, locks records when they are being updated to ensure data integrity. If User A is editing a record, and User B (or another process) tries to edit that same record at the same time, B will have to wait until A's transaction completes. Salesforce will queue B's action for up to 10 seconds; if the lock isn't free by then, B's action fails with an error "Unable to obtain exclusive access to this record". This is usually seen as the dreaded *"UNABLE_TO_LOCK_ROW"* error. In short, only one transaction at a time can hold a lock on a given record.

In smaller orgs, lock conflicts are rare, but as you scale up, you might have many users or automated processes touching the same records or related records. Common examples:

- Many child records updating the same parent (which causes a parent record lock).
- Integration jobs mass updating records that users also might be editing.
- One "hot" record (like a single config record) that lots of things update (e.g. a single "settings" record toggling statuses, etc.).

One notorious scenario is **data skew**. Data skew refers to an imbalance or concentration of records that can cause performance issues. Salesforce specifically points out having over 10,000 records associated to a single thing (one user or one account) as a threshold to watch. For example, if one **user** owns 50,000 contacts (ownership skew), updates to that user's role or sharing can lock up lots of contacts at once. Or if one **account** has 20,000 contacts (account skew), and you update that account, Salesforce will lock the account and possibly some contacts while it recalculates sharing, potentially causing other processes that need those contacts to wait/fail. The rule of thumb: avoid any design where one "parent" (be it an account, owner, or lookup parent) has an extremely large number of children. Salesforce recommends no more than 10,000 child records per parent in general.

So how do we manage concurrency and avoid locks? Some strategies:

Ownership and Data Partitioning: Spread out record ownership and parent-child relationships to avoid giant single bottlenecks. In an **enterprise nonprofit** setting, suppose you load all donations under a single master "Annual Fund" Account – that account will have hundreds of thousands of child donation records. If someone updates that account's name or ownership, it could lock all those donations (not to mention cause a huge sharing recalculation). A better approach: partition the data, e.g. create multiple accounts (perhaps one per year or per region) to distribute the child records. Or use a different logical data model to avoid one super-busy parent. Similarly for ownership skew: if you have an integration user that creates all records, that user will own, say, 1,000,000 records. That's ownership skew. To solve, you might use a **pool of users** (like 5 integration users each owning some portion) or periodically transfer ownership to relevant actual users or queues. In a **distributed education environment**, imagine a central admin user uploading all student records – instead, assign ownership to individual school coordinators so no one user is overloaded.

Avoiding Locking via Process Design: If you have automations that might collide, consider their order and triggers. For example, if you have two flows that both update the same parent record when children change, that's a lock contention risk if those children are updated in parallel. Maybe combine them into one flow or ensure they run serially via a controlling process. If an external integration is bulk updating records that users also edit, try scheduling that integration during off-peak hours or lower its batch size. You can also leverage **Granular Locking** (Salesforce automatically does some of this for certain objects like Campaigns) – for Master-Detail relationships, Salesforce can sometimes lock at a more granular level (the parent only, not all related). But you as an admin mostly influence locking by reducing simultaneous touches on the same records.

Record Access Design: A bit out of pure declarative scope, but how you configure roles and sharing can affect contention. For example, if a record's ownership changes and that triggers role hierarchy sharing recalculations for 50,000 child records, that's heavy. In large concurrency scenarios, some architectures even opt to remove roles from heavy-transaction users (so that sharing rules are simpler – this is noted in some data skew best practices). As an admin, you can ensure big data volume owners are lower in hierarchy or isolated.

To illustrate, consider an **enterprise-scale nonprofit** using Salesforce for program management. They have a "Volunteer Hours" custom object tracking volunteer time logs, with millions of records. Initially, they had one record type for all programs and a single user in charge of data entry – meaning that user owned all volunteer hour records (bad skew!). They also related all hours to one master Program record. This led to frequent lock errors when two staff tried to approve hours concurrently. The solution: they partitioned the data by program and region. Each Program got its own parent record, and volunteers log hours under the specific program. They also created separate owner queues for each region so ownership is distributed. This reduced the chance that two transactions hit the exact same parent or owner. In effect, they **partitioned the workload** both by record relationship and by ownership.

Another example in a **university** (education): student records were experiencing locks because multiple departments (admissions, registrar, housing) tried updating the same Student record concurrently (like during term start, many updates happen at once). To alleviate this, the university created child objects to store department-specific data so that, say, housing updates a Housing record (child of Student) while registrar updates a separate Academic record, reducing direct colliding edits on the Student parent. They also implemented a platform event: if a Student core field needs to be updated by multiple sources, instead of all directly updating Student (causing collisions), each source publishes an event and a single subscriber orchestrates sequential updates.

How do you detect if you have a locking/concurrency problem? Often, users will complain of random errors when saving ("unable to lock row") or flows failing with that message. You might notice this especially during peak usage or batch jobs. Tools:

- **Event Monitoring** has an event type for Locking errors (if you have that feature, you can see frequency of locks).
- Debug logs will show if a transaction was waiting on a lock.
- Salesforce also provides a "Concurrent Async Apex Limit" error if too many parallel operations happen; if you see those, you might need to stagger jobs.
- And as always, **Salesforce Optimizer** might highlight data skew situations (like a user owning too many records).

ChatGPT can help by analyzing patterns or log snippets:

User (typed): "I got a lot of errors 'UNABLE_TO_LOCK_ROW' when importing data. How can I find what's causing it and fix it?"

ChatGPT: "The 'UNABLE_TO_LOCK_ROW' errors indicate record locking issues. Common causes: you might be inserting or updating child records that tie to the same parent, and another process (or another batch in the import) is doing the same simultaneously. First, identify which object/record is locked – the error message often includes an Id of the record that couldn't be locked. If it's the same parent Id repeating, that parent is a bottleneck. To fix: If doing a data import, sort or chunk the data by parent to avoid hitting the same parent in parallel. If it's happening with users editing records concurrently, investigate if there's a design issue (like one account with many children). Possibly implement partitioning – e.g., split that one account into multiple. Another angle: If using flows, see if two flows might be fighting over the same record updates (like both trying to update the parent when a child changes). Merging such flows could help. Essentially, reduce contention by restructuring data or timing so that one transaction locks a given record at a time. You can also catch and retry logic in Apex if necessary, but declaratively the focus is on prevention by design."

Platform Cache and Caching Strategies: The prompt mentioned "effective use of caching mechanisms." Salesforce has a Platform Cache feature that lets you store data in memory for quick retrieval. How does this relate to concurrency? One use case: if you have reference data that many users or processes query (e.g., a conversion rate or a config setting), storing it in cache can reduce repeated queries and maybe reduce chances of locks on that reference record. For instance, if everyone reads a "Settings" custom object, that's fine (reads don't lock), but if many update it (like writing a log count to it), that's bad design – better to avoid frequent updates to a single settings record (that's a lock magnet). Instead, maybe use cache or a big object or some other way to track that info.

In high concurrency environments, sometimes a pattern is to redirect loads to "child" records rather than one parent. Example: instead of everyone updating one "Counter" record, have them insert child records and have a nightly job aggregate them. This avoids lock contention on the single counter record.

Minimizing DML Collisions: DML collisions basically mean two operations trying to change the same data. Minimizing collisions might mean queuing certain operations. If you have control over an external integration, you might ensure it doesn't overload a single record. If multiple systems need to update a record, consider having one system own it and others send requests to that system (hub and spoke).

Another trick: **skew mitigation via sharing settings** – if a skewed owner must exist, keep them out of roles or in a role where no one is above them (so no cascading sharing calc). Salesforce documentation suggests for heavy skew owners, put them in their own role at top of hierarchy with no subordinates , or even remove them from role hierarchy if possible, to limit impact of their record changes on others.

For concurrency, also consider **Gov. limits on concurrent threads:** Salesforce limits the number of long-running transactions and concurrent batch jobs. If you plan many async processes, make sure to schedule them thoughtfully (don't have 50 scheduled flows all at the same minute – stagger them).

In summary, handling high concurrency comes down to:

- Designing data model to avoid single points of contention (big fan-out relationships or one-user-owns-all).
- Staggering and decoupling processes so they don't all hit the same record at once.
- Using asynchronous design (from previous section) to separate transactions (which also helps reduce lock conflicts because operations happen sequentially or in smaller chunks).
- Monitoring and adjusting based on where locks occur.

By employing these strategies, you ensure that adding more users or parallel processes scales linearly rather than causing a traffic jam. A large nonprofit with volunteers worldwide or a university with simultaneous logins for course registration can maintain smooth operations because their Salesforce org is architected to distribute the load gracefully.

Now that we've covered how to design for performance and scalability – from queries to concurrency – our last step is learning how to **measure and monitor** it on an ongoing basis. After all, you can't improve what you don't measure. In the next section, we'll look at the tools Salesforce provides (Optimizer, Event Monitoring, logs, Lightning usage metrics, Health Check, etc.) and how to use them (plus ChatGPT) to keep a pulse on your org's performance health and continuously tune it.

Next, we'll focus on monitoring, measuring, and improving performance – leveraging Salesforce's built-in tools and analytics to keep your org running at peak performance.

MONITORING, MEASURING, AND IMPROVING PERFORMANCE

Performance management is not a one-and-done task – it requires continuous monitoring and tuning. Salesforce provides several tools to help you gauge how your org is doing and identify bottlenecks, from the

Salesforce Optimizer app to Event Monitoring logs and the Lightning Usage App. In this final section of the chapter, we'll cover how to use these tools step-by-step, and how to interpret their output to drive improvements. We'll also discuss how to set up a routine for org health checks (performance-focused), and of course, how ChatGPT can assist you in analyzing results or comparing environments. Think of this as establishing a "performance DevOps" mindset: regularly measuring key metrics and acting on them to keep your org healthy.

Salesforce Optimizer: The Optimizer used to be a PDF report; now it's an interactive app in Setup. When you run it, it scans your org configuration against a set of best practices. For performance, what does it highlight? A few examples:

Profiles & Permissions: too many of them can slow admin tasks, though not directly performance for end user, but complexity.

Fields on Page Layouts: It will warn if a page layout has an excessive number of fields (as we discussed) or if you have many required fields (affecting edit performance perhaps).

Apex and Flow: It might highlight flows with many elements or suboptimal configurations, or Apex issues (like too many future calls).

Lightning Pages: Newer versions can flag pages with long load times or suggest using the Analyze button.

After running Optimizer, you get a list of items with recommendations. For instance, it might say "Object X has 352 fields, consider reducing or auditing usage" or "Lightning Page Y takes longer than 5s to load for some users, consider reducing components." Each item has an explanation. This is a goldmine for quick wins. As a practice, run Optimizer after every major release or org change. It's like a diagnostics scan.

Lightning Usage App: This app (accessible from the App Launcher in Lightning) provides metrics on Lightning Experience usage and performance. Key things to look at:

Slowest Desktop Record Pages: It literally lists which record pages are slow (over 4 second EPT) and how often they're viewed. This helps you prioritize which pages to optimize first – maybe your Case page is the slowest, affecting 200 users daily, whereas an Account page might be slower but only used by 5 execs occasionally.

Browser and Network metrics: It shows if certain browsers are slower. For example, maybe a lot of users on an old version of Internet Explorer (hopefully not by 2025!) have poor performance. It might prompt an IT policy to update browsers.

Most Viewed Pages: If a page is very heavily used but borderline in performance, it's worth focusing on – small improvements there yield big overall gains.

The Lightning Usage App fosters a data-driven approach. You might discover, for instance, that a certain custom home page is quite slow – perhaps due to an embedded Wave (Einstein Analytics) dashboard. With that insight, you could modify that page for performance or train users on using a different view.

Event Monitoring (Performance Edition or add-on): This feature gives you detailed logs of various events, including page loads, API calls, reports run, etc. Of particular interest for performance:

ApexExecution and FlowExecution events: These can show CPU time and SOQL counts for transactions, which help find inefficient processes.

LightningPageView events: These record front-end performance data for Lightning pages (similar to what Lightning Usage shows, but raw data per load).

ReportExport or ReportRun events: If reports are timing out or slow, these logs can reveal which reports and how long.

URI events: Overall web requests, could surface if certain Visualforce pages or custom endpoints are slow.

An admin can use the built-in **Event Monitoring Analytics App** (if purchased) which has dashboards for some of this. Or you can manually download CSV logs for specific event types from Setup (Event Monitoring) or via API. These are rather granular; to make sense of them, ChatGPT can help. For example, you could paste a snippet of an ApexExecution event log to ChatGPT and ask it to explain which operation took longest.

Debug Logs & Analysis: When troubleshooting a specific slow operation, capturing a debug log is invaluable. If a user says "Saving Contact takes forever," you can log their save and see maybe a flow ran and took 8 seconds in loop, or an Apex trigger did heavy queries. The log has timestamps for each step. It's a lot of text, and this is where ChatGPT can be a hero. You could feed relevant portions of the log (being careful to omit sensitive info) and ask "Where are the delays?" or "Which SOQL query ran too many times?" ChatGPT can parse the patterns and point out, e.g., "It appears the flow 'Update Related Accounts' executed an UPDATE 150 times, which suggests it's doing one update per record in a loop."

Lightning "Analyze" Button: When editing a Lightning record page in the App Builder, there's an "Analyze" button that can predict page load performance (it uses something called Predicted Page Load Time) and gives suggestions. For example, it might say "This page has 8 related lists, consider reducing to 4 or fewer" or "The component X is taking a long time to load." Use this whenever you make a complex page. It's like having a little performance consultant built into the builder.

Health Check: Salesforce Health Check is mostly about security settings (password policies, etc.), but a "healthy org" performance-wise might mean also reviewing things like Indexes and ensuring up-to-date statistics. While not directly in Health Check, consider a periodic check of things like:

Indexes in use: Maybe use a query or ask ChatGPT to list fields that might benefit from an index (ChatGPT could suggest, but actual index creation requires Salesforce Support for some).

Selective vs non-selective query stats: If you have access to something like Query Analytics (some have used the Tooling API to find queries that used full scans), that can be advanced though.

Comparing Environments: If you have a Full sandbox or Staging org, it's good to compare performance with Production. Sometimes performance issues don't show until production (due to data volume differences). Using Sandbox Performance testing (maybe via Load testing tools) can help. ChatGPT can generate test data or test scenarios if needed ("simulate 100 users editing records at once").

Monitoring Routine: A best practice for enterprise thinking is to incorporate performance monitoring into your regular org maintenance:

- After each major deployment, do a quick performance test on key pages and reports.
- Monthly or quarterly, review Lightning Usage App stats and Optimizer again.
- If new large data sets are added (say you onboard a new business unit with 1 million records), watch how queries and page loads behave with that injection.
- Use sandbox testing plus the optimizer to pre-empt issues.

User Feedback Loops: Sometimes users spot issues before tools do. Encourage users to report slowness specifics ("This dashboard takes 2 minutes to run" or "Whenever I open Opportunity X it spins"). Those are canaries in the coal mine.

Let's illustrate improvement with a scenario: A **global nonprofit** noticed that during their annual fundraiser event, the system bogged down for staff entering donations. By looking at event logs and Lightning usage, they saw spikes in certain operations and locking. They realized the donation entry form triggered a cascade of flow updates (some of which could be deferred). By adjusting those flows (some actions moved to async paths) and scaling up their integrations to handle bursts (maybe temporarily turning off an on-save confirmation email to batch send later), they improved throughput. They monitored the next event in real-time with Event Monitoring dashboards to ensure response times stayed acceptable.

Finally, how can ChatGPT assist in ongoing improvement?

- If you get an Optimizer report (you can still export details), you could ask ChatGPT to summarize top issues.
- For a Lightning Usage CSV (custom report), ChatGPT could highlight which pages consistently slow.
- For comparing two sets of metrics (pre- and post-change), you can feed both and have it identify what improved or regressed.

- Use ChatGPT as a sounding board: *"I see that our Case page median load time is 5.2s, what are the likely causes?"* and it will likely list the usual suspects (too many fields, etc., which you can cross-check with the Analyze suggestions).

User (typed): "Here is an excerpt from Salesforce Optimizer results: 'Accounts Object: 55 fields on page layout (Recommended < 60). 12 report types with no records. 3 flows on Account object.' What should I do with this info?"

ChatGPT: "The Optimizer is indicating a few things. For Accounts, 55 fields on the layout is within recommended range (under 60) but close; it might be fine, but if users find the Account page slow, consider reducing fields a bit or using Dynamic Forms to hide seldom-used fields. The '12 report types with no records' is an admin cleanup thing – unused report types don't directly hurt performance, but cleaning them up reduces clutter. The '3 flows on Account object' suggests you have three separate automation processes on Accounts; it might be worth reviewing if all three are needed or if they can be consolidated. Multiple flows could mean slightly more overhead on each Account save. Focus first on user-facing performance items (fields on layout), then on org cleanliness (unused report types), and keep an eye on flows (maybe they're fine, but ensure they are efficient and not overlapping in functionality)."

Such assistance can help prioritize and clarify actions from these tools.

As we close this section, the theme is clear: **Measure, then improve.** Use Salesforce's tools to get facts about your org's performance, not just hunches. Every release, every significant data growth, make it a habit to assess performance metrics. A culture of performance tuning means you catch issues early (perhaps in a sandbox or small scale) and keep users happy with a snappy experience.

We've journeyed through a comprehensive set of best practices for performance and scalability. From understanding platform limits, to writing efficient queries, avoiding configuration mistakes, archiving old data, using async processing, preventing lock conflicts, and continuously monitoring – these practices ensure your declarative Salesforce solutions can grow with your business. In the chapter summary next, we'll synthesize these key points and prepare you for the next chapter, which will extend into maintaining org health over time (a related and continuing effort).

Finally, we'll summarize the key performance and scalability best practices from this chapter and set the stage for Chapter 13, which focuses on ongoing org health and maintenance.

CHAPTER SUMMARY

In this chapter, we tackled the critical challenge of keeping a Salesforce org performant and scalable as it grows. We began by establishing **Salesforce performance fundamentals:** recognizing that governor limits and multi-tenant resource sharing are not obstacles but guidelines that, when respected, keep your org (and everyone else's) running smoothly. By understanding transaction limits and platform architecture, you can design declarative solutions that work within these boundaries, avoiding runtime errors and slowdowns. We highlighted the importance of knowing limits (like the 50k query row cap and CPU time limits) and showed how ChatGPT can help explore those constraints in your context.

Next, we dove into **optimizing queries for large data volumes.** The key takeaway: use selective filters and indexes to make data retrieval efficient. Whether you're building a list view for a finance team's loan accounts or a report on hospital appointments, crafting queries that use indexed fields and return a modest subset of records is crucial. We saw that non-selective queries (e.g. filtering on unindexed fields or very common values) are blocked or slow in Salesforce. Practical tips included avoiding null/not null filters, breaking up pages with too many results, and using features like Skinny Tables or SOSL search for better performance in specific cases. Cross-industry examples (retail order lookups, healthcare record filtering, etc.) illustrated how these principles apply universally where data is big. By implementing these query optimizations, you lay the groundwork for an LDV-friendly org.

We then shone a light on **performance anti-patterns** – those innocent-looking configurations that can undermine performance. From overusing formula fields (which compute at read time and can slow record access), to building flows that iterate with excessive DML or SOQL, to loading pages with an overload of components, we identified common pitfalls. For each, we discussed remedies: e.g. minimize complex formulas or calculate values in advance, bulkify flow logic to avoid "N+1" query issues, limit roll-up summaries and use them wisely, and trim down Lightning pages by leveraging tabs, limiting fields (Salesforce's guidance ~60 or fewer on a layout) , and removing unnecessary components. We demonstrated how tools like Salesforce Optimizer and the Lightning App Builder's Analyze can catch these issues, and even how ChatGPT can assist by reviewing flow designs or formula text to suggest improvements. The overarching message: simplicity and focus in configuration lead to better performance – every extra calculation or component has a cost, so include only what provides clear value.

The chapter also stressed data management through **effective archival strategies**. We learned that keeping every record forever in Salesforce can degrade performance over time (large data sets slow down searches, reports, and even operations like deployments). To combat this, we explored how to archive or purge data that is no longer actively used but must be retained. Salesforce Big Objects emerged as a powerful declarative tool for on-platform archiving of massive data volumes. We saw how a scheduled flow or other declarative mechanism can migrate old records into a Big Object or external database, thereby freeing up the primary org. Examples from education (archiving old student records) and nonprofit (archiving donations beyond retention period) showed the real-world applicability. The key best practice is to establish criteria for archiving (e.g. age or status of records) and implement a routine process for it. By doing so, you ensure your active org remains lean and fast, while still meeting data retention needs.

In **asynchronous processing and platform events**, we addressed how to handle heavy processing without impacting users. The strategy of "fire and forget" via async means users don't have to wait for long operations to complete. We explained how scheduled flows can offload nightly or periodic tasks, and how Platform Events allow decoupling of processes – a user action triggers an event, and one or multiple subscribers handle the follow-up work in separate transactions. This pattern improves UI responsiveness and can increase throughput (since asynchronous contexts often allow more generous limits and parallelism). Retail and manufacturing scenarios demonstrated how complex workflows (order fulfillment, inventory updates, IoT data handling) can be scaled via events and background processing. The guidance here: identify processes that don't need to be synchronous and convert them to async – whether through standard platform features (like scheduled paths in flows or event-triggered flows) or with a bit of Apex (invocable actions for queueing jobs). The result is a more resilient system that can handle spikes and large workloads without timing out or frustrating the user.

We also covered **high concurrency and contention** management. As orgs grow in user count and integration processes, the risk of record locking conflicts rises. We learned why data skew (like one user owning >10k records or one account with >10k children) is harmful – it creates choke points that can lead to lock errors and slow performance. Best practices to avoid these issues include partitioning data (both logically and by ownership) to spread load. We discussed minimizing scenarios where multiple processes hit the same record at once – for instance, avoiding designs where a single configuration record is updated by everyone frequently. Through examples, we saw how nonprofits and universities can distribute record ownership or break apart monolithic record relationships to reduce collisions. The advice is to proactively structure data and automation in a way that, even under heavy simultaneous use, processes operate in parallel as much as possible without stepping on each other's toes. And if locks do occur, we now have strategies to identify and remediate them (monitoring for lock errors, using smaller transaction batches, etc.).

Finally, we emphasized **monitoring and continuous improvement**. A scalable org isn't set-it-and-forget-it; it requires ongoing attention. Salesforce's toolbox for this includes the Optimizer app (for config best practice checks), Lightning Usage metrics (for page performance and usage patterns), Event Monitoring logs (for deep technical metrics on queries, CPU time, etc.), and even simple tools like debug logs and the "Analyze" button. We laid out how to use these tools, and how to interpret their output to guide performance tuning – for example, using Lightning Usage to find the slowest pages and then simplifying those pages for a quick win. We also illustrated how ChatGPT can partner in this analysis, e.g. summarizing an Optimizer report or suggesting fixes for a flagged issue. The key habit encouraged is regular health checks: incorporate performance review into each release cycle and periodically audit the org as data and usage patterns evolve. By catching issues early (maybe a report starting to slow down or a flow nearing limits) you can address them before they impact users or become incidents.

Throughout the chapter, we included **ChatGPT prompt callouts** demonstrating how a Salesforce professional might use conversational AI in various modes – typed text for quick Q&A on limits, voice for hands-free problem explanation, file upload for analyzing things like flow diagrams or logs, SORA (video output) for generating stakeholder-friendly explainers, and even image generation for visualizing architecture. These examples underscore the theme of the book: that AI like ChatGPT can enhance a declarative developer's toolkit, providing guidance, validation, and creativity in tackling performance challenges.

In summary, ensuring performance and scalability in declarative Salesforce solutions comes down to mindful design and proactive management:

Mind the fundamentals: Always consider governor limits and multi-tenancy – design within those guardrails.

Efficient data access: Use selective queries and indexes; don't let data growth surprise you into slow queries.

Avoid known pitfalls: Simplify complex automations and pages; eliminate patterns known to cause slowness or limits errors (e.g. loops with DML, too many formulas).

Manage data growth: Implement archival and purging strategies so that your org's operational data set stays efficient.

Leverage async: Offload what you can to asynchronous processes to keep user interactions snappy and allow for larger scale processing in the background.

Design for concurrency: Partition and distribute to prevent lock contention; be aware of data skew and mitigate it in your data model.

Monitor and iterate: Continuously measure performance, get insights from Salesforce tools (and even AI analysis), and iterate on your configurations to address bottlenecks.

By following these best practices, you build an org that not only works well today but can handle the demands of tomorrow. The principles are cross-industry – whether it's a hospital system or a global sales CRM, the approach to performance is largely the same: know the system's constraints, design smartly, and keep a vigilant eye as things evolve.

As we transition to Chapter 13, we'll extend this mindset into overall org maintenance and health. Performance and scalability are one aspect of org health; the next chapter will cover how to maintain a clean, optimized, and low-technical-debt Salesforce org over time. You'll learn about governance, ongoing monitoring, and leveraging new features to continuously improve. Essentially, where this chapter armed you with tactics to keep things running fast, Chapter 13 will ensure they keep running *smoothly* and reliably in the long run.

With a firm grasp on performance best practices, you're now prepared to not only build powerful declarative solutions, but also to ensure they stand up to real-world scale – all with the help of thoughtful design and a bit of AI assistance along the way. Let's carry this proactive, enterprise-minded approach forward as we look at maintaining and monitoring org health in the next chapter.

13

MAINTENANCE, MONITORING, AND ORG HEALTH

It's tempting to think of a Salesforce implementation as "one and done" – build the solution and let it run. In reality, successful organizations treat a Salesforce org as a living system that needs care and feeding over time. Even the most elegant declarative build will accumulate some clutter and technical debt if left unattended. Consider a bustling kitchen after a long dinner service: without regular cleaning and organizing, the workspace becomes chaotic. In the same way, a Salesforce org requires periodic attention to stay efficient, secure, and aligned with business needs. This chapter embraces the philosophy of continuous improvement over a "set-and-forget" mindset, showing experienced admins and developers how to keep their org in top shape day after day.

Maintaining org health begins with knowing where you stand. We'll start by exploring how to **conduct regular Salesforce org health checks**. Just as an annual physical can catch issues early, routine org checkups can reveal unused features, performance bottlenecks, or security gaps before they grow into bigger problems. From built-in tools like Salesforce Optimizer to admin-savvy strategies, you'll learn how to systematically assess your org's wellbeing and prioritize fixes. This proactive habit not only prevents headaches but also sets the stage for everything else – after all, it's much easier to document and improve a system that you understand clearly.

Good documentation is often the unsung hero of a healthy Salesforce org. **In the second section, we'll discuss the importance of thorough documentation and best practices to achieve it.** Experienced professionals know that today's quick fix can become tomorrow's mystery if it isn't well-documented. We'll discuss techniques for capturing the "why" behind each customization, from field descriptions to process notes, and how doing so streamlines onboarding and troubleshooting. You'll even see how tools like ChatGPT can assist in generating and updating documentation – whether by voice dictation or summarizing an uploaded configuration file – making the process less tedious and more collaborative.

Of course, knowledge on paper (or wiki) only goes so far. Equally critical is **active monitoring of your org's usage and performance, the focus of our third section.** You'll learn how to set up monitoring dashboards and leverage Salesforce's own analytics (such as the Lightning Usage App and Event Monitoring) to keep an eye on user activity, adoption rates, and system performance. Think of this as the org's fitness tracker: with the right metrics, you can detect anomalies (like a sudden drop in login rates or spikes in data errors) and act before they impact the business. We'll walk through examples – like tracking which pages are slowest or which features go under-used – and discuss how ChatGPT can help interpret these signals. For instance, you might upload log data and ask the AI to highlight usage trends or even generate a quick visualization for a weekly report. By the end, you'll have a toolkit for staying ahead of issues through smart monitoring.

Another pillar of org health is managing change deliberately. In **section four, we turn to change management and configuration versioning.** Every experienced admin or developer has seen the chaos that ad-hoc changes can introduce – conflicting automations, forgotten workflow rules, or multiple teams unknowingly working on the same object. We'll cover strategies to bring order to this process: maintaining version control for declarative changes (yes, even Flows benefit from a changelog), deciding when to retire or consolidate older components, and using source control tools or DevOps Center to track modifications. This isn't just theory; we'll share practical tips such as naming conventions for versioned Flows and how to document changes so that nothing slips through the cracks. You'll also find out how you can enlist ChatGPT as a virtual release assistant – from summarizing differences between Flow versions to brainstorming deployment checklists – to ensure smooth, low-risk changes every time.

Hand in hand with change management is governance. Think of governance as the rulebook and referee that keep your org from descending into the Wild West. We'll discuss processes like evaluating new field requests through a governance lens: Do we really need this field, or does an existing one serve the purpose? Who approves its addition? How do we ensure it follows naming standards? By institutionalizing such practices, you prevent unnecessary complexity from creeping in. Real examples will show how a governance board or Center of Excellence can continuously review and refine standards (for data, security, release processes, and more) rather than treating governance as a one-time project. The payoff is huge – a cleaner org, less redundancy, and a team that's always on the same page. And if you're wondering how to make this less bureaucratic, we'll explore ways to use AI assistance for routine governance tasks, like using ChatGPT to draft a "new field request" questionnaire or even employing an Operator-style agent to flag rule violations automatically.

No Salesforce org lives in a static bubble of requirements. Business needs evolve, and so must your solutions. That's why **section six tackles the balance between flexibility and maintainability.** Salesforce's declarative power makes it easy to spin up new features or customizations on the fly – sometimes too easy. We'll examine the trade-offs between delivering exactly what users ask for versus maintaining a simpler, more supportable design. You'll read about scenarios where an ultra-flexible configuration (say, dozens of record types or nested automation rules) became a maintenance nightmare, and how to avoid those traps. Strategies like using modular design, setting boundaries on one-off customizations, and regularly refactoring overly complex setups will be highlighted. We'll show how you can even use ChatGPT as a design sounding board: describe an intended solution and have the AI critique its complexity, or perhaps generate a wireframe to visualize a simpler approach. Through these examples, you'll develop an intuition for when to say "yes, we can do that" versus "maybe there's a simpler way to achieve the core goal."

Finally, no chapter on org health would be complete without looking ahead. **In our seventh section, we discuss future-proofing your Salesforce implementations.** Salesforce evolves with three major releases each year, bringing new features and retiring old ones – and a top-notch admin/dev team stays ahead of that curve. We'll provide guidance on how to anticipate and embrace these changes: from migrating legacy automations to the latest tech (for example, moving Workflow Rules to Flow Builder as Salesforce phases out the former) to adopting features like Permission Set Groups that simplify user management and reduce complexity. You'll learn practical ways to keep your org modern and minimize technical debt accumulation – whether it's scheduling a "release review" every season, running pilots in a sandbox, or using ChatGPT to summarize long release notes into a targeted impact analysis. We'll even touch on new AI capabilities on the horizon (imagine leveraging OpenAI's Sora for auto-generating tutorial videos, or an Operator agent to handle routine org checks) that could become part of your toolkit for staying current. By cultivating a forward-looking approach, you'll ensure your Salesforce org isn't just healthy today, but primed to take on tomorrow's challenges.

By the end of this chapter, you will have a 360-degree view of Salesforce maintenance and optimization. From daily upkeep to strategic upgrades, from hands-on troubleshooting to AI-assisted foresight, we cover the best practices that keep enterprise orgs running like a well-oiled machine. Ready to begin? Let's start with the foundation of it all – knowing the state of your org. In the next section, we'll dive into **regular org health checks**, your first line of defense in maintaining a robust Salesforce environment.

CONDUCTING REGULAR ORG HEALTH CHECKS

Every seasoned Salesforce professional has experienced the unease of inheriting an org that's been neglected for too long. You log in and find dozens of unused fields, automation rules firing with no one sure why, and inconsistent configurations layered over the years. Regular org health checks are how we prevent getting to that point. This section explores how to systematically evaluate the health of your Salesforce org on a routine basis. Think of it as a proactive medical exam for your CRM: the goal is to catch issues early, clean up clutter, and ensure everything is running as intended.

Salesforce Optimizer: Your Org's X-Ray Machine. One of the best places to start an org health assessment is with Salesforce Optimizer, a native tool built to scan your instance for potential problems and improvement areas. Optimizer generates a report (now often accessible as an interactive app in Setup) that highlights things like fields with no data, overly permissive profiles, long-running Apex queries, and other red flags. In a matter of minutes, you get a bird's-eye view of your org's configuration hygiene. As an example, Optimizer might reveal that you have 20 custom fields on an object that haven't been populated in months, or perhaps several profiles that haven't been assigned to any users. These findings are pure gold for an admin focused on cleanup. The tool even provides recommendations for many items – if it flags an unused field, it might suggest removing it or at least hiding it from layouts. Salesforce Optimizer covers a broad range of areas, helping identify "problem areas" in your build such as unused fields, stale reports, or roles with no users. It's not perfect (you'll occasionally see false positives or minor suggestions that might not apply to your org), but it's a solid starting point.

Establish a Regular Cadence. Running a health check isn't a one-time event – it's a habit. Many organizations find that a quarterly review works well, aligning with Salesforce's seasonal release cycle. Others might do a lighter check monthly and a deep dive annually. The key is consistency. By scheduling regular org health checks (and sticking to the schedule), you create a feedback loop of continuous improvement. For instance, you might reserve the first Friday of each quarter for running Salesforce Optimizer and reviewing its findings, along with other checks we'll discuss shortly. Over time, you'll likely notice the reports get "cleaner" as you address recurring issues, which is a satisfying way to measure progress. Remember, **proactive maintenance saves time –**

fixing five small issues every quarter is much easier than confronting fifty issues at year-end.

Beyond Optimizer – What to Check Manually. While Salesforce Optimizer is comprehensive, it doesn't cover every aspect of org health. Seasoned admins often supplement it with their own checklist. Here are some additional areas to include in your org check-up:

Profile and Permission Creep: Review profiles, permission sets, and permission set groups to ensure they follow the principle of least privilege. Perhaps over the last few projects, certain profiles were given extra permissions "just to make things work" – now is the time to dial those back if they're not needed. Also, check for profiles with zero users and consider retiring them to simplify your security model.

Role Hierarchy and Sharing: Ensure your role hierarchy still reflects the org's reporting structure and that sharing rules make sense. Changes in company structure or team roles might necessitate updates. An outdated hierarchy can lead to either data not being visible when it should be, or worse, visible when it shouldn't.

Data Quality and Storage: Look at data volumes on key objects. Are there tens of thousands of old records that could be archived? How about files and email attachments consuming storage? Salesforce provides a **Storage Usage** page and **Report** tools (like Big Objects or reporting snapshots) to monitor growth. Identifying large data accumulations can prompt archiving strategies that keep performance snappy.

Automation Audit: Examine active automation – Flows, Process Builders (if any remain active), Workflow Rules, and Apex triggers. Check for overlap or contradictory logic. For example, you might discover two record-triggered Flows on the same object that could conflict or an old Workflow Rule that's technically inactive but not yet removed. It's best practice to keep only the automations you need and ensure they are up-to-date (especially given that older Workflow Rules and Processes should be migrated to Flow as part of your org's evolution).

By running through a list like the above in addition to Optimizer, you'll cover both the automatically detectable issues and the more context-specific ones.

Prioritizing and Remediating Findings. Once you've gathered findings from the health check, the next step is figuring out what to tackle first. Not all issues flagged are equally urgent. A missing description on a field is a mild annoyance, whereas an overly permissive profile or an automation error could pose immediate risks. One effective approach is to categorize findings into buckets such as "Quick Wins," "Medium-Term Fixes," and "Long-Term Refactoring." Quick wins might include deleting an unused report or adding help text to a misunderstood field – tasks that are low effort and low risk. Medium-term fixes could be things like deactivating an obsolete Flow (after confirming it truly isn't needed) or merging duplicate picklist values. Long-term refactoring might involve larger projects like a role hierarchy overhaul or re-building an overly complex automation in a cleaner way. By categorizing this way, you ensure that you capitalize on easy improvements immediately while scheduling the heavier lifts for when you can allocate proper time and resources.

Pro Tip: **Document your findings and actions.** Create a simple "Org Health Log" – a spreadsheet or a document – to record each health check cycle's results and what was done. This not only helps track progress over time but also serves as communication material to leadership about the ongoing investment in org quality. You might note, for example, that in Q1 you identified 15 unused fields and deleted 10 after stakeholder sign-off, in Q2 you optimized page load times by cleaning up a slow Lightning component, and so on. This log becomes a narrative of continuous improvement, demonstrating the value of proactive maintenance.

Leverage ChatGPT for Health Check Analysis. Handling the output of a thorough health check can be a bit like sifting through a medical report – there's a lot of information. This is where an AI assistant can lighten the load. For example, Salesforce Optimizer's PDF report might be lengthy, listing dozens of items. Instead of manually summarizing them, you can copy the relevant sections or export the data and ask ChatGPT to summarize or prioritize issues for you. *Imagine you have an Optimizer result showing 25 fields with missing data and 5 with questionable* *permissions settings.* You could feed that list (with sensitive details removed) into ChatGPT and prompt: *"Group these issues by theme and suggest which ones to address first and why."* The AI might respond with a helpful summary, e.g., "**Field Usage:** 15 fields have 0% fill rate – consider removing or repurposing these first as they add no value. **Security:** 5 permission sets grant admin-level access – review these immediately for compliance risk. **Performance:** 3 long-running Apex queries were flagged – these can be optimized in the next development cycle." This kind of AI-generated triage can save you time and ensure you don't overlook something critical.

ChatGPT Prompt Example: "I've run Salesforce Optimizer and identified 10 unused fields, 2 unused profiles, and 3 flows with best practice warnings. Suggest a prioritized action plan to clean up these items." – By asking ChatGPT in this way, you'll get a structured response that you can compare against your own plan. It's like having a second pair of eyes to validate your thinking. You might even discover a rationale you hadn't considered for prioritizing a certain fix. Always remember to double-check AI suggestions against your real-world knowledge, but often it will echo sound reasoning.

Incorporating Health Checks into Team Culture. The most effective org health programs are those that become part of the team's DNA. Encourage your admin and development team to view maintenance as an integral part of the development lifecycle, not an afterthought. This could mean including a "health impact" assessment whenever a new feature is built (e.g., will it add a lot of fields or complexity, and how will we mitigate that?), or making org health a standing agenda item in team meetings. Some organizations even gamify it – perhaps rewarding the "cleanup hero" who resolves the most Optimizer findings each quarter, or running a short "org cleanup sprint" with a cross-functional team. By making org health a shared responsibility, you avoid the scenario where everyone assumes someone else is taking care of it. Instead, much like a DevOps culture shift, maintenance becomes embedded in everyday work.

As you wrap up your org health check, you should have a clear understanding of the current state of your environment. You've identified what's working well and what needs attention, and you've planned next steps to remediate any issues. Crucially, you're doing this on a regular basis, which means problems won't accumulate unnoticed. This continuous care approach pays dividends in stability and agility – your org stays **lean, clean, and ready** for whatever comes next. In fact, one of the big takeaways from running these health checks is often a realization: "Wow, I need to write this stuff down so others (and future me) understand it." That epiphany leads neatly into our next topic. After all, findings and fixes are great, but if they aren't documented, you risk repeating mistakes or losing knowledge. In the following section, we'll dive into the art of documentation – how keeping good records of your org's configuration and decisions is itself a form of maintenance that yields long-term benefits.

IMPORTANCE AND BEST PRACTICES FOR DOCUMENTATION

Imagine walking into a library where none of the books have titles or labels. You could perhaps still find what you need by reading everything, but it would be painfully slow and prone to error. An undocumented Salesforce org is a bit like that library – all the knowledge is *somewhere* in the configuration, but without guideposts and context, it can take an inordinate amount of time to decipher. That's why documentation is so vital to org health. In this section, we'll highlight how clear, up-to-date documentation acts as both a compass and a safety net for admins, developers, and analysts. It ensures that knowledge doesn't leave when people do, and it helps even your future self remember why certain choices were made.

Documentation as Org Memory. One of the simplest yet most powerful forms of documentation in Salesforce is the description field on metadata components (objects, fields, flows, etc.). Every custom field you create has a description box begging to be filled in – but too often it's left blank. A best practice (which experienced practitioners swear by) is to always populate these descriptions with the *purpose* of the component and any important context. Don't just say "Field used in Project X"; say "Field used to capture Project X compliance score, needed for regulatory reporting. Updated via nightly Flow. (Added 2025-01, requested by Compliance team)". In other words, capture the **why**, not just the what.

Well-written descriptions explain the rationale behind a customization and how it fits into the bigger picture. This might feel like extra work in the moment, but it pays off immensely down the line. When someone new joins the team or when you yourself revisit the org after a year, those little nuggets of information save hours of digging and guesswork. In fact, with the rise of AI tools, having good metadata descriptions can even boost AI performance – one report noted that AI recommendations improve dramatically ("off the charts") when metadata descriptions are well-maintained. It turns out that documenting your org not only helps humans but machines too, which is a neat incentive in today's world.

What to Document – Beyond the Basics. We all know we should document fields and processes, but what does **comprehensive** documentation look like in practice? Here's a checklist of areas to consider documenting in your Salesforce org.

Business Processes and Requirements: For each major process (e.g. Lead to Opportunity conversion, Case escalation, etc.), maintain a brief description of how Salesforce supports it. This can be in a business requirements document or within a tool like Salesforce's built-in **Path** guidance or even an external wiki. The key is to tie configurations to business meaning.

Data Model and Key Fields: Keep a data dictionary that lists custom objects and fields, their meanings, owners, and any dependencies. If a field is calculated by a formula or automation, note that. If certain picklist values carry special significance, write it down. Modern org documentation apps (and even something like an **Elements.cloud** repository or a simple spreadsheet) can serve this purpose.

Automation Flows and Logic: Maintain a diagram or at least a bullet list of your active automations. For example: "Account Cleanup Flow (Screen Flow) – launched by button on Account page, guides users to merge duplicate accounts." Mention version numbers if multiple are in use. Documenting flows could be as simple as exporting a screenshot of the Flow and storing it with a few notes. There are also AI tools emerging that can ingest your flow XML or screenshots and produce an outline – consider leveraging them.

Integration Points: If Salesforce is integrated with other systems (ERP, marketing tools, etc.), document what data flows in/out and how (API, middleware, etc.). Note any scheduled jobs or critical external IDs. This helps troubleshoot when data doesn't sync as expected.

Security Model: Document roles, profiles, permission set usage, and sharing rules at a high level. For instance: "Profile: Sales User – baseline CRM access without export permissions. Field-level security locked down for pricing fields; extended via 'Pricing Access' permission set to those who need it." This kind of summary helps anyone quickly grok who should have access to what.

Remember that effective documentation is **concise and clear**. It's not about writing a novel for each item; it's about providing enough context that someone else (or future you) immediately understands the intent and impact of a configuration.

Pro Tip: **Adopt a documentation standard and stick to it.** Whether it's a naming convention or a template for notes, consistency is king. For example, you might decide that every Flow description will start with a one-line purpose, followed by "Created: [Date], by [Author]" and if relevant "Replaced [Old Process Builder X]". For fields, you might use a format: "Purpose: ...; Used in: [Processes/Flows]; Owner: [Dept or Contact]". By standardizing, you make it easier to parse documentation quickly. Plus, if you ever use automated tools to scan documentation quality, consistency will yield better results (as AI finds patterns easier when you use a consistent format).

Keeping Documentation Up-to-Date. Out-of-date documentation can be as misleading as no documentation. We've all opened a wiki page that says one thing, only to find the org behaves differently because the doc wasn't updated after a change. To avoid this, integrate documentation into your change process. Whenever a new feature is built or an existing one is modified, updating the related documentation should be a required step in your deployment checklist. If you're using agile methodologies, make "Documentation updated" a definition of done for each user story. Some teams even turn this into a quality gate: a build isn't considered complete until the associated documentation in Confluence, SharePoint, or wherever is updated. Another practical tip is to conduct periodic "docathons" (documentation hackathons). For instance, allocate one day every six months for the team to collectively review and polish the org documentation – fill in gaps, remove outdated sections, and improve clarity. This can catch those bits that slipped through the cracks during busy project work.

Speaking of keeping things current, ChatGPT and similar tools can be a lifesaver for documentation maintenance. Imagine you have a lengthy technical design document from a past project – rather than reading it end-to-end to update it, you could feed it into ChatGPT and ask for a summary or ask specific questions like, "Which parts of this process changed after implementing feature X?" You can even upload configuration files (like a profile XML or object metadata file) to ChatGPT and prompt it to generate documentation. Naturally, do this carefully – **ensure no sensitive data is shared** unless you're in a secure environment and have privacy settings locked down. Many organizations use ChatGPT's integration features or on-premise versions to analyze code and metadata safely. The benefit is that you can quickly produce draft documentation and then refine it.

ChatGPT Prompt Example: "Summarize the purpose and main steps of the 'Deal Approval' Flow from this JSON export." – If you export a Flow's definition or even copy the key elements into ChatGPT, the AI can outline what the Flow does (e.g., "This Flow checks if Deal Size > $100k, then routes for Finance approval, else auto-approves. It sends an email to the owner upon final approval."). This draft summary can then be used in your documentation repository. It's a clever way to document complex automations without starting from scratch.

Voice Interaction: Don't overlook the convenience of voice for documentation tasks. If you're more of a talker than a typer, try using ChatGPT's voice input (for example, via a mobile app or voice assistant integration) to **dictate** a description of a new feature or object, then have it transcribed and refined. You could say, *"We've added a custom object called 'Vendor Assessment' that tracks supplier evaluations. It's linked to Accounts and has an approval process for finalizing scores."* ChatGPT can take that spoken explanation and structure it into a neat paragraph or list. Later, you or a colleague can polish it and add it to the official documentation. This approach adds a bit of Elmore Leonard's easygoing conversational style to your writing process – sometimes speaking out loud helps make the description clearer and more straightforward.

Documentation and Onboarding/Transition Support. A huge payoff of diligent documentation is smoother transitions – whether onboarding new team members or handing over a project. Experienced Salesforce folks know the pain of reverse-engineering a predecessor's work. By maintaining good docs, you're doing your future colleagues (and possibly consultants or auditors) a favor. For instance, a new hire can get up to speed much faster if they have an architecture diagram of the system and a glossary of terms used in the org. Consider including a "Getting to know our Org" section in your documentation that highlights the most critical things a newcomer should read first (like core object model and key processes). It's akin to the introduction chapter of a book – it frames the story and prevents newcomers from getting lost in details prematurely.

Ensuring Accessibility and Usage. The best documentation is the one that's actually used. Make sure your documentation is stored in a place that everyone on the team can easily access. This might be an internal wiki, a shared drive, or even within Salesforce itself (for example, using the Lightning Experience help menu to link to internal pages, or a custom app that surfaces documentation). Tag or organize content by topic so people can find what they need with a quick search. And periodically, remind the team that the documentation exists! It could be as simple as referencing it in meetings ("According to our design doc for that integration, the data flows nightly...") or encouraging a culture where people check the docs before pinging a coworker with a question. Over time, as people trust and rely on the documentation, it creates a virtuous cycle: they'll be more likely to keep it updated since they see its value in their daily work.

Maintaining robust documentation might not be the flashiest part of Salesforce administration, but it's undeniably one of the most crucial for long-term org health. A well-documented org is like a well-charted map of a complex territory – anyone venturing in can navigate with confidence. You reduce the "it's all in my head" risk, decrease ramp-up time for new projects, and even improve collaboration with non-technical stakeholders (when they can read process documentation in plain language). Before we close this section, take a moment to reflect on your own org: if you left tomorrow, would those inheriting it have a guidebook or a scavenger hunt? With good documentation, we aim for the former.

Having emphasized the written (and spoken) word, let's shift our focus to actionable oversight. Documentation provides knowledge, but we also need continuous awareness of how the system is behaving in real-time. In the next section, we'll explore **org monitoring techniques** – the dashboards, logs, and tools that help you keep a finger on the pulse of your Salesforce implementation day by day.

EFFECTIVE ORG MONITORING TECHNIQUES

Knowledge is power, and in the context of Salesforce maintenance, that knowledge comes from monitoring the right things. Even a well-built and well-documented org can run into issues if usage patterns change or if something unexpected occurs. Think of monitoring as the nervous system of your Salesforce org – it alerts you to pain points (like performance slowdowns), unusual behavior (like a spike in data exports), or indicators of success (like high user adoption of a new feature). In this section, we'll discuss practical ways to monitor your org's health in real-time or near-real-time, using Salesforce's native tools and some clever AI assistance to stay informed and proactive.

User Activity and Adoption Dashboards. One of the first areas to monitor is how users are interacting with Salesforce. After all, an org is only as healthy as it is useful – if users aren't logging in or completing key activities, that's a red flag for adoption or training issues. Salesforce provides a great built-in tool called the **Lightning Usage App**. If you haven't used it before, it's an app (accessible from the App Launcher) that gives you statistics on daily and monthly active users in Lightning Experience, how often users switch back to Classic, what browsers they use, and even which pages are slow or most viewed. For example, you might discover that **5% of your users account for 50% of the switches to Classic** – indicating maybe a certain role isn't fully comfortable in Lightning and needs some attention or features not yet migrated. Or you might see that a particular Lightning page is consistently one of the slowest to load, prompting you to optimize that page or component. Adoption metrics such as daily active users, most visited pages, and feature usage tell a story about user engagement. By monitoring these, you can catch early signs of user frustration (like heavy Classic usage or low logins in a sales team) and address them with training or tweaks. Dashboards summarizing these metrics can be placed on an Admin Home page for ongoing visibility.

System Performance Monitoring. Performance issues in Salesforce can be subtle or sudden. While Salesforce as a platform handles a lot of heavy lifting, the way you configure things (complex formulas, many concurrent automations, huge reports, etc.) can impact the responsiveness for end users. Monitoring system performance involves keeping an eye on things like **page load times, Apex execution times, SOQL query counts**, and integration throughput. Some of this is available in the Lightning Usage App (e.g., identifying the slowest pages), and more detailed info can come from **Event Monitoring logs.** If your org has Salesforce Shield or Event Monitoring enabled, you gain access to nearly 80 different event types logging various aspects of the system's operation. For instance, there are event logs for page performance (tracking front-end load times), Apex executions (tracking CPU time and heap size), and API calls. By analyzing these, you might identify a specific Flow that often consumes near the CPU limit or a custom Lightning component that loads slowly due to heavy client-side processing. Setting up alerts or at least regular reports on key performance indicators is wise. A simple approach is to use Salesforce's **Report and Dashboard** on the Login History and Apex Execution objects for basic monitoring. For example, create a report of "Average page load time by application" and watch for upward trends. If you spot one, drill deeper to find the culprit.

Security and Audit Monitoring. Salesforce provides an array of logs that are invaluable for security monitoring. The **Setup Audit Trail** is a basic one – it logs all changes made in Setup (who changed what configuration and when). Regularly review this, especially in orgs with multiple admins or developers, to ensure no unauthorized or mistaken changes slipped through. More comprehensively, **Event Monitoring's security-related events** (like Login events, Logout, Export, etc.) can reveal patterns such as login attempts at odd hours or data exports larger than usual. As highlighted by security experts, proactive monitoring can catch things like an internal user suddenly exporting a large report of contacts (could be legitimate, could be a data leak) or multiple failed login attempts indicating someone's trying to guess a password. For organizations in regulated industries, tracking these events isn't just good practice – it might be required by compliance. You can use Salesforce's built-in **Health Check** (for security settings compliance) and set up notifications for when certain thresholds are crossed (for example, using a tool or custom code to alert if a user grants a high-risk OAuth token or if a field policy is changed).

Pro Tip: **Automate your monitoring where possible.** Instead of manually poring over logs every day, set up automated alerts for key conditions. Salesforce's Platform Events and Transaction Security (part of Shield) let you define policies, such as "if more than 10 records are exported by a user in a short time, alert Security" or "if a user logs in from a new country, notify the admin team." There are also third-party monitoring tools that can send email or Slack alerts for Salesforce events. By automating, you ensure that critical signals don't rely on someone remembering to check a dashboard – they proactively come to you.

Leveraging Reporting Snapshots for Trends. Some metrics are only meaningful in context over time. For instance, 100 open Cases might be fine if it was 150 last week (trend improving), but alarming if it was 20 last week (trend worsening). Salesforce's reporting snapshots allow you to take regular "pictures" of data and store them for historical trend analysis. Consider using a snapshot to track things like the number of open opportunities, backlog of approvals, or even config counts (e.g., number of active workflow rules vs flows). These historical metrics can be charted on dashboards to visually show trends. If you see a sudden jump or dip, it prompts further investigation. Maybe a big data load happened, or perhaps an automation failed to run. One could say this crosses into the realm of **operations analytics** – using data to ensure your processes are flowing smoothly.

AI in Monitoring: Interpreting the Data. With so many sources of data (dashboards, logs, events), one challenge is interpreting it all. This is where ChatGPT or similar AI tools can add value as a kind of intelligent assistant that helps make sense of raw numbers. Suppose you export a month's worth of Event Log Files for user logins and see thousands of rows of data. Instead of manually crunching that, you could ask Chat-GPT: *"Here is our login data for April (paste anonymized data or summaries): can you identify any unusual patterns or outliers?"* The AI might highlight that *"User A logged in from 5 different countries within two days, which is unusual,"* or *"There's a consistent drop in logins every Friday, which could indicate something about user behavior or system availability."* This doesn't replace formal analysis, but it gives you quick insights or at least starting hypotheses.

Another scenario: say you notice a performance issue intermittently. You gather logs and see no obvious single cause. You might describe the symptoms to ChatGPT: *"We experience occasional slow page loads around 3 PM daily for our Sales app. What could be some reasons or where should I look?"* A well-trained AI might respond with plausible explanations: *"It could be a scheduled job (like a large data export or batch) kicking off at that time, consuming resources. Or perhaps that's when many users are logging activities simultaneously causing record locking or sharing recalculations."* It might then suggest checking the Apex Job schedule or the sharing recalculation queue. Essentially, the AI can act as a brainstorming partner, ensuring you don't overlook less obvious angles.

ChatGPT Prompt Example: "I have an event monitoring log of all data exports in the last week. Please list any users who exported significantly more data than others and any days that stand out." – With a prompt like this, after you feed in the relevant log data (perhaps aggregated), ChatGPT could produce a neat little report: "User X exported 5,000 records on Tuesday (others averaged 200). User Y had high exports on Friday. Overall exports spiked on Tuesday." You can then take that info and verify it, maybe cross-checking if User X had a legitimate reason (like a scheduled report) or if it's something to investigate.

Monitoring for Feature Adoption. Beyond raw usage and performance, consider monitoring how new features or changes are adopted. For example, if you roll out a new custom page or process, you can monitor its usage via custom logging (perhaps a record is created each time the process runs) or by simple metrics like record counts. If you introduced a new "Customer Feedback" custom object and expect each Sales rep to create one record per week, you can quickly report on how many feedback records each rep created this month. This isn't about catching people out, but about seeing if the process is working or if additional enablement is needed. If some teams have zero records, maybe they didn't receive adequate training or the feature doesn't fit well in their workflow. Monitoring adoption helps close the loop between design and actual usage.

Visualization and Wireframes for Analysis. Sometimes, describing a system issue or usage pattern in words is hard – a picture might be more effective. While Salesforce provides charts in dashboards, you might occasionally need a custom visualization (like a specialized graph or a UI wireframe to debug a layout issue). ChatGPT's advanced features (especially with Code Interpreter or integration with tools) can potentially help create quick charts. For instance, if you have CSV data of API call counts per hour, you could use ChatGPT to generate a line chart image to spot surges. On the UI side, if users complain about a confusing layout, you might sketch a proposed new layout in text and ask ChatGPT (with an image generation capability or via a design plugin) to produce a wireframe. While this is a bit experimental, it's increasingly feasible. The idea is to leverage AI not just for text, but for visual aids in monitoring. A wireframe could help communicate a UI improvement idea to the team, and an AI-generated chart might reveal correlations in monitoring data that aren't obvious in tables.

In summary, effective org monitoring is about being **proactive and informed.** Rather than waiting for a user to report "Salesforce feels slow today" or "I can't find data I need," you have systems in place to catch those situations early. By combining Salesforce's native tools (Lightning Usage App, Event Monitoring, dashboards) with smart analysis – potentially boosted by AI – you create a safety net for your org. This keeps you in a preventative stance rather than a reactive one.

Armed with the insights from monitoring, you can ensure your org not only stays healthy but also evolves in the right direction. For instance, monitoring might reveal that a particular process is heavily used and slow – which could feed into a decision to optimize or refactor it. And that brings us to our next topic: how to manage changes and versions of your configuration to continuously improve without breaking what already works. In the following section, we'll tackle **managing change and configuration versioning**, where the focus is on maintaining order in the midst of evolution.

MANAGING CHANGE AND CONFIGURATION VERSIONING

Change in a Salesforce org is constant – new business requirements, process improvements, and Salesforce releases all drive the need to modify your configuration. However, uncontrolled change can be more damaging than no change at all. In this section, we explore strategies to manage changes methodically and maintain version control over your declarative configurations. The goal is to avoid the "wild west" scenario where multiple people deploy updates haphazardly, and instead institute a disciplined approach akin to what software developers use for code. For declarative development, this means tracking the evolution of things like Flows, validation rules, and page layouts, and having a clear process for making, reviewing, and backing out changes if needed.

Treat Declarative Config Like Versioned Code. One mindset shift that has gained momentum in recent years is to handle your metadata (declarative configs) with the same care that developers handle source code. This doesn't mean admins need to become Git experts overnight, but it does mean establishing a system of versioning and backups. Salesforce has provided tools like **Change Sets** historically, but now more modern tools like **Salesforce DevOps Center** and third-party solutions allow tracking changes in a source control repository. If you're not already using something like DevOps Center, consider it – it's a free Salesforce-provided app that streamlines change tracking and integrates with Git behind the scenes. By moving to a source-driven approach, you automatically gain version history. That means if someone accidentally deletes a key component or a change introduces a bug, you can compare with a previous version and even roll back if necessary. It's peace of mind in a complex org.

However, even without sophisticated tools, you can implement basic version control practices. For example, maintain a manual changelog (a simple spreadsheet or document) where every admin or dev logs what they changed, when, and why. It sounds onerous, but it's incredibly useful for troubleshooting. "Oh, Leads stopped syncing to Marketing Cloud on Tuesday – what changed on Tuesday? Let's check the changelog. Ah, someone modified the Lead mapping." This beats scouring Setup trying to guess what might have impacted the behavior.

Flow Version Management. If there's one declarative component that absolutely needs version control, it's **Salesforce Flow**. Flows are powerful and often mission-critical, and Salesforce allows you to keep multiple versions of a Flow. Best practice is to always create a new version when modifying a Flow (rather than overriding the active one immediately) so you can test it and potentially roll back by re-activating an older version if something goes wrong. However, orgs can accumulate a dizzying number of Flow versions over time – sometimes dozens per Flow. Salesforce currently permits up to 50 versions of a single Flow. It's wise not to let it max out; otherwise you'll be forced to delete old versions to make new ones. A strategy here is to periodically review and prune Flow versions. Keep perhaps the last few major versions for safety, but delete truly obsolete ones that you'll never revert to (especially if they're inactive and the business logic has significantly moved on). Also, **naming conventions** help a ton. Consider including version numbers or dates in the Flow label (not API name, just label) for quick reference, e.g., "Opportunity Discount Approval v3 (2025-05)" as the flow name. In descriptions, note changes per version: "v3 – added region-based approval steps; v2 – initial release with basic approval". This way, even within Salesforce, someone can quickly see the progression.

When should you retire an old version? A good rule of thumb is: if a Flow version is no longer in use (no pending paused interviews, and you've run a couple cycles of business with the new version with no issues), it's a candidate for cleanup. Also, if you've cloned a Flow or replaced it with a new Flow entirely, consider deactivating and later uninstalling the old one after a safe period. Keeping a bunch of inactive flows around clutters your org and can confuse new team members.

Other Declarative Versions to Track. Flows are the most obvious, but think about other things that change: validation rules (did we change the logic? when? who asked for it?), approval processes, even page layouts ("we added these 3 fields to the Contact layout in May to support a new initiative"). For complex orgs, a lot of these changes are captured in project management tools or release notes, but it's helpful to maintain an internal history too. Something as simple as adding a text note on a page layout's description, like "Apr 2025: added Region and Tier fields – requested by Sales Ops" can save confusion later if someone wonders, "Why do we even have this field on the layout? Who uses it?" Essentially, don't be afraid to **annotate your configuration** with version info and rationale. These annotations act like breadcrumbs for future maintainers.

Pro Tip: **Sandbox and Deployment Discipline.** One of the best ways to manage change is to have a clear sandbox strategy and deployment process. For example, you might use a dedicated **UAT sandbox** where all changes for the next release are assembled and tested by users before going to production. In that sandbox, encourage testers and builders to note down any issues or adjustments in a shared log. By the time you deploy to production, you have both the final change set and a record of what changed. Additionally, always deploy using a metadata deployment tool (Change Sets, DevOps Center, CI/CD pipeline) rather than making ad-hoc prod changes for anything significant. This ensures changes are tracked and can be repeated in another org if needed. Ad-hoc changes are sometimes necessary (a quick field edit for an urgent fix), but treat them as exceptions and document them immediately when they happen.

Utilizing Source Control for Declarative Config. Modern Salesforce development encourages the concept of **Source of Truth = Source Control** (not the org itself). If you have developers on the team, you might already use Git for code. It's increasingly feasible to use it for declarative parts too, especially with Salesforce DX. Admins can get involved by learning to retrieve and deploy metadata using tools like VS Code or Salesforce CLI, which isn't as scary as it sounds for key items. Alternatively, a middle-ground approach is using a GUI tool or a service like Gearset, Copado, or Flosum that offers version tracking. The benefit cannot be overstated: you get a timeline of every change to your metadata. For example, if a profile's permissions mysteriously change, you could diff the profile metadata in Git and pinpoint exactly when and who changed it. In a fast-moving environment with multiple contributors, this traceability is a lifesaver for auditing and accountability.

Now, how can ChatGPT assist in change management? A few ideas:

Change Summaries: Let's say you have two versions of a flow's XML (one from last month and one from this month). You could feed the textual differences to ChatGPT and ask for a summary: *"Summarize what changed between Flow version 2 and 3."* It might reply, "Version 3 adds a new decision branch for high-value deals and an email alert; removes the obsolete discount approval step." That's a quick way to generate release notes content or documentation of changes.

Naming Suggestions: If you're struggling with naming conventions or consistent labels, ask: *"What's a clear naming scheme for multiple Price Book layouts for different user profiles?"* ChatGPT could propose something like "Price Book Layout Sales, Price Book Layout Admin" etc., following a pattern.

Backup Verification: You might even script or use an API to extract a list of recent changes (via Setup Audit Trail or metadata) and then use ChatGPT to explain them in plain language for a non-technical stakeholder report.

ChatGPT Prompt Example: "We plan to replace our old 'Deal Approval' Process Builder with a new Flow. Draft a deployment plan ensuring we don't disrupt business operations." – The AI might enumerate steps: "1. Create the new Flow with all required logic. 2. Test in sandbox with sample records. 3. Schedule a deployment during off-peak hours. 4. Before deployment, deactivate the Process Builder in production (to avoid parallel logic) then activate the new Flow. 5. Post-deployment: monitor approvals closely for one business day and have a rollback plan (i.e., re-activate Process Builder) if issues arise." This kind of high-level plan can serve as a starting checklist to refine with your team. It's especially useful for junior admins to validate their thinking or for anyone to ensure they didn't miss a step.

Preventing Configuration Conflicts. In multi-admin or multi-team orgs, one headache is conflicting changes – for example, two people edit the Opportunity page layout at the same time, and one overwrites the other's changes unknowingly. Good communication is one solution (like change request tickets and clear assignment of tasks), but technical safeguards help too. Source control, as mentioned, will flag conflicts if two changes collide in the metadata. If you're not at that stage, a simpler approach is to "section off" areas of work: e.g., Admin Alice handles Sales Cloud configs this sprint, Admin Bob handles Service Cloud – or one works in one sandbox, another in a different sandbox, and you merge via deployment. Also consider enabling **DevOps Center or Change Sets with a pipeline** where changes are collected and merged deliberately rather than everyone clicking in production.

Retiring Old Configurations. A critical aspect of change management is knowing when to deprecate or remove what's no longer needed. Hanging on to every field "just in case" or leaving old automation deactivated but present can cause confusion and even performance issues. A best practice is to retire configs in stages: first deactivate (but leave in place with a note "to be deleted after X date if no issues"), then after a safe interval, completely delete it from the org. If you worry about hard deleting (for example, fields with data), consider taking an extract of the data or metadata for archive before deletion, so you have a backup if you ever needed to reference it. When Salesforce announced workflows and process builders will be retired, it signaled to many orgs that it was time to migrate and clean house. Taking action on such guidance is a hallmark of an org with good change management – it means you're planning ahead, not scrambling at the last minute.

As we refine our change processes, we naturally bump into the broader concept of governance – ensuring that changes align with a controlled vision and compliance. We've touched on internal team process here, but how do we ensure all this is happening under a unified strategy? That's where governance comes back into play. In the next section, we will examine **integrating governance with ongoing maintenance**, effectively tying together the proactive maintenance, documentation, monitoring, and change management practices under a coherent framework of rules and guidelines. Governance will provide the guiding hand to all these efforts, ensuring that as changes happen, they happen in a responsible, strategic manner.

INTEGRATING GOVERNANCE WITH ONGOING MAINTENANCE

The term "governance" might conjure images of committees and red tape, but in a Salesforce context it's really about *structured decision-making and oversight* to keep your org sustainable. Earlier in the book, we discussed the concept of a governance framework or even a Center of Excellence – now it's time to see how that framework operates continuously, not just as a one-time setup. Integrating governance with ongoing maintenance means your org's rules and best practices are actively enforced and revisited as living principles, not shelved documents. This ensures that every new field, every process change, every integration is evaluated not only on its immediate merits but on its long-term implications for org health.

Start with Guiding Principles. A well-governed Salesforce org typically has a set of guiding principles or a charter. For example, a principle might be "No duplicate fields for the same purpose – data is entered once and reused" or "All automation must be documented and have an owner." Such high-level rules set the tone. When maintenance tasks come up, like cleaning fields or approving a new app, these principles guide decisions. Let's say a sales manager requests five new checkbox fields because they think it'll help track something. A principle-driven governance approach would ask: do these represent new data points or ones we could capture with existing fields? Is there a way to achieve this with configuration (perhaps a child object or a single picklist) rather than adding five separate checkboxes? It's almost like having a little architect on your shoulder asking, "Does this align with our standards?" By filtering requests through agreed principles, you reduce knee-jerk additions that later contribute to bloat and confusion.

Governance Board and Change Approval. If your org is large enough, you might have a formal body (a Steering Committee or Center of Excellence team) that reviews major changes. Integrating this with ongoing maintenance means that even the routine things (like adding a picklist value or installing an AppExchange package) follow a lightweight approval process. This doesn't have to be heavy. It could be as simple as a weekly meeting where admins/devs bring up upcoming changes and get a quick thumbs-up, or a Jira ticket that needs a "review by architect" before implementation. The idea is to avoid the scenario of one team implementing something that inadvertently conflicts with another team's domain. For example, adding a field to an object that another department was already using differently. A governance review can catch such overlaps: "We already have a field called 'Client ID' over here, maybe you should reuse or at least coordinate naming." It's far better to have that conversation beforehand than to discover it later when reconciling reports.

Speaking of requests, one practical governance element is a **New Field Request form or process.** Many organizations require that anyone asking for a new field (especially on key objects) fills out some details: Why do you need it? What options were considered? How critical is it? Will it be used in reports? This can be a simple form that the admin team evaluates. It forces a bit of due diligence and often uncovers that the field might duplicate something or isn't truly needed. Research has shown that up to 50% of built features in Salesforce (fields, objects, etc.) are never used – a statistic that underscores how important it is to scrutinize "Can we add a field?" requests. By ensuring new fields (and other components) go through a review, you significantly reduce the chance of adding unused clutter.

ChatGPT Prompt Example: "Draft a checklist for evaluating a new custom field request in Salesforce." – An AI could generate items like: "1. Check if an existing field serves this purpose. 2. Identify the data type and ensure it aligns with the intended use. 3. Determine who will populate and use this field (user roles). 4. Assess impact on page layouts and reports. 5. Consider data volume and integration impacts. 6. Define clear field name and description per standards." Such a checklist, once refined, becomes part of your governance toolkit and can be shared with any stakeholder to set expectations.

Maintaining Standards and Best Practices. Governance isn't just about saying no; it's about setting standards that make saying yes safer. One example: a naming convention standard (for fields, picklist values, etc.). During maintenance, someone might notice inconsistent names (like some API names have prefixes, others don't). The governance role would be to establish a convention (e.g., all integration-related fields start with

"INT_" or all text fields end with "_txt" – whatever works for your org) and then maintain it. Periodic audits as part of maintenance can catch deviations. If you find things that don't adhere, decide whether to rename them (which might involve refactoring code or reports) or at least note them. Similarly, coding best practices (for those orgs using Apex) or Flow design patterns could be documented by governance and reviewed regularly.

Another key area is **data governance.** Who "owns" each major data entity? Governance ensures that if Marketing wants to purge a bunch of Leads, there's a process to check with Sales or archive the data instead of losing it outright. Or if a compliance rule says "We must encrypt all customer PII fields," governance makes sure those platform encryption or field-level security settings remain in place and are updated when new fields are added. Ongoing maintenance under governance means routinely checking that governance policies (like required fields, validation rules for data quality, etc.) are working and updated when business policies change.

Feedback Loop and Continuous Improvement. A mature governance practice will also gather feedback from users and admins about what's working or not. For instance, maybe a certain approval process has too many steps and people are finding workarounds. The governance team should hear about that and consider streamlining it – that's maintenance too: not just cleaning up unused things, but optimizing what's heavily used. Regular meetings between IT and business stakeholders as part of governance can surface these insights. It's better to proactively refine processes than to let them become pain points that degrade user experience or data quality. Also, when Salesforce releases new features (like a standard feature that could replace a custom one you built), governance would champion evaluating and possibly adopting that (this is a crossover into future-proofing, which we'll hit in the next section). For example, suppose you governed a standard that "We use Process Builder for approvals" back in 2018, but now in 2025, Flow Orchestration might be a superior standard – governance should update the playbook and oversee the transition.

Educating and Empowering Through Governance. Good governance also has an educational component. It's not just top-down control; it's also about empowering team members with knowledge of the rules so they can self-govern in a sense. If your developers and admins clearly know the design patterns to follow, they'll implement changes that require minimal oversight. For instance, governance might produce a "Declarative Development Handbook" for your company – a concise guide covering things like how to design automations, how to name components, how to handle common requests. New team members get up to speed faster with such a guide, and existing members have a reference to align their work. This reduces the need for constant intervention. Part of integrating with maintenance is to keep that guide updated as the org evolves. Perhaps every quarter, review it: did we learn something new? Did we decide to ban a certain practice (like no more Workflow Rules, only Flows)? Add that in.

Pro Tip: **Governance as a Service, not a Gatekeeper.** Shift the perspective of governance from "the team that says no" to "the team that helps everyone do it right." If you position your governance committee or architect group as advisors and enablers, teams will involve them early and value their input. For example, a business unit wanting a new Salesforce app should feel comfortable approaching the governance folks for guidance on how to do it within org standards, rather than trying to go around them. This cultural aspect is crucial – when governance is collaborative, maintenance becomes much easier because everyone's pulling in the same direction (preventing messes rather than cleaning up after them). In practice, this could mean the governance team holds open "office hours" for any admin/dev to consult on a tough design issue or does periodic roadshows to highlight new best practices or common mistakes to avoid.

Governance and Technical Debt Management. We touched on technical debt in earlier chapters; governance plays a big role in keeping debt under control. A governance group should maintain a backlog of "debt items" – things that aren't urgent but should be fixed eventually (like "too many duplicate fields capturing similar data" or "some hard-coded picklist values in code that should be dynamic"). By tracking these and periodically addressing them (maybe in each release cycle allocate some time for debt cleanup), you prevent the gradual decay of the org. It's analogous to infrastructure maintenance in a city – you fix the roads and clean the pipes regularly, not only when there's a crisis. Governance ensures those maintenance tasks are recognized and prioritized amid new feature development.

Metric-Driven Governance. How do you know your governance and maintenance efforts are effective? Consider establishing metrics. For example, measure the total custom field count and see if its growth aligns with genuine needs or if it's ballooning. Track how many of those fields are filled (Optimizer can help here) – if only 50% of fields are populated regularly, that's a sign of potential bloat. Another metric: user adoption rates or satisfaction surveys (if governance is too strict, maybe people are unhappy; if it's effective, users might notice more consistency). Even measuring the number of org health issues found each quarter – hopefully, it trends down or stays low due to proactive management.

By weaving governance into the fabric of daily and weekly operations, you create an environment where maintenance isn't an afterthought or a chore – it becomes a natural part of the development lifecycle. This means fewer emergencies, fewer regressions, and a more harmonious evolution of your Salesforce org. A well-governed org can adapt to change smoothly because the guardrails are in place to prevent derailment.

We've now covered how governance provides the umbrella under which all maintenance activities thrive. With a solid governance structure, you can confidently allow flexibility for business needs while safeguarding maintainability. This sets the stage for a nuanced discussion in our next section: **balancing flexibility and maintainability**. No matter how good your governance is, there will be times you must choose between building a highly flexible solution or a simpler, more maintainable one. How do we strike that balance? Read on, as we discuss strategies to get the best of both worlds without falling victim to the pitfalls of over-engineering or under-delivering.

BALANCING FLEXIBILITY AND MAINTAINABILITY

Salesforce's declarative tools are amazingly flexible – you can often achieve the same goal in ten different ways. But as the saying goes, *just because you can, doesn't mean you should.* This section is all about walking the fine line between giving the business what it asks for (flexibility to handle every use case, nuance, and exception) and keeping the system straightforward enough to maintain and evolve. In other words, how do we avoid turning our org into a Frankenstein of convoluted logic while still meeting requirements and allowing growth?

The Allure of Flexibility. Let's acknowledge why we sometimes end up with overly complex configurations: business needs can be very complex! Stakeholders might request that a form's behavior changes based on 10 different conditions, or that an automation covers every edge case they can imagine. As Salesforce professionals, we pride ourselves on being able to solve problems, so the natural impulse is to say "yes" and figure out a way to build it declaratively. Salesforce's platform often obliges – you can, for instance, create a formula field with nested IF statements to handle a ton of scenarios, or build a mega-flow that branches out to handle each special case. In the short term, this delivers exactly what was asked. But what's the long-term cost? Typically, maintainability suffers. The formula becomes so complex that no one wants to touch it (or it hits limits). The mega-flow works initially, but a year later when someone needs to modify it, they spend hours trying to understand its labyrinthine logic.

Spotting Over-Engineering. How can you tell if a solution is too flexible for its own good? One red flag is if you find yourself adding "just one more" tweak repeatedly. For example, a validation rule that started simple now has additional OR conditions for five different profiles and a checkbox to exclude special records. Or a set of record types proliferating because each department wanted its slight variation of a process. If you see proliferation – of conditions, of components – step back and reassess. Another sign is if the business users themselves are confused by the solution ("We have 7 different case forms depending on scenario and even we can't remember which to use"). Consistent user confusion or frequent admin errors when updating a flow are indications that the design might be too clever or too bespoke.

Designing for Simplicity. One of the core principles of Salesforce's Well-Architected framework is "Easy" – solutions should be easy to use, and easy to change. Achieving that often means intentionally limiting flexibility to what is truly necessary. For instance, suppose you have a multi-step approval process that varies slightly by region. You could build one giant approval process with conditional logic for regions, but an easier-to-maintain approach might be to have separate, simpler approval processes per region (if the processes are distinct enough). Yes, that's a bit of duplication, but each is straightforward. The complexity is segmented rather than intertwined. Conversely, if you have many duplicate automations, perhaps consolidation is the simplicity you need. It's a balancing act: sometimes combining things reduces maintenance (fewer places to update logic), other times separating them does (each piece is cleaner).

A useful strategy is to apply the **80/20 rule**. Design for the 80% standard scenarios in a clean way, and handle the 20% exceptions separately. Maybe the 80% can be done with a simple automation, and the few outliers are managed with a manual process or a smaller secondary flow. Not every single corner case must be fully automated – if doing so adds significant complexity, weigh the cost/benefit. Often stakeholders are okay with handling rare exceptions manually if you explain that automating them would make the whole system harder to maintain or slower. It's our job to articulate that trade-off.

The Role of Experience and Governance. This balance often comes down to experience – knowing when to push back on requirements or propose an alternative that's simpler. Governance, from the previous section, can support you by giving you the authority to say "Actually, our standards avoid formulas that reference more than X fields; let's find another way." Experienced architects on a governance board can also review designs and spot potential over-engineering. They might suggest using out-of-the-box features instead of custom solutions. For instance, rather than creating a crazy custom sharing setup for a special use case, maybe leverage Salesforce's Territory Management or Einstein Activity Capture if it fits – standard tools that are supported and updated by Salesforce, hence lower maintenance for you.

Documentation of Complex Solutions. If you do end up building something inherently complex (sometimes it's unavoidable for genuinely complex business processes), mitigate the pain by documenting it thoroughly (as we highlighted earlier). Map out the flowchart for a complex automation and store it in a place everyone can find. Comment your formulas (Salesforce formula fields don't allow comments, but you can write out logic in a text attachment or description field). The point is, acknowledge "this is a bit intricate" and leave breadcrumbs. That way maintainability is bolstered even when flexibility made things complicated. Also, include the *why* – why it had to be this way. Perhaps compliance requirements forced a detailed solution; note that, so future maintainers know it wasn't just overzealous design.

Simplifying an Existing Tangled Org. Many readers might not be building fresh solutions but dealing with legacy ones. How do you introduce maintainability to something already overly complex? Gradually and carefully. Identify hotspots that cause frequent issues or that no one understands. Refactor one at a time. For instance, if you have a gargantuan Process Builder with 30 nodes, maybe plan to split some logic into reusable Flows or move some conditions into formulas on the object to simplify the flow. Use tools like Salesforce Optimizer and the Apex PMD (for code complexity) to pinpoint pain points. Another effective tactic is to convene a brainstorming session (with the team or even with ChatGPT as a virtual participant) to ask, "If we were building this today from scratch, how would we do it differently?" You might find that Salesforce now has a feature that handles it out-of-the-box, or that what seemed necessary years ago no longer is due to process changes.

ChatGPT Prompt Example: "We have a formula field with 8 nested IF statements and several cross-object references. It's hard to maintain. Suggest ways to simplify or alternative approaches." – An AI response might suggest using a helper object to hold some of that logic, or breaking it into multiple smaller fields, or using a Flow to populate a value instead of an ultra-complex formula. These suggestions can inspire a redesign that you then validate. Maybe it says, "Could any of these conditions be moved to validation rules or picklist dependencies?" prompting you to think out-of-the-formula. In essence, use AI as a rubber duck (an analogy from programming where explaining a problem to a rubber duck can help you solve it) – except the duck talks back with ideas.

Balancing Config vs. Code. Sometimes, maintainability might mean deciding to write a bit of Apex or use an invocable method rather than a convoluted declarative hack. This is a delicate point, as our book focuses on declarative best practices, but the best practice is *not to be dogmatic*. If a small snippet of code can perform a calculation or enforce a rule far more cleanly than a combo of validation rules, formula fields, and flows, it could be the more maintainable route (provided you have the resources to support that code). The key is to make that decision consciously: weigh the pros and cons. Code can carry its own technical debt if not managed, but sometimes it's the appropriate tool. The ideal is to use "clicks not code" for the majority of cases, and "code with purpose" when needed, and even then, try to encapsulate complexity. For example, one could create a simple Apex @AuraEnabled method to handle a complex calculation and call it from a Flow, rather than contorting the Flow to do something it isn't good at.

Future Flexibility – Designing for Change. Another dimension of flexibility vs. maintainability is designing *today* for *tomorrow's* changes. A very rigid solution might work now but be hard to adapt when things inevitably shift. On the other hand, a solution too generic might be hard for users to grasp. Consider using Custom Metadata Types or Custom Settings as a way to inject flexibility without hardcoding logic. For instance, if you have criteria that might change (like thresholds or approvers), store them in a custom metadata record that a Flow or formula reads. That way if the policy changes, an admin can just update the record, not redesign the Flow. This approach can significantly improve maintainability because updates become data changes, not structural changes. It adds a layer of complexity (the indirection of referencing a metadata record), but if done for the right reasons, it's a net win. It's all about *intentional complexity* – adding a bit to save a lot later.

Pro Tip: **Use Modular Design in Declarative Solutions.** Break big problems into smaller ones. Instead of one Flow that does ten things, perhaps have it invoke sub-flows (using the "Subflow" element) each handling a portion of the task. This is analogous to how developers write functions. It makes each piece more digestible, testable, and reusable. Similarly, instead of one mega validation rule with 20 OR conditions, you might have a few targeted validation rules each with a clear message and purpose. Salesforce has limits on how many rules can fire, etc., so you don't want to go overboard, but as long as it's within limits, multiple simpler rules can be easier to maintain than one super rule that tries to do everything. The trade-off is performance (100 little rules vs 10 bigger ones – you have to balance that), but clarity often trumps a tiny performance difference, especially if user error prevention is at stake.

Communicating Trade-offs to Stakeholders. Achieving balance often requires negotiation with the business. It's important to communicate the *why* of a simpler approach. If you tell a department, "We won't build a custom object for this; we'll use tags and a report instead," explain the maintenance burden that the custom object would introduce (new fields, more data to manage, etc.) versus the simplicity of leveraging existing functionality. Often, non-technical stakeholders appreciate that there is a lower-cost, lower-risk solution even if it's not 100% tailored. They just need to understand the rationale. Using analogies can help (e.g., "We can either add a new wing to the house for a home office, or we can repurpose an existing room – the second option won't require new construction and we can do it faster, even if it means you share the space a bit").

In summary, balancing flexibility with maintainability is about conscious design decisions. It's about considering not just "Can I build this?" but "Should I build it this way? What happens next month or next year when something changes?" By thinking ahead, involving governance, and not being afraid to simplify or even say no to unnecessary complexity, you ensure your Salesforce org remains nimble and robust. The beauty of Salesforce is that it can bend in many directions – but you always want to avoid bending it to the point of breaking.

With this balanced mindset, you are equipped to build solutions that stand the test of time and shifting requirements. This also naturally prepares you for the future, because a maintainable org can evolve more easily. And speaking of the future – our final section is all about **future-proofing your Salesforce implementations.** The platform is ever-changing, and staying ahead is the ultimate way to ensure longevity and health. Let's explore how to keep your org not just current, but ready for what's next.

FUTURE-PROOFING SALESFORCE IMPLEMENTATIONS

The only constant in the Salesforce world is change. Three times a year, Salesforce pushes out new releases with features that can redefine best practices. Additionally, the broader technology landscape (think AI, integrations, evolving security standards) continuously influences how we use Salesforce. Future-proofing is the practice of anticipating these changes and structuring your org and team to adapt with minimal friction. It's about building not just for what you need today, but with an eye on what you'll likely need tomorrow. In this final section, we'll cover strategies to ensure your Salesforce implementation remains modern, leverages the latest improvements, and avoids accumulating obsolete baggage.

Stay Educated and Embrace the Release Cycle. Salesforce releases (Spring, Summer, Winter) are packed with enhancements – some minor, some major. A cornerstone of future-proofing is being proactive about these releases rather than reactive. Allocate time in your team's schedule for **release readiness.** This could mean assigning someone to be the "release captain" for each cycle, who will review the release notes (which can be hundreds of pages long) and distill the relevant changes for your org. What new features could replace customizations you've built? For example, if you have a custom lightning component for a dynamic form and Salesforce announces Dynamic Forms for your object, that's a sign you can eventually retire the custom work and use the standard approach. Similarly, Salesforce's introduction of things like **Permission Set Groups** was a game changer for orgs with too many profiles – by grouping permission sets, you could simplify user access management. If you jumped on that, you likely reduced complexity and prepared for Salesforce's stated direction of potentially minimizing profile usage. Always align with Salesforce's roadmap; they often telegraph their intentions (like the end-of-life announcement for Workflow Rules and Process Builder in favor of Flow). When Salesforce says "We're deprecating X," plan your move to Y sooner rather than later.

Pro Tip: **Use Preview Sandboxes and Pilot Programs.** Salesforce offers preview sandboxes before each release (usually if your sandbox is on a certain instance, it gets early access to the new release). Utilize this! Have a sandbox where you can test critical use cases against the upcoming release. This not only catches any breaking changes early (rare, but it happens, especially if you use some less common features or unmanaged packages) but also lets you play with new toys. If a new feature looks promising, you can experiment and even demo it to stakeholders. Additionally, Salesforce has pilot/beta programs for new features – if a feature could give you a competitive edge or simplify something, consider joining those. You'll get direct experience and can influence the feature's development.

Build with Extensibility in Mind. Earlier we touched on not over-engineering, but part of future-proofing is making sure your solutions aren't closed boxes. For instance, if you create a custom object to track a business process, think about whether that data might someday need to integrate with another system or be reported on in combination with other data. That might influence your choice of naming, or ensuring you include integration-friendly fields (like external IDs). If you design a Flow, consider if it might be reused – perhaps build it as a subflow that can be called from multiple places. When creating Lightning web components or other custom code, try to make them configurable (e.g., attributes that allow an admin to tweak behavior without code). This way, when business inevitably says "that's great, but now we need a slightly different variation," you're not starting from scratch.

Keep an Eye on Technical Debt Retirement. We've talked a lot about cleaning up and retiring old stuff. Future-proofing also means phasing out things that won't stand the test of the next platform upgrades. A prime example is the effort to migrate away from Workflow Rules and Process Builder to Flow (which by now, hopefully, you've planned or done, since after 2025 they're no longer supported). Another example is formulas vs. Lightning Web Components or screen flows: If you have a Visualforce page that's been around for 10 years, maybe plan to replace it with an LWC for a better UI and forward compatibility. If you have legacy Apex that doesn't use bulk patterns or is on an older API version, schedule time to refactor it. Salesforce may not force you to upgrade such things often, but you don't want to be caught on older tech when the world (and talent pool) has moved on. The more your org uses current platform features, the easier it is to find help, whether that's hiring new team members or consulting the Trailblazer Community.

Adopt New Features Judiciously. Future-proofing doesn't mean jumping on every shiny new beta feature, but you should definitely leverage improvements that reduce your maintenance overhead. A classic case: the introduction of **Dynamic Forms and Dynamic Actions.** If you migrated some classic page layouts to use Dynamic Forms, you probably made life easier by controlling field visibility in the app builder instead of multiple page layouts or record types. Or consider **Orchestrator** for complex multi-step approvals which previously might have required a lot of custom flows and custom objects. As another example, Salesforce in recent releases has introduced tools like **Salesforce Optimizer App** (interactive) and **Translation Studio** for managing translations better. If you operate in multiple languages and you were maintaining translations manually in spreadsheets, using the new Translation Studio can future-proof that part of your org by centralizing and easing translations.

However, it's wise to evaluate new features in a sandbox thoroughly. Not every new feature is mature on day one. Sometimes waiting a release or two is prudent. A good approach is to pilot a feature with a small subset of users or a small dataset. For instance, if Salesforce releases a new AI-driven field prediction and you think it could replace a custom prediction you built, test it in parallel to see if it's reliable before deprecating your solution.

Leverage AI and Automation for Maintenance. As we look to the future, AI (like our friend ChatGPT) is becoming a regular part of the admin toolkit. We've already discussed use cases of AI throughout earlier sections, but thinking ahead: OpenAI's **Sora** (text-to-video model) could one day help generate training videos or user guides from your process documentation automatically. Imagine releasing a new feature and instantly getting a draft of a how-to video for users – that addresses the user adoption side of future-proofing, making sure people use what you build. Similarly, the **Operator** tool, as it matures, might allow you to automate more org management tasks. For instance, an Operator agent could conceivably handle repetitive tasks like creating users, resetting passwords, or even populating test data on command.

Keeping abreast of these AI developments and thinking creatively about offloading grunt work to them will be increasingly important. This doesn't replace the strategic role of an admin, but it augments your capacity, allowing you to focus on higher-level design and user collaboration.

Community and Ecosystem Engagement. No one can predict the future alone, but fortunately we have the vast Salesforce community and ecosystem. Future-proofing includes staying connected: attend local user groups, join webinars on new releases, follow Salesforce influencers or product managers on social media who often hint at things to come. Platforms like IdeaExchange let you see what features the community is pushing for – which often end up on the roadmap. By knowing these trends, you might avoid spending lots of effort on a custom solution that Salesforce might deliver natively next year. For example, say a lot of people are asking for a certain type of component or integration, and you see Salesforce has it under consideration – maybe you hold off on a big custom build to see if it gets delivered. Or at least design your solution in a way that it could be swapped out.

Cross-Training the Team. Future-proofing isn't just tech – it's people. Ensure your team is continuously learning. If only one person knows a certain area (say, CPQ or Experience Cloud), that's a risk. Encourage skill sharing and backup plans for key knowledge domains. The more versatile your team, the more resilient your org is to changes, including team changes (people leaving) or new project types. Trailhead, of course, is a great way to keep up to date on new features (Salesforce releases modules on new stuff each release). Allocate some team time for training, it pays off.

Plan for Scale and Org Strategy. Lastly, consider the growth of your org. Future-proofing means thinking about higher data volumes, more users, possibly multiple orgs (if your company acquires another with Salesforce, how will you integrate?). Laying groundwork like a solid **data archive strategy** is future-looking: you might not need it now, but if you know that in 2 years you'll have double the data, better to implement archiving or cleanup processes early. Similarly, if multi-org is a possibility, start aligning configurations so that merging later is easier (consistent record names, avoiding hardcoded references to things that might differ). If you foresee needing Salesforce's newer clouds or products (maybe Marketing Cloud or Industries), try to keep your core data model in Salesforce CRM aligned with those (for example, using standard objects where possible, since other Salesforce products often sync with them more easily than with custom objects).

In a nutshell, future-proofing is about being proactive, staying flexible, and continuously aligning your org with where the Salesforce platform and your business are headed. It's the capstone of everything we discussed: regular health checks, documentation, monitoring, change management, governance, and balanced design all feed into a future-proof org. If you practice all those diligently, you're naturally poised to handle new requirements and platform shifts.

ChatGPT Prompt Example: "Summarize the key Salesforce Winter '26 features that could replace custom solutions in our org (I use Service Cloud, Sales Cloud heavily)." – This kind of query could help you quickly identify things like, "Winter '26 introduces native appointment scheduling – could replace your custom scheduling app; new AI-driven case classification – could reduce your custom case assignment rules; etc." It's like having a personal analyst comb through the release with your org in mind. Incorporating that into your release readiness can highlight future-proofing opportunities.

As we conclude this section and the chapter, reflect on how far we've come: from checking the pulse of your current org, to methodically cleaning and documenting, to watching it in action, to managing change deliberately, to enforcing good practice through governance, to designing smartly, and finally to gazing at the horizon. It's a lot to take in, but together these practices truly embody a Salesforce org that is *healthy, agile, and prepared.* In the next chapter, we will build on these maintenance and strategy insights as we move into even more advanced considerations, ensuring that your declarative expertise continues to expand alongside the ever-evolving Salesforce platform.

SUMMARY

In this chapter, we journeyed through the lifecycle of keeping a Salesforce org at its best. We began with **regular org health checks**, learning how tools like Optimizer and structured review cycles help catch issues early and keep our environment clean. From there, we stressed the **importance of documentation** – treating it as the org's memory – so that every field, flow, and business process has context and clarity for anyone who comes across it. We explored **monitoring techniques**, setting up dashboards and utilizing event logs to stay alert to usage patterns, performance, and security signals, effectively giving our org a real-time health monitor. We then tackled **change management and versioning**, implementing practices to control and track modifications, ensuring that each update is intentional and reversible if needed. Governance emerged as the glue in **integrating maintenance efforts**, providing standards and oversight so that as we maintain and enhance the org, we do so cohesively and responsibly. We then balanced the scales between **flexibility and maintainability**, learning to deliver robust solutions without over-complicating them – an art of saying "yes" to requirements in a smart way. Finally, we set our sights forward with **future-proofing strategies**, embracing Salesforce's rapid evolution and positioning our org to leverage new features while minimizing technical debt.

Together, these facets form a holistic approach to Salesforce org success: one where continuous improvement is a mindset, not an afterthought. With a healthy, well-monitored, and well-governed org, you are not only solving today's problems but also building a foundation for tomorrow's innovation. As we move into the next chapter, which covers advanced declarative techniques and strategic considerations, you'll carry forward this strong maintenance ethos – because the best Salesforce solutions aren't just powerful when delivered, they stay powerful and reliable in the long run, evolving gracefully as needs change. In the world of Salesforce Enterprise Thinking, that is the ultimate best practice: to create systems that endure and excel, release after release, year after year.

14

ADVANCED DECLARATIVE TECHNIQUES AND EXTENSIBILITY

Salesforce professionals often reach a point where "point-and-click" configuration needs to stretch beyond its usual limits. This chapter explores how experienced admins, developers, and architects can push declarative development further using advanced techniques and emerging tools. We'll demonstrate that many challenges traditionally solved with code can now be addressed with clever configuration – all while maintaining the agility and clarity that declarative solutions provide. The focus here is on extensibility: extending Salesforce's **no-code** toolkit to do more, integrate wider, and solve tougher problems, all without writing large amounts of Apex.

Over the next sections, we consider ways to connect Salesforce to external systems **declaratively**, design loosely-coupled architectures inside Salesforce with events, and make your automations more scalable by externalizing logic and data. You'll learn how features like External Services enable integrations via clicks (bringing in outside API capabilities), and how Platform Events let you decouple complex processes for greater flexibility. We also look at Custom Metadata Types, a powerful mechanism to store and manage configuration data that drives your automation – allowing admins and business analysts to adjust business rules without editing flow formulas or Apex code.

As we explore these tools, we'll highlight how to **extend Flow Builder's capabilities.** This includes using invocable actions (small Apex or packaged components that plug into flows) to perform tasks that flows can't do out-of-the-box, and leveraging the ecosystem of pre-built flow extensions available on the AppExchange and in the community. An underlying theme is understanding the boundaries of declarative tools: knowing when you're better off writing a snippet of code versus contorting a no-code solution. Paradoxically, mastering "clicks, not code" also means knowing when **to** use code for the sake of maintainability or performance. We'll discuss how that boundary is continually shifting outward as Salesforce releases new features – what required Apex last year might be doable with a Flow this year.

Throughout, we'll maintain a practical, forward-looking lens. The tone is friendly and conversational, but also analytical: think of it as an experienced colleague walking you through advanced Salesforce techniques with a mix of hard-earned wisdom and enthusiasm for what's next. You'll find **ChatGPT prompts** in call-out boxes, suggesting ways to use an AI assistant to further explore each topic or assist in real-world tasks. We encourage you to try these prompts – whether by typing or even speaking to ChatGPT – to see how an AI helper can augment your Salesforce work. For example, you might **ask ChatGPT to draft an external API schema**, brainstorm an event-driven design, or generate a testing checklist for a complex flow. As always, apply critical thinking to the AI's suggestions, but you may be surprised at how it can accelerate your work or inspire new solutions.

By the end of this chapter, you will have an expanded toolkit of declarative techniques and a deeper understanding of when and how to use them. More importantly, you'll be equipped to combine these approaches – integration, events, metadata-driven logic, and slight touches of code – into sophisticated solutions that deliver business value with minimal custom code. This is the art of Salesforce declarative extensibility: **assembling powerful, maintainable solutions by clicking cleverly.** Let's jump in by looking at how Salesforce can connect to the wider world without a single **executeAnonymous** or **HttpRequest.invoke** in sight, setting the stage for truly integrated, extensible architecture.

DECLARATIVE INTEGRATION USING EXTERNAL SERVICES

Integration has long been a domain where many Salesforce teams assumed Apex or middleware was the only answer. Calling external APIs, connecting to third-party platforms, or syncing data with other systems traditionally required custom code or tools like MuleSoft. Today, that's changing. Salesforce's declarative integration options have matured – we can now connect Salesforce to external services **with clicks, not code**. In this section, we explore External Services, Flow HTTP callouts, and even no-code integration tools like MuleSoft Composer and Slack integrations. These tools empower admins to build integrations declaratively, expanding Salesforce's reach without sacrificing the simplicity of a no-code approach.

External Services: Salesforce External Services is a feature that lets you integrate RESTful APIs by importing their specifications. In plain terms, if you have an API "blueprint" (an OpenAPI schema or a definition of endpoints and data), Salesforce can ingest that and turn it into point-and-click building blocks. Imagine you want to retrieve real-time exchange rates from an external currency service within a Flow. Instead of writing an Apex callout, you can register the API with External Services. You'd obtain the service's OpenAPI schema (a JSON or YAML file describing the API's methods and data structures) and create a Named Credential in Salesforce for authentication (to securely store the endpoint URL and credentials). With those in place, the External Services wizard in Setup will generate **invocable actions** for each API operation. Now, your Flow can simply include an "Action" element that calls "CurrencyService.getLatestRates" (for example), map input values, and receive the output – all without writing code. When the flow runs, Salesforce takes care of making the HTTP callout and parsing the response.

This might sound complex, especially preparing the OpenAPI schema. In practice, many API providers publish OpenAPI specs. But if you don't have one, this is where ChatGPT can assist. You can feed ChatGPT the API's documentation or a sample response and **ask it to draft an OpenAPI schema**. For instance, if you have a sample JSON response from a weather service, you could provide it to ChatGPT and request an OpenAPI 3.0 spec covering the needed endpoints. Once ChatGPT suggests a schema, you'd review and tweak it (ensuring it meets Salesforce's support requirements), then import it as an External Service. Always double-check that the schema is correct and complete – a small error could cause your External Service registration to fail.

ChatGPT Prompt: "I have a REST API that provides currency exchange rates (in JSON). Can you help me create an OpenAPI schema I can use to set up a Salesforce External Service for this API?"

Pro Tip: If you plan to upload an API spec or any sensitive data to ChatGPT for help, make sure your OpenAI settings are configured to keep your uploads private. For example, you might disable chat history or opt out of model training with your data. This ensures you're not unintentionally sharing proprietary information when getting AI assistance. Always follow your company's data security policies when using tools like ChatGPT with real integration details.

External Services work best for well-defined, RESTful APIs. They currently support OpenAPI 2.0 and 3.0 schemas for JSON-based services. There are some limitations to be mindful of: for example, complex nested data structures might be tricky to handle in Flow, and certain authentication flows (like OAuth 2.0 with multiple handshake steps) aren't handled automatically by External Services. Typically, simple API keys or basic auth can be managed via Named Credentials, but for more complex auth you might still need a bit of Apex or an external integration tool. Nonetheless, the scope of what you can do declaratively is impressive. Admins have used External Services to do things like **validate addresses** against an external service during lead conversion, **fetch credit scores** from a financial system in real time, or **send messages to a collaboration tool** when a record changes. All of this can be orchestrated in Flow Builder, alongside other logic, without writing a custom HTTP callout class.

Salesforce has recently gone even further by introducing **Flow HTTP Callouts.** This feature (available in modern releases) allows you to make a callout to an external API directly within a Flow, even if you don't have a formal OpenAPI schema. In Flow Builder, you can add an "HTTP

Callout" element, configure the method (GET, POST, etc.), the endpoint URL, headers, and body, and then use the response in subsequent steps. Behind the scenes, Salesforce actually creates an "External Service" for you when you use this, but it saves you from hand-crafting the full schema upfront. For example, you could set up a Flow HTTP Callout to a simple REST endpoint (maybe a weather API URL) by entering the URL and mapping the expected JSON response to a Salesforce Apex-defined type automatically. This lowers the entry barrier for quick integrations – ideal for one-off calls or simple services. The trade-off is that you bypass the formal schema, so maintaining or updating that integration might be less structured. Think of Flow HTTP Callouts as a quick, declarative way to do what you might have done with a few lines of Apex in the past. They're great for straightforward needs, but for a large or mission-critical integration, you'd likely still prefer the robustness of a full External Service or a middleware solution if complexity grows.

Declarative integration isn't limited to calling APIs – it also encompasses connecting Salesforce with other platforms in user-centric ways. A prime example is **Salesforce's Slack integration** capabilities. Since Salesforce's acquisition of Slack, a number of point-and-click integration options have emerged. For instance, using **Flow Builder's Slack actions**, an admin can post a message to Slack from a Salesforce Flow. If a high-priority case is updated, a flow could automatically send a formatted Slack message to a support channel, alerting the team – no Apex needed, just a Slack app configuration and a Flow action. Conversely, Slack itself can be a trigger for Salesforce actions: with the right setup, a user in Slack could invoke a Salesforce Flow (e.g. typing a slash command in Slack that kicks off a Salesforce approval process or queries an account record). These integrations are powered by a combination of Salesforce's Slack app, permissions, and Flow's built-in support for Slack actions. The key point is that even for what used to require a custom Slack bot and code, we now have *declarative glue* available. Admins can incorporate Slack in their automation toolbox, keeping users informed in their preferred channels without writing a custom integration service.

Another powerful tool in the no-code integration arsenal is **MuleSoft Composer**. Whereas External Services and Flow callouts handle connecting Salesforce to external APIs one call at a time, MuleSoft

Composer (a separate product, but built for admins) provides a higher-level integration workflow capability. Composer allows you to create integrations between Salesforce and other systems (think SAP, Workday, databases, or even Google Sheets) through a UI, using pre-built connectors and templates. It's essentially Flow Builder but across systems: you configure triggers (e.g., "a new record in Salesforce" or "a new row in Excel") and then actions in another system ("create a record in Netsuite" or "send an email via Gmail"), all with mappings and logic configured by clicks. While Composer is outside of core Salesforce Setup (and requires an additional license), it exemplifies the "clicks, not code" philosophy for integration at scale. An admin or integration specialist can build multi-step, cross-system automations without writing scripts or managing servers. For example, with Composer you might synchronize Salesforce opportunities to an ERP system once they reach a certain stage, or automate a workflow where a new customer in Salesforce triggers creating a project in a project management tool and a welcome message in Slack. Historically, such orchestration would be the realm of IT integrators or custom code; now it's accessible to the declarative builder.

It's important to approach these integration options with an architectural mindset. Just because you *can* integrate without code doesn't mean complexity disappears. You'll need to consider error handling (e.g., what if the external service is down—does your flow handle that gracefully?), transaction boundaries (external callouts happen after Salesforce commits, so design accordingly to avoid data mismatches), and limits (Salesforce enforces limits on callouts per transaction and total time—declarative callouts are no exception). Always test integrations thoroughly in a sandbox. Use tools like Named Credentials not only for security but also for flexibility—if an endpoint changes, you update it in one place.

Also, remember governance: document these integrations just as you would code. It should be clear to others how data flows in and out of your Salesforce org. If you've set up an External Service, keep a copy of the API schema in a repository or documentation portal. If using MuleSoft Composer, build monitoring and logging into your flows so you know when something fails.

Finally, don't forget that ChatGPT can be a teammate in integration work beyond just generating schemas. You might use it to **troubleshoot** an integration by pasting an error message and asking for likely causes. Or use it to brainstorm integration mapping: for instance, feeding it field names from two systems and asking for suggestions on how to map or transform the data. With the new ChatGPT **voice capabilities**, you could even talk through an integration plan while commuting, using ChatGPT as a rubber duck to catch any blind spots in your logic. AI tools can't replace thorough testing and human insight, but they can accelerate the path to a working solution.

As you become comfortable with declarative integrations, you'll find yourself able to build an integrated Salesforce experience that feels seamless to users. They click a button and data comes back from a third-party service in seconds, or an event in Salesforce instantly notifies a user in Slack. This level of connectivity – achieved with configuration – is empowering. It broadens what "Salesforce admin" work entails, inching into areas that used to be exclusive to integration specialists or developers. It's a perfect illustration of Salesforce's Enterprise Thinking: using high-productivity tools to deliver business value faster, while still respecting best practices and maintainability.

Having explored how to reach *outside* of Salesforce declaratively, let's now turn to connecting processes *within* Salesforce in a more flexible way. In complex orgs, it's not just external systems that need integration – often our internal business processes need to talk to each other without becoming tangled in a web of triggers and flows. This is where **Platform Events** come into play, enabling a pub/sub model inside Salesforce that decouples processes for resiliency and scalability.

UTILIZING PLATFORM EVENTS FOR DECOUPLED PROCESSES

In a traditional Salesforce setup, when something important happens (say a record is created or updated), we often handle related business logic with triggers or record-triggered flows. These fire immediately and perform necessary follow-up actions. However, as an org grows, this approach can lead to tightly coupled processes. One save operation might kick off a cascade of automations that all run in the same transaction, potentially causing performance issues or making it hard to isolate failures. **Platform Events** offer a sophisticated declarative (and programmatic) solution to this: they allow different parts of your system to communicate through an asynchronous, decoupled mechanism. By publishing and subscribing to events, you can design Salesforce applications where processes are **loosely connected** – improving scalability and making the overall architecture more modular.

Think of platform events as a messaging system within (and even beyond) Salesforce. You define a custom event much like you'd define a custom object – specifying fields that the event message will carry. But unlike a record, an event is ephemeral; it's published, flows through the system to any subscribers, and isn't stored in a standard table (though you can retain event messages for a period if you need replay or auditing). The power comes from the **publish/subscribe model** (pub/sub): one component publishes an event, and one or many other components react to it, without the publisher needing to know about those subscribers. This decoupling is gold for complex environments. For example, when an Opportunity is closed, you might need to kick off a bunch of things: update some summary data, send an invoice to an ERP, notify the sales team in Slack, and perhaps do some AI analysis on win/loss reasons. Rather than one monolithic flow or trigger trying to do all that, you could have the Opportunity "publish" an Opportunity_-Won__e platform event (perhaps via a little record-triggered flow or Apex trigger that just does that one thing). Downstream, you create separate platform event–triggered flows: one listens for Opportunity_-Won__e and handles the Slack notification, another listens and does the ERP integration (maybe via an External Service), another does the AI analysis. Each flow runs independently, and if one fails, the others still proceed. They're also easier to maintain because each has a single responsibility.

Setting up platform events declaratively involves a few steps, but it's well within an admin or architect's reach. In Setup, you create a new Platform Event definition, giving it a name and adding fields just as you would for a custom object (e.g., Case_Update__e with fields like CaseId, UpdatedBy, Priority, etc.). Once deployed, you can publish an event using a

Flow or **Process Builder** (Process Builder is older, but could publish events; now Flow is the go-to). In Flow, there's an action "Publish Platform Event" where you select your event type and set field values from context. For example, in a record-triggered flow on Case (after save), you might add an action to publish Case_Update__e, mapping CaseId to the event's CaseId field, and maybe including certain details like the new Status.

On the subscriber side, you can create a **platform event–triggered flow**. When building a new flow, one of the options for trigger is "Platform Event" – you then choose the event type. This type of flow will run whenever an event of that type is received by Salesforce. Inside that flow, you can do whatever is needed for that event's occurrence. In our case update example, perhaps one subscriber flow creates a chatter post to notify a group if the case's priority is high and status changed to "Escalated." Another subscriber flow (separate) could take the same event and log an entry in a custom "Case History" log object for auditing. Because these flows run asynchronously and independently, you avoid slowing down the original operation (the case update) and avoid interdependencies between the post to Chatter and the logging logic. Each is its own mini-process that just happens to be triggered by the same event message.

It's not just flows that can subscribe to events. Apex triggers can listen for platform events too (useful if you need to do complex logic in code upon receiving an event), and even external systems can subscribe to Salesforce platform events through CometD or the newer Pub/Sub API. This means platform events can bridge Salesforce and external apps in real-time. For instance, your org could publish an event when a big deal closes, and a separate Node.js application listening could receive that and trigger a celebration animation on a big screen in the office, or maybe update a legacy system. All without the Opportunity directly invoking those external actions – it just emits the event and moves on.

From an **architectural perspective**, platform events enforce a clean separation. The publisher doesn't have to worry about what happens after publishing – it just fires the event. Subscribers don't interfere with the original transaction; they do their work in their own context. This yields a more robust system. If one subscriber flow fails (say, the ERP is down and the integration flow errors out), that error doesn't roll back the Opportunity save or affect the Slack notification flow. Salesforce will log the error for that flow and even automatically try to replay the event if you've configured a durable subscription with a replay window. In high-volume scenarios, you might use **High Volume Platform Events** (which allow more throughput). Admins should be aware of limits, though: there are daily event delivery limits and a concept of "event bus" usage. Usually, Salesforce gives generous allocations, but if you plan to fire tens of thousands of events per day, you'll want to check your license allowances and perhaps purchase add-ons for high-volume events.

One potential challenge with platform events is **debugging** and visibility. Because everything is decoupled, tracking the end-to-end flow of what happened can be tricky. Salesforce provides some tools: the **Event Monitoring** and **Transaction Finalizers** can help in certain cases, and you can always log some info in subscriber flows (e.g., create a log record when an event is received). For critical processes, design some monitoring – maybe an email alert if a subscriber flow fails, or a daily report of important events processed. It's wise to also use the **Platform Events Trail** (available in Setup) to see if events are being published successfully and if subscribers are connected.

Let's consider a scenario that ties these ideas together: You have a complex order processing system. When an order is placed (in Salesforce), you need to do a bunch of things: reserve inventory (in an external inventory system), send a confirmation email, generate an invoice in a finance system, and update a dashboard. Rather than one monster flow or trigger, you design it event-driven. The Order object has a record-triggered flow that, after saving a new Order, publishes an Order_Placed__e event with all the relevant info (order Id, customer Id, total amount, etc.). Then you have three separate flows: one is triggered by Order_Placed__e and calls an External Service (without code) to the inventory API – updating stock and getting a confirmation back, which might update the Order record. Another flow listens for the event and uses the Email Alert action to send a nicely formatted confirmation email to the customer (or perhaps it invokes a Marketing Cloud Journey – whatever suits, the point is it's separate). A third subscriber could be a small Apex trigger or an invocable action that takes the event data and creates an invoice in the finance system (maybe via another integration tool or an API). Each piece can be developed and tested independently, and if one piece needs to change (say you switch finance systems), you update that subscriber or event publisher without overhauling the whole chain.

Using platform events effectively requires a bit of an adjustment in mindset. It's about designing for **asynchronicity** and separation of concerns. This is a design principle often seen in large-scale enterprise architecture, and Salesforce provides it natively so you can apply those patterns declaratively. Platform events also encourage thinking about eventual consistency – acknowledging that not everything has to happen in one transaction. A slight delay (usually a few seconds or less) between an action and a downstream reaction can be acceptable in exchange for a more reliable and scalable system. Users typically won't notice that your Slack notification or inventory update happened asynchronously, as long as it's timely enough and the process is well-communicated.

As you design event-based architectures, consider using ChatGPT as a brainstorming partner. You can describe your current tightly-coupled process to ChatGPT and **ask for ways to decouple it using events or other patterns.** It might suggest which parts make good event publishers vs subscribers. Sometimes explaining your design to an AI (or anyone who will listen) surfaces new ideas – for example, you might realize you can simplify a process by having two events instead of one, or combining a couple of steps into one subscriber. You could even have ChatGPT help draft a sequence diagram of your event-driven process for documentation purposes.

ChatGPT Prompt: "I want to decouple our case management process using platform events. Currently, when a case is closed, the same process updates related records, sends emails, and posts to Chatter. How can I use a platform event to make this more modular, and what would the subscriber flows do?"

Using voice, you could literally talk through a flow: "When a case closes, I need A, B, and C to happen" and let ChatGPT's voice companion confirm the event architecture. Salesforce's emerging tools like the **AI-**

driven "Agent" (Operator) could in the future even automate parts of this setup for you – imagine telling an AI agent, "Set up an event for Case Closed and corresponding flows for each follow-up action." We're not quite there yet, but it's on the horizon. In fact, OpenAI's new **Operator** tool promises to help with web and file actions; one day such agents might assist in configuring Salesforce or at least retrieving the necessary information (for instance, pulling a list of all current triggers to suggest which could be converted to events).

By adopting platform events for decoupling, you're adding a powerful pattern to your arsenal of declarative techniques. It will make your org more **Adaptable** (one of Salesforce's well-architected pillars), since you can change pieces of processes without breaking others. It also naturally supports a **composable architecture** – you can plug new subscribers into events later without touching the original logic. This is a forward-looking way to design Salesforce apps, aligning with modern microservice and event-driven trends in the wider tech world.

We've now looked at reaching outside Salesforce and structuring inside Salesforce. Next, we'll explore how to make your declarative solutions themselves more adaptable and scalable. One key strategy is to move hard-coded logic out of automation and into data that can be easily maintained. This brings us to a deceptively simple but game-changing tool: **Custom Metadata Types**, which let you store and manage configuration data declaratively.

LEVERAGING CUSTOM METADATA TYPES FOR DECLARATIVE SCALABILITY

One hallmark of an advanced Salesforce solution is that it's **data-driven** and easily adjustable without changing the logic. In the context of declarative automation (Flows, validation rules, etc.), this often means avoiding hard-coded values or one-off branching that requires editing a Flow every time a business rule changes. Enter **Custom Metadata Types** – a feature that allows you to define custom data structures (similar to custom objects) whose records are metadata rather than regular data. These records can store configuration: think of thresholds, business rules, mappings, or settings that your automations refer to. By using custom metadata, you make your declarative solutions more scalable and maintainable, because administrators or even business analysts (with the right permissions) can update a business rule by tweaking a metadata record instead of modifying a Flow or Apex code.

Let's break that down with an example. Suppose your sales process has various tiers of discount approval: 10% discounts can be auto-approved, 10–20% require manager OK, above 20% require finance approval. A novice admin might implement this with a bunch of Decision elements in a Flow or multiple Approval Processes, each one checking a specific percentage value. That works, but what if those thresholds change? Or you introduce a new tier? You'd have to edit the flow or approvals (which means testing, deploying, maybe even downtime for changes). With custom metadata, you could instead create a custom metadata type called "Discount Threshold" with fields like Max_Discount__c and Approval_Level__c. You'd then create records: one record says "Max 10 -> Auto-approve", another "Max 20 -> Manager", another "Max 100 -> Finance". Your Flow can be simplified to a loop or lookup: when a discount is requested, the Flow **queries the custom metadata** to find which band the discount falls into, and then it knows who needs to approve. If tomorrow the business says "change the manager threshold to 15%", you just update the metadata record (from 20 to 15) – no flow changes needed. The logic was effectively externalized as data.

Custom Metadata Types are defined in Setup (just like objects). You add custom fields to them to structure your config data. The records of custom metadata can be packaged and deployed through metadata API just like object definitions, which is a huge advantage over the older approach, **Custom Settings.** (Custom settings also stored config data, but were org-specific and not easily deployed between environments with their values.) With custom metadata, your configuration values become part of your deployment artifacts, ensuring consistency across sandboxes and production. Also, custom metadata records can be referenced in formulas via the $CustomMetadata global variable, enabling dynamic formulas. For example, a validation rule could compare an Opportunity's discount to a value stored in a custom metadata record (e.g., $CustomMetadata.Discount_Threshold__c.Manager.Max_Discount__c) rather than a hardcoded percentage, meaning if the threshold changes, the validation rule automatically uses the new value.

From a declarative automation perspective, you'll typically leverage custom metadata in **Flows** by using the **Get Records** element to query the custom metadata records, or by using formulas that refer to them. One pattern is to load a set of custom metadata records and loop through them in a Flow. Let's say you have a custom metadata type for "Data Quality Rules" that defines a set of criteria and messages for records that don't meet standards. A single after-save Flow on, say, a Contact could retrieve all active "Data Quality Rule" records and then loop, checking each rule against the Contact's data (maybe using formula fields or decision logic), and accumulate any issues found. If a new rule is needed, you add a new custom metadata record – the Flow automatically picks it up. This is a very scalable design: one flow to handle many rules, driven by data that non-developers can adjust.

Another real-world use is mapping configurations. Imagine you integrate with multiple fulfillment vendors, and depending on a product, a different vendor should handle the order. You could store a mapping in custom metadata: Product Category -> Vendor. Your flow, when creating an order, does a quick lookup to find which vendor applies, then routes the logic accordingly. If a mapping changes (new vendor or product line), update the metadata, and the flow's behavior updates instantly.

Custom metadata types can also serve as feature toggles or settings for your org. Have an advanced Flow that you only want active under certain conditions? Instead of hardcoding a check, you might have a custom metadata "Org Settings" with a field "Enable_NewTriageFlow__c = true/false". The flow can reference that. This way, if you need to quickly disable the flow's functionality (perhaps something went wrong in production), you flip the switch in metadata. This idea of a "kill switch" or feature flag is a best practice in software development and can be applied declaratively via custom metadata.

Pro Tip: If you use custom metadata for feature flags or toggles, consider building a simple Lightning App Page for admin users to manage those settings, or at least document them well. Since custom metadata is typically accessed via Setup (which only certain users might have access to), you want to ensure the people who might need to toggle something can do so quickly. You could even use a Screen Flow or a Lightning component to expose those settings in a controlled way (maybe only for System Admins or through a custom permission).

Let's address how an AI like ChatGPT can assist with custom metadata planning and usage. First, designing the structure: you might describe your business rules to ChatGPT and ask, "How can I model this using custom metadata types?" The AI might help you outline what fields you need and how many types. For instance, say you have a complex pricing logic – ChatGPT could suggest using one custom metadata type for product base prices and another for regional price adjustments. It's like brainstorming with a colleague. You can also have ChatGPT draft sample records in CSV format, which you could then upload via Data Loader to create the metadata records (since custom metadata records can be loaded like any object, though you often deploy them via deployment tools). If you have a spreadsheet of rules from business users, you might even paste a snippet into ChatGPT and ask it to convert that into a set of SOQL insert statements or a formatted JSON that matches your custom metadata – basically using it as a translator.

Another helpful use is generating references for formulas. The syntax for $CustomMetadata in formulas can be a bit finicky (it requires the type and record developer names). You could give ChatGPT the names and ask for the exact formula syntax to retrieve a value. Or if you're writing an Apex invocable action that interacts with custom metadata, ChatGPT can stub out that Apex code. We'll cover invocables more soon, but note that reading custom metadata in Apex is straightforward and ChatGPT likely "knows" the pattern (just a SOQL query on the metadata type, which is accessible in Apex).

ChatGPT Prompt: "We have a set of loan approval rules that vary by country and loan amount. I want to use custom metadata to avoid hard-coding these rules in flows. How might I design the custom metadata types and records? And how would a Flow use them to apply the right rule?"

When using ChatGPT for these purposes, remember to abstract any sensitive business data. You can say "Country A, threshold X" rather than actual internal numbers if needed, or use dummy data to illustrate. If you do need to share something like an exported CSV of custom metadata records for analysis, ensure it doesn't contain secrets and leverage your privacy settings as mentioned earlier.

Custom Metadata vs. Custom Settings vs. Custom Objects – a quick note for clarity. **Custom Settings** (particularly List type settings) were the old way to store config data accessible in formulas and code. They are still around and have the advantage of allowing org-specific values (Hierarchy settings can vary by profile or user). But they are not easily deployable with values (you often had to script the creation of the settings in each environment) and they aren't as powerful in formulas (you can reference only one custom setting record in a formula easily, whereas you can reference multiple custom metadata records if needed). **Custom Objects** could also be used to store config, but they are meant for business data primarily – using them for config means those records count against data storage, and you have to ensure they're not modified by users inadvertently, etc. Custom metadata records don't count against data storage and can be hidden from typical users (they're usually only accessible via API or Setup). So, for most cases where you want "constants" or "rules" that govern logic, custom metadata is the recommended approach today.

From a scalability standpoint, one of the biggest benefits of custom metadata is consistency across environments. When you deploy your changes (say from a sandbox to production), your flows, validation rules, **and the referenced data** (the metadata records) all go together. This avoids those infamous "works in UAT but not in prod" issues because someone forgot to set up the same custom setting values in prod. With metadata, what you test is what you deploy. Seasoned architects love this because it aligns with proper DevOps processes.

Another advanced use of custom metadata is to drive **dynamic Apex or dynamic Lightning components**, but that veers into code. For example, you might have a generic flow that calls an Apex action which reads a custom metadata to decide which of several actions to execute – effectively using metadata to reconfigure code behavior. This is an example of composing programmatic and declarative solutions. Even if you're not writing such code yourself, be aware that some managed packages use custom metadata to allow admin users to toggle features or provide inputs without the package having to be custom-coded for each org. It's an architecture pattern you'll see often: "configuration over code."

As you incorporate custom metadata into your declarative solutions, you start achieving something powerful: **scalability in maintenance.** Imagine a year from now, your org has twice as many business rules and processes. If you've built them with metadata-driven flows and event-driven architecture, adding new rules or processes doesn't mean exponentially more flows or a tangle of logic. It might just mean adding rows to a table (metadata) or hooking a new subscriber to an existing event. This is akin to treating parts of your application like a mini database of logic that the declarative engine reads from. It's a very "enterprise" way of thinking – externalizing variability and keeping the engine (flows) as generic as possible. It's also very friendly to collaboration between admins and business analysts: a BA could be empowered to adjust a threshold in a custom metadata record given the proper access, rather than asking IT to modify automation.

We've now seen how to integrate externally, decouple internally, and make our logic data-driven. Next, we'll look at how to **extend** declarative automation itself when its native capabilities aren't quite enough. This involves stepping slightly into the world of code, but in a controlled way that complements our no-code solutions. In the following section, we'll discuss invocable actions and other ways to expand Flow Builder's reach without abandoning declarative approach.

EXPANDING FLOW CAPABILITIES WITH INVOCABLE ACTIONS

Flow Builder is incredibly powerful out-of-the-box – it can create, update, delete records, post to Chatter, send emails, submit approvals, make decisions, loop, and even call external services as we covered. But every experienced Salesforce professional eventually encounters a requirement where a Flow's native features fall just short. Perhaps you need to perform a complex calculation, manipulate text in a way formulas can't, or integrate with something in a manner that External Services doesn't support. Rather than switching wholesale to Apex and writing an entire trigger or batch job, a best practice is to fill these gaps with **invocable actions**. Invocable actions are like **plugins for Flow**: small chunks of Apex (or even External Service callouts or other processes) that can be dropped into a Flow as an "Action". By leveraging invocable actions, you extend the declarative toolkit while keeping the overall orchestration in the no-code realm.

An invocable action in Apex is simply a class with a method annotated with @InvocableMethod (and optionally an @InvocableVariable for structured inputs/outputs). These Apex methods can then be exposed to Flow Builder. For example, suppose your users want a flow that, given an address, finds the nearest Salesforce Account (perhaps representing a store or partner) and returns it. There's no out-of-the-box Flow feature to do geolocation queries or complex distance calculations. However, a developer could write a small Apex method findNearestAccount(latitude, longitude) that queries Account records by location and returns the closest one. With the invocable annotation, that method becomes available in Flow: the admin building the flow simply sees an action "Find Nearest Account" (perhaps under a custom category), plugs in the coordinates (which the flow might have gotten via some address-to-coordinates service), and then uses the result downstream in the flow (maybe to display info to the user or assign the case to that nearest office). The heavy lifting (spatial math and SOQL) is done in a few lines of Apex, but the process and UI remain declarative.

What's beautiful about invocable actions is that they encourage a **modular approach**. Instead of writing one big Apex class to do everything, you write small, focused actions that do one thing well and can be reused across flows. Over time, you might build or accumulate a library of such actions – or better yet, leverage the community and AppExchange. Salesforce and third-party vendors have released many pre-built invocable actions. Some are official (like Salesforce has actions for posting to Slack, which we discussed, or some basic math operations). Others come from partners – for example, a CTI (telephony) vendor might provide an invocable action to send an SMS or initiate a phone call. You can find packages for actions to generate PDFs, perform advanced string manipulations (e.g., a Regex matcher), do calendar calculations, and more. A very popular community source is **UnofficialSF.com**, where Salesforce community contributors share free, open-source Flow extensions, including both invocable Apex and Flow Screen Components (which allow custom UI in flows). As an advanced declarative builder, it's worth exploring these resources before you decide to write your own Apex. Why reinvent the wheel if someone already packaged a "Create Google Calendar Event" invocable action or a "Sort Collection" action?

To use an AppExchange or open-source action, you'll typically install a package into your org. Once installed (and after maybe a permission set assignment if required), those actions just appear in your Flow's Action palette. From the admin user's perspective, it's as if Salesforce magically gained a new feature. For instance, after installing a Twilio SMS action (a hypothetical example), your Flow can now have a step "Send SMS" where you configure phone number and message, all without writing a single line of code yourself – the code is embedded in the package. When using third-party actions, always vet them: ensure they come from a reputable source, check if the code is visible or if the provider is a known Salesforce partner, and test them in a sandbox. They run in your org, so you want to be sure they behave and don't hit limits or have side effects.

Now, if you do need to create a custom invocable action (maybe your requirement is too niche for any pre-built solution), ChatGPT can be a surprisingly helpful assistant. Many Salesforce developers have turned to ChatGPT to generate Apex code snippets. You could prompt: "Write an Apex invocable method that takes a list of record Ids and returns a map of Id to record Name for a given sObject type" – and it will likely produce a decent starting point. You'd still need to know how to deploy Apex (e.g., using Salesforce DX or change sets), and you must write a test class for it (Apex code needs test coverage to be deployed to production). ChatGPT can even help with the test class generation if you ask. However, you should thoroughly review and tweak any AI-generated code. Ensure it follows best practices (bulkification, security with sharing, etc.). The advantage is that ChatGPT can do the boilerplate for you and save time, especially if writing Apex isn't your primary skill set. Some admins use it to bridge the gap: they describe what they want in natural language and get a code template that they can hand off to a developer or refine themselves.

ChatGPT Prompt: "Generate an Apex invocable action that takes a text string and returns the text in reverse order. I want to use this in a Flow to display a reversed string."

The above prompt would result in a simple class, which you could then deploy. It's a trivial example (you probably won't need a reverse string in real life), but it illustrates the point: ChatGPT can handle from trivial to fairly complex logic in Apex generation. If your invocable needs to call an external API (maybe a unique integration scenario that External Services can't handle), ChatGPT could draft the callout code including the HTTP request and JSON parsing. Again – review for security (use Named Credentials, avoid hardcoding endpoints, etc.).

Another emerging capability is the ChatGPT Operator tool, which, in the context of development, might help manage files and even perform deployments in the future. We're seeing the early days of AI-assisted software development. One can imagine down the line saying: "Operator, create an Apex class for me with this specification and deploy it to my Dev sandbox." Some of this is possible today with scripting, but having a conversational agent manage it would be new. As we stand now, you still have to take the code from ChatGPT and put it into Salesforce (via Developer Console or VS Code), run tests, deploy, etc. That's a healthy process because it forces a human in the loop to validate. Always remember that deploying anything to Salesforce – be it a flow or Apex – should follow your org's change management and testing protocols.

Beyond Apex, note that Salesforce has invocable actions for things like Process Builder (legacy), and even you can create invocable actions via Flow (subflows can be invoked as actions in other flows). However, here we're focusing on Apex invocables as the key to extensibility. Also, keep in mind that Lightning Web Components (LWCs) can be used in Flow Screens for custom UI – that's another form of extending Flow's capability, by adding new screen components (for example, an LWC that shows a graphical chart or a multi-select dropdown). Creating those requires code as well (JavaScript/HTML), and while not invocable actions, they similarly expand what you can do in a Flow screen beyond text boxes and picks. An admin might collaborate with a web developer to build a rich component, say a map display, which can then be dragged into a Flow Screen. Together, invocable Apex actions and custom screen components allow a nearly unlimited expansion of Flow's powers while keeping the overall logic in Flow.

Pro Tip: Maintain a registry or documentation of all custom invocable actions in your org, especially those that are home-grown or installed via packages. Because these actions run behind the scenes, other admins might not immediately know what they do or where they come from. A simple way is to name them clearly (the @InvocableMethod annotation has a label attribute you can set, e.g., label='Find Nearest Account'). You might also comment your flows to indicate which actions are custom and where their code lives. Down the line, if something needs updating (say the API version of that Twilio action or a bug fix in a custom action), you'll thank yourself for having that info readily available.

Let's illustrate how invocable actions can combine with the earlier topics: Suppose we want to implement a robust **"Case Escalation" process.** When a support Case is escalated, we want to: (a) evaluate a complex set of criteria to assign a priority score, (b) send notifications to various teams (with some logic on who to notify), and (c) integrate with an external incident management system by creating a ticket there. A possible solution:

- **Flow**: a record-triggered flow on Case runs when Escalation checkbox = true.
- Flow first calls an invocable action "Calculate Escalation Score" (Apex) which takes in case details and returns a numeric score (the logic might involve text analysis of description or past history – easier to do in Apex or even via an AI call).
- Flow then makes decisions based on the returned score and perhaps case product area (maybe using custom metadata to decide thresholds as in previous section).
- For notifications: flow uses an invocable action "Get Escalation Contacts" (could be Apex or External Service to query a contacts database) to figure out who should be notified.

- Then the flow posts to Slack (using Slack invocable action or built-in Slack action) and sends an email (standard action).
- For external incident system: maybe we have no direct External Service, but we do have an Apex invocable action "Create Incident in ServiceNow" that handles that via a callout.
- Additionally, the flow could publish a Platform Event "CaseEscalatedEvent__e" for any other systems or processes internally that care (maybe for analytics or to trigger a customer SMS via another system).

In this scenario, the Flow orchestrates everything declaratively. We sprinkled in invocable actions where needed for specialized tasks. Each action can be developed and maintained separately. If tomorrow the way we calculate the escalation score changes (say we incorporate an Einstein AI prediction instead), we can adjust that one action (or even replace the action in the flow with a different one) without rewriting the rest. If the external incident system changes its API, we only update the "Create Incident" action. The Flow doesn't need to change unless inputs/outputs change drastically.

Invocable actions remind us that "declarative vs. programmatic" isn't a black-and-white dichotomy. The best solutions often use a blend. It's akin to cooking: flows are the recipe (the orchestration), invocable actions are like pre-made ingredients or spices that you add for flavor that you can't get otherwise. As an architect or advanced admin, you strive to keep as much logic declarative as makes sense (for clarity, admin ownership, and agility), and insert code in small doses where it genuinely adds value or efficiency. This is the essence of an extensible, composable approach.

We've covered a lot of ground on extending flows and avoiding unnecessary code by using just a bit of code smartly. This naturally raises the question: how do we know when we've stretched declarative to its appropriate limit? In other words, how do we discern the line where a problem is better solved with Apex or a programmatic solution rather than with flows and clicks? In the next section, we'll tackle that decision-making process and discuss the limits of declarative tools, ensuring we maintain a balanced approach.

UNDERSTANDING DECLARATIVE LIMITS AND WHEN TO USE APEX

No matter how powerful Salesforce's declarative tools become, there will always be edge cases and scenarios where writing code is either more efficient or even necessary. A savvy Salesforce professional isn't one who insists on "no code ever," but one who knows **when to lean on code** to complement declarative solutions. In this section, we examine the limits of declarative development: technical limits (governors and capabilities) and practical limits (maintainability and complexity). We'll provide guidance on identifying when you've reached those limits and how to decide if a requirement should be addressed with Apex (or another programmatic approach) instead of a purely declarative design. The goal is to preserve the "clicks-first" mindset while acknowledging that Apex and code have their place – often as an enabler of declarative, or as a more suitable tool for certain jobs.

First, let's talk about **technical limits.** Salesforce imposes governor limits on all operations to ensure one user's automation can't hog resources in a multitenant environment. Flows, being on the platform, share many of the same limits as Apex: SOQL queries, DML statements, CPU time, heap size, etc. If you design a Flow that loops through thousands of records performing complex logic, you might hit those limits just as an Apex developer would if they wrote inefficient code. Some limits are harder to manage in declarative world: for example, if you have a Flow with multiple "Get Records" elements inside a loop, you might inadvertently run more queries than the limit allows (100 queries per transaction typically). An Apex developer might catch that pattern and optimize by bulk querying outside the loop; an admin might not realize the Flow is doing something similar under the hood. Salesforce has improved Flow's engine over time to handle bulk records better (e.g., auto batching in before-save flows), but you still need to be aware of designs that don't scale.

Another area is **performance.** Complex flows, especially those with many elements and complicated branching, can become slower to save records or execute. A rule of thumb: if a single transaction (like a record save) triggers more than, say, 4-5 flows (between before-save, after-save flows on the same object, plus any related record flows via things like platform events or subflows), you should evaluate the design. It might be fine, or it might be getting hard to manage and approach a point where debugging or ensuring correct order is tough. Apex gives you finer control over order of execution and transaction handling. For instance, if two after-save flows both update the same record again (to do different things), they each cause a second round of saves (potentially causing recursion or just extra processing). An Apex trigger could handle those two things in one context, more efficiently. We're not advocating to immediately collapse everything into triggers – but being aware of when flows might step on each other's toes is key. Sometimes the answer is still declarative (maybe merge the flows or use a decision to avoid conflicts), but sometimes the answer might be to replace a tangle of flows with a well-structured Apex trigger for clarity and reliability.

Capability gaps can also necessitate Apex. For example, certain Salesforce APIs or advanced platform features are only accessible via Apex. If you need to, say, perform a complex SOSL text search across multiple objects and parse that to find some data – there's no Flow equivalent for SOSL queries at the moment. Or if you want to generate a PDF from HTML, you might use Visualforce/Apex because Flow cannot do HTML rendering. Another example: calling a SOAP web service. External Services supports REST APIs with OpenAPI, but SOAP (XML) services might require using the old WSDL2Apex (which generates an Apex class). If your project heavily relies on a SOAP integration, you'll likely involve Apex. In such cases, you often still use flows to orchestrate (maybe the flow calls an invocable method that wraps the SOAP call), so you keep the integration "declarative-first" in design but leverage code where needed to bridge the gap.

There are also **soft limits**, like complexity and readability. A Flow that handles a very complex set of conditions might become hard for humans to follow. Picture a flow with 10+ Decision diamonds, cross-connected lines everywhere, maybe a dozen variables – it can be the equivalent of spaghetti code. At that point, writing a clear Apex method with if/else logic and comments might actually be more maintainable and less error-prone. One way to gauge this is to ask: if a new team member looked at this automation, would it be faster for them to understand a well-written piece of code or the flow diagram? If the flow has turned into a labyrinth, it's no longer serving the goal of declarative (which is to be easily understood and managed without deep technical knowledge). You might refactor the flow into smaller subflows (decluttering), or decide to implement some parts in Apex for simplicity. A general best practice: flows should ideally implement one logical chunk of work – if you find one flow trying to do everything, consider breaking it up or rethinking it.

Testing and governance considerations also influence the declarative vs. code decision. Apex requires writing test classes, which is extra work but also ensures you think about edge cases and outcomes. Flows don't require tests to deploy; that speed can be a double-edged sword. On one hand, it's great for agility. On the other, it means without discipline you might deploy automation that hasn't been thoroughly vetted in all scenarios. If you have a business-critical process with lots of logic, you might prefer Apex simply so you can cover it with unit tests that run on every deployment. Salesforce is working on tools for automating Flow tests (there have been pilots for a Flow testing framework), but it's not as mature as Apex testing. Some teams mitigate this by writing Apex tests that indirectly invoke flows (like calling a small bit of Apex that triggers a flow, or verifying outcomes after a record-triggered flow runs in a test context), but that requires some coding anyway. In a regulated environment, you might have policies that any logic affecting financial data must have test coverage – which leans toward using Apex or at least carefully designed flows with a manual test script.

From a **forward-looking perspective**, each Salesforce release erodes some reasons to use Apex. For example, historically one of the reasons to use code was to create complex reports or roll-ups – then along came Declarative Lookup Rollup Summaries (a community tool), and later Salesforce's own Roll-Up Summary fields for Master-Detail, and now new features like Summary formulas in reports or even the possibility of doing multi-object roll-ups with tools. Another example: before Winter '20, you needed triggers to run automation on record save; then after-save flows replaced that need in most cases. We used to code for sending custom notifications or for doing scheduled jobs, and now we have Scheduled Flows. So the line is moving. We expect that trend to continue. It's quite plausible that what is an Apex-only task today (say, handling complex JSON parsing) might be offered in Flow tomorrow (maybe a "Parse JSON" element could appear, who knows?). As someone building for the long term, it's smart to keep an eye on upcoming features. It might influence whether you invest time in a coded solution or wait for a declarative alternative. That said, waiting isn't always an option when a requirement is on the table – so you do your best with what's available, and perhaps architect things in a way that a future swap-out is easy (for instance, if you write an invocable action because Flow can't do something, keep it small and focused; the day Flow can do it natively, you can remove that action and re-wire the flow).

It's worth mentioning the concept of **"polyglot" Salesforce development** – using the right mix of tools. A well-architected Salesforce solution might have 70% of logic in declarative (flows, formulas, etc.), 20% in Apex, and 10% in other things (maybe external scripts or integrations), or some ratio that fits the scenario. There's no pride lost in writing Apex where needed. The real measure is: are we delivering a solution that is reliable, maintainable, and meets requirements? Sometimes that means writing a 50-line Apex class instead of building a 300-element flow. Other times, it means the opposite: deleting 500 lines of legacy Apex and replacing with a configuration that the admin team can own going forward.

If you find yourself unsure whether a use case can be solved declaratively or if it's pushing too far, that's a great moment to consult colleagues or even ChatGPT. The community is full of discussions on "Flow vs Apex for X scenario". ChatGPT can help summarize pros and cons if you ask something like, "Should I use a Flow or an Apex trigger for a scenario where I need to update 10,000 records in batches?" It might mention that Apex with batchable interface could handle 10,000 records efficiently in the background, whereas a Flow might hit limits or be manual to chunk. Or consider using ChatGPT's Code Interpreter or Advanced Data Analysis (if available to you) by feeding it some pseudo-metrics (like "I have 50 flows on one object, is that too many?") – it might not give a perfect answer, but could outline concerns to consider.

ChatGPT Prompt: "I have a Flow that updates related records and sends emails when an Opportunity is closed. However, we have very high volume (hundreds of opportunities closing daily) and complex logic. Would it be better to use Apex for this, and if so, why? What would a hybrid approach look like?"

Another dimension to consider is **when to use other Salesforce features** instead of bending flows or code. For example, sometimes admins implement scheduled flows to do daily jobs like recalculating fields – but could that have been done with a report and analytic snapshot? Or a Big Object for data archiving instead of trying to store tons of records and process them in flows? Knowing the entire platform's capabilities (from Einstein bots to Scheduled Jobs to formula limits) helps make informed decisions. Earlier chapters in this book likely covered choosing the right tool for the job – here in the advanced stage, it's about recognizing when you're stretching a tool past its comfort zone.

Pro Tip: When writing Apex to complement declarative, keep the Apex logic as separated from the database as possible if you want admins to trigger it easily. For example, rather than embedding SOQL inside a large trigger logic, you might expose an invocable method that accepts data as parameters. That way a Flow could gather what it needs and call the action, and the action purely performs logic on the given input and returns output. This makes your Apex more reusable and testable. It also means if Salesforce introduces a declarative way to do part of it, you can gradually migrate pieces without rewriting everything.

To illustrate this balance: consider a **data archival** requirement. You have an object with millions of records, and you need to move older ones out periodically to a big object or an external storage. You could attempt a monstrous scheduled Flow to find and delete or archive records, but that's likely not efficient. A better approach is an Apex batch job – something flows can't do (except calling an invocable to run a batch maybe). So you schedule an Apex batch via Apex or even via a very small flow that invokes an Apex action to start the batch. The heavy lifting is in code because it's a heavy-duty operation. Meanwhile, the decision of *when* to run it and maybe the conditions (which could be stored in custom metadata) are still declarative. That's a pragmatic split.

It's also important to know when *not* to use Apex. A common error in judgment is writing code out of habit for things that flows can now do. If someone on your team suggests writing a trigger for something straightforward like "when field X is this, update field Y," pause and consider: a record-triggered flow might handle that elegantly with zero code – plus it's easier for an admin to tweak if the logic changes. Code has ongoing ownership costs; if you can achieve the same with clicks and it's within the bounds of good design, do it with clicks. Part of an architect's role is guarding against over-engineering with code when a simpler solution exists.

As Salesforce continues to evolve, the line between code and clicks will keep moving. We also see the emergence of *"low-code Apex"* in a sense – features like Apex defined calculated insights, or AI-assisted code generation blur the line. But fundamentally, always align with this principle: **Use the highest-level tool that accomplishes the task without undue compromise.** High-level means more declarative and abstracted if possible – because it's generally faster to build and easier to maintain – but without undue compromise means you won't twist a tool to do something it's not meant for at the cost of reliability or clarity.

Having a clear grasp of when to step outside pure declarative ensures that you deliver solutions that work well and can be maintained by your team's skill set. Often it's not an all-or-nothing choice; a blend can yield the best result. This hybrid mindset is key to Salesforce Enterprise Architecture: optimize for clarity, performance, and agility, whatever combination of tools that requires.

Now that we have a framework for deciding when to use declarative vs. code, let's appreciate how far we can actually go with declarative before hitting that boundary. In fact, using the advanced techniques covered so far, we can build surprisingly sophisticated solutions almost entirely declaratively. In the next section, we'll showcase an example of such a solution – combining integration, events, metadata, invocable actions, and more – to solve a complex business problem. This will demonstrate the art of assembling these pieces into a cohesive, advanced declarative implementation.

BUILDING SOPHISTICATED DECLARATIVE SOLUTIONS

By combining the advanced techniques we've discussed, you can create solutions that would have seemed nearly impossible without code just a few years ago. Let's paint a picture of one such **sophisticated declarative solution** to illustrate how far "clicks, not code" can really go. Along the way, we'll highlight which tools are used and how they interplay. The example we'll explore is a **Case Triage Wizard** for a support center – a guided intake and resolution assistant that leverages dynamic screens, external AI services, platform events, and custom metadata-driven logic, all orchestrated with Flow and a sprinkle of Apex invocables. It's the kind of solution that delivers a lot of business value (faster, smarter case handling) and is built with minimal code, showcasing extensibility and forward-thinking design.

Scenario: A company wants to improve how customer support cases are triaged and resolved. When a new case comes in (or when a customer calls in), support agents currently navigate multiple systems: they check troubleshooting guides, search for similar past cases, possibly invoke an external knowledge base or AI tool for suggestions, and then decide how to route the case (which team, what priority). This is time-consuming and inconsistent. The goal is to provide a one-stop guided experience for agents that helps them gather the right info and provides intelligent recommendations for solutions or next steps.

Solution Overview: We create a **Screen Flow** in Salesforce that acts as an interactive wizard. When a case is created or escalated, an agent launches this flow (possibly automatically via a quick action or even embedded on the case page as a dynamic component). The flow goes through several screens: asking the agent structured questions (or auto-filling from case data), presenting relevant information, and offering suggestions. Behind the scenes, it uses **Dynamic Forms** and conditional logic to adapt to the situation – for example, if the case is about a billing issue, it might show different fields or steps than a technical issue. It also calls an **AI service** (using External Services or an Apex action) to analyze the case description and suggest likely solutions (like Knowledge Articles or troubleshooting steps). Throughout the process, the flow might log key events (like "triage started" or "recommendation shown") via platform events for audit or metric gathering. At the end, the flow could even automatically escalate or route the case based on what happened (perhaps invoking an invocable action to update a field or call another system).

Now, let's break down the components and techniques:

Dynamic Screen Design: Using the power of Flow's screen components and conditional visibility (a feature where you can show/hide fields based on criteria), the wizard tailors itself. For example, the first screen could ask "What is the nature of the issue?" (with options like Technical, Billing, General). The second screen might branch: if Technical, show a component that collects product version and error details (maybe using a custom Lightning web component for file upload or screenshot). If Billing instead show fields for invoice number or billing date. This avoids a one-size-fits-all form and makes the agent's job easier. Dynamic Forms on Lightning Record Pages could also complement this – for instance, the Case record page might have a dynamic region that surfaces a "Run Triage Wizard" button or highlights certain fields if the case is of certain type. The combination of dynamic page layouts and the flow means the UI is highly responsive to context without needing hardcoded logic for each scenario.

Custom Metadata–Driven Logic: Suppose the company has a list of key questions or checks for different case categories. Instead of building each of those into the flow as literal screens, we could store them in custom metadata. For instance, a custom metadata type "Triage Checklist" with fields: Category (Tech/Billing/etc), Question_Text, Field_API (to map to a case field or a flow variable), and maybe a sequence number. We populate records like ("Tech", "Did you reboot the device?", "Reboot_Confirmed__c", order 1) and a few more. The flow, when it knows the category, can **retrieve all matching Triage Checklist metadata records** (Get Records where Category = "Tech") and then loop through them, dynamically generating a screen or a set of inputs for each question. This could be done with a bit of creativity (for instance, using a single generic screen that references a variable and having the loop update the variable with each question text, or using a screen component that can render a list of questions). Even if some Apex is needed to facilitate dynamic screens (like creating a custom LWC that takes a list of questions and displays a checklist), the metadata approach means adding a new question is as simple as adding a new custom metadata record – no flow changes. This is **declarative scalability** in action.

AI Integration via External Services or Invocable: One of the coolest extensibility points is calling an AI service to get recommendations. Salesforce has introduced Einstein GPT and other AI assist features, but let's say we want to use a generative AI (like an OpenAI API) that takes the case description and returns a summary or potential solution. We can do this by creating either an External Service (if OpenAI or an intermediary provides an OpenAPI spec) or by writing a small Apex invocable action that calls the AI API. For the sake of argument, imagine we have an Apex invocable action GetAISuggestions that sends the case description to an AI endpoint and gets back, say, three suggested Knowledge Article titles or steps. The flow can call this action after the initial details are gathered. It then receives the suggestions and on the next screen, presents them to the agent: e.g., "The AI Assistant suggests these possible solutions – (1) Check firewall settings, (2) Reset user password, (3) Refer to KB-101 article. Do any of these resolve the issue?" The agent can choose one if applicable, or skip if not. If one is chosen, perhaps the flow can automatically send the corresponding Knowledge Article link to the customer or log that suggestion was provided.

Platform Events and Orchestration: As the wizard progresses, it can publish events to keep other parts of the org in the loop or to trigger additional asynchronous processes. For example, if during triage it's discovered this issue is actually caused by a product bug, the agent might check a box "Potential Product Bug". The flow could publish a Potential_Bug__e platform event with the case info. There might be a subscriber flow that listens for this event and alerts the product team or creates a work item in a project management system (via another integration). This way, the triage flow doesn't directly integrate with the product team's system; it just signals an event. Similarly, at the end of the flow, if the case is indeed escalated, it could publish a Case_Triaged__e event which could be used for dashboards (maybe a Lightning web component listening via CometD to update a real-time dashboard of how many cases have been triaged today, etc.). Platform events ensure these side processes are decoupled from the main flow. The agent's interaction stays fast; anything extra happens in the background.

Flow Orchestration (where applicable): Salesforce offers a feature called Flow Orchestrator which could string together multi-step, multi-user processes. If our triage involves multiple people (maybe one step for front-line agent, another for a tier-2 specialist), orchestrator could manage assignments and hand-offs declaratively. This might be beyond our single wizard example, but it's worth noting as an advanced declarative tool: you can have something akin to a state machine or workflow engine coordinating complex sequences – all configured with clicks.

User Experience Enhancements: This advanced solution might also incorporate things like **screen flow embedded components** – for example, an LWC that displays a timeline of past similar cases on one of the flow screens (by querying cases with similar keywords via an invocable or External Service to an AI search). It could also use **file upload** components for the agent to attach screenshots directly in the flow steps, or a **progress indicator** (Flow has a built-in progress indicator for screen flows to show which stage you're in, improving usability).

The result of this solution: A support agent launches the triage wizard on a case. They see a tailored experience guiding them through relevant questions, possibly auto-filling known info. The system fetches intelligence (AI suggestions, related cases) in real-time and offers them within the flow. The agent, empowered with this info, either resolves the case immediately with the provided solution or properly categorizes and escalates it with a single click. Throughout this, the agent doesn't leave Salesforce, doesn't juggle multiple windows – everything is integrated. And if no solution is found and escalation is needed, the flow might automatically create a high-priority task for Tier 2 support, attach all the gathered info to it, and even inform the customer that it's been escalated with certain expected next steps (maybe via a quick email template sent).

From the development side: we achieved this with configuration plus targeted extensions. We used custom metadata for flexibility, platform events for decoupling, an external AI service for smarts, and a few invocable actions for glue. If the business wants to change the triage questions or add a new category of cases, it's a metadata and admin task, not a coding project. If they want to switch out the AI provider (say move to Salesforce's native Einstein GPT), we can replace the invocable action but keep the flow structure – minimal disturbance. If they realize they need to collect an extra piece of info, maybe it's as easy as dragging a new field onto a screen or adding a custom metadata record and field.

This example underscores how **forward-looking design** allows declarative tech to handle complexity. It's like constructing with Lego blocks: flows, metadata, events, and invocables are pieces that, when assembled thoughtfully, create something much grander than each piece alone. It does require an architect's mindset – you have to envision the end-to-end and ensure all the parts will cooperate. But Salesforce has reached a point where that vision is quite attainable without writing a full custom application from scratch.

One might ask: what about performance and limits in such a big solution? True, you should always consider those. In our triage wizard example, we likely want to keep the number of screens reasonable (not 20 screens, or agents will get fatigued – maybe use logical grouping, possibly even let them jump to sections). For AI calls, ensure you handle timeouts or errors (the invocable could catch exceptions and return a message like "AI suggestions not available now"). The flows and events should be designed so that if any piece fails, it's not catastrophic (e.g., if the product team integration event fails, it shouldn't break the agent's flow – just log it and proceed). We also should measure user adoption and experience: fancy flows are useless if agents find them cumbersome. So in a rollout, we'd gather feedback and possibly tune the screens or logic.

Now, let's consider how ChatGPT and similar tools can assist in creating such a solution:

- During design, you could **brainstorm with ChatGPT** on the flow structure. For example, feed it the scenario and ask it to outline steps or suggest what data to collect for each case category. It might list common troubleshooting steps or key fields which you might incorporate.

- For building the External Service schema or Apex for the AI call, as earlier mentioned, ChatGPT can help generate those.
- You might even use ChatGPT to create sample data for testing: e.g., "give me 5 example case descriptions and a likely solution for each" to seed your testing of the AI suggestion feature.
- If you're creating custom Lightning components for the flow (like that similar cases viewer), you could ask ChatGPT to outline a Lightning Web Component that takes a case ID and queries similar cases (it can't run SOQL, but it might pseudo-code it for you).
- Documentation is another aspect: ChatGPT could help draft a quick user guide for agents using this new wizard, summarizing how it works and the benefits. This guide could be refined and distributed as part of enablement.

One thing not to forget: **ChatGPT can generate visuals too (indirectly)**. If you want a flow diagram or wireframe of the wizard to communicate with stakeholders, you could describe it and see if ChatGPT (potentially with integration to image generation) can produce a mock-up of the screens or a diagram. With features like SORA (OpenAI's tool for generating videos from prompts, as hinted by new ChatGPT features), you could even imagine generating a short explainer video. For instance, SORA could potentially create a simple video of an agent interacting with a conceptual interface of the triage wizard. That might be too bleeding edge, but it's interesting to note how AI could assist in UX design communication. At the very least, a static wireframe or flowchart is doable – there are tools where you feed a description and get a draft UI. ChatGPT could output a pseudo-HTML or a textual description that is then converted by a design tool.

ChatGPT Prompt: "Create a step-by-step outline for a Case Triage Wizard flow that includes dynamic questions for different case types, an AI recommendation step, and an escalation step. Include which Salesforce features (metadata, events, etc.) would be used at each step."

Using such a prompt, you'd get a narrative you can compare with your plan, maybe catch something you missed. It's like having a second opinion in the design phase.

As you implement complex flows, a **Pro Tip** is to leverage the new **Flow Debug and Runtime options** Salesforce provides. For example, you can debug a screen flow with sample data without actually publishing it to users, to see the path it takes. Additionally, monitor the flow's performance in dev – if a screen is taking noticeably long to load (perhaps due to an AI call), consider UX tweaks like adding a spinner or asynchronous loading of that part. Some invocable actions can be made to run asynchronously (via the Queueable interface) if they don't need to return data immediately – something to consider for long processes triggered by the flow.

We should also mention **Einstein Bots or Next Best Action**, as they border on this kind of solution. Einstein Bots (chatbots) could handle some initial Q&A with a customer before a case is made, but in our scenario we focused on agent-side. Next Best Action is a declarative tool that can pop up recommended actions in the UI based on rules or predictive models. One could use Next Best Action to show a recommendation when viewing a case (like "Suggest Knowledge Article X to the customer"). That might actually be a simpler alternative to building an entire flow for suggestions. In our wizard, we essentially built that logic ourselves. It's a reminder that Salesforce often has multiple ways to tackle a problem – an advanced solution might combine many, but we should avoid redundancy. If NBA (Next Best Action) can do part of it, maybe use that instead of a custom AI call. The reason to go custom might be needing more control or integration of steps.

The bottom line is, sophisticated declarative solutions are very much achievable. They require thoughtful combination of features and sometimes a dash of code, but far less code than an equivalent custom app off-platform or even an all-Apex solution on Salesforce from 10 years ago. This signifies a broader trend: **low-code doesn't mean low capability**. We're in an era where you can solve complex, cross-system business challenges with mostly declarative configuration, given the right know-how. It's incredibly empowering for organizations because it means faster development times, easier changes, and more of the solution in the hands of those who understand the business problem (often admins or consultants) rather than exclusively with developers.

Now, as we construct ever-more complex declarative systems, we must also acknowledge the new challenges they bring. Complexity is complexity, whether in code or in config. There are potential pitfalls – performance bottlenecks, maintainability issues, security considerations, and more – when pushing the envelope. In the final section of this chapter, we'll discuss how to manage the risks that come with advanced declarative configurations, ensuring that our powerful solutions remain reliable and governed. After all, with great (declarative) power comes great responsibility.

MANAGING RISKS IN ADVANCED DECLARATIVE CONFIGURATIONS

The more sophisticated our declarative solutions become, the more important it is to manage the **risks** that accompany them. A complex flow or a web of automation can misfire just as dramatically as a buggy piece of code if not designed and overseen carefully. In this section, we'll cover strategies for keeping advanced declarative configurations under control: maintaining clarity, preventing performance issues, securing and governing processes, and having contingency plans for when things go wrong. We'll also touch on the unique considerations when using AI-assisted development and external integrations, to ensure that leveraging these cutting-edge tools doesn't inadvertently introduce vulnerabilities or compliance issues.

Complexity & Maintainability: One risk of building very advanced solutions with clicks is that they can become "black boxes" of logic that are hard to follow. A flow that chains to subflows and platform events and invocables might be tricky for a new team member to grok. Documentation and visualization are your friends here. Make liberal use of descriptions on your flow elements, input/output labels, and comments (Flow now allows adding text template elements that don't execute but can hold notes). Create diagrams that map out the process: for example, a Lucidchart or Visio diagram that shows "Case Triage Wizard" steps, or a schematic of how a platform event flows to various subscribers. Keep these docs updated as the solution evolves. It's similar to commenting code, but since the "code" is visuals and config, you may need external docs to capture the big picture. If your org has a lot of complex automation, consider maintaining an **automation catalog** – a living document or spreadsheet that lists all flows, what they do, when they run, and contacts for each. This helps when multiple admins or devs are collaborating.

For complicated flows, another tactic is modularization (just as we do in code). Break big flows into subflows or segments. Use consistent naming conventions so it's clear how things relate (e.g., prefix subflows with the parent process name). This way each piece is digestible. Yes, there's a balance: too many tiny flows can also be confusing, but generally smaller, purposeful flows are easier to test and debug.

Testing & Quality Assurance: As noted earlier, flows don't require automated tests like Apex does, so we must impose our own QA discipline. In advanced scenarios, it's wise to create a set of **test case scenarios** for your flows and automation. For example, for the case triage wizard, enumerate scenarios: technical issue solved by AI suggestion, billing issue requiring escalation, etc., and test each thoroughly in a full sandbox. Have end-users (like support agents) participate in UAT for user-facing flows – they'll identify usability issues or edge cases you didn't think of. When possible, test not just the "happy path" but also failure modes: disconnect the external API and see how the flow behaves when the AI call fails – does it show an error, and is that user-friendly? If a platform event subscriber flow fails, do you get notified (maybe via an error email from Salesforce)? How will you catch and fix such an error in production?

Consider simulating load if relevant. For instance, if your flow could be run by 50 users simultaneously (e.g., in a call center), does it hold up? Salesforce generally handles concurrency well, but if all those flows call the same external API, could that API rate-limit you? These are things to check in advance. If needed, implement throttling or graceful degradation – e.g., maybe limit how often the flow calls the AI for the same case, or cache results in a custom object for reuse within a short window.

Governor Limits & Performance: We touched on technical limits earlier; managing them in advanced configurations is critical. Keep an eye on Flow interviews (runs) that are long or that bulk process large data sets. Salesforce's "Flow Debug Details" and the **"Pause on Fault"**

option are useful for pinpointing where a flow fails. If a flow is hitting limits, refactor it – perhaps by splitting transactions (e.g., using platform events or scheduled paths to do some work later). There's also a relatively new feature: **Asynchronous paths in record-triggered flows** (formerly called "After Save – Run Asynchronously"). Use those for non-urgent actions to reduce transaction load. Also, prefer "before-save" flows for simple field updates (they are much more efficient) instead of after-save plus an update element.

Be mindful of **cascading automation**: for example, a record-triggered flow updates a record which triggers another flow on the same object. Salesforce tries to prevent infinite loops with a recursion flag on flows, but you could still get ping-pong effects or simply unnecessary extra processing. Try to consolidate where sensible, or use a flag field that flows set to indicate "don't re-run this logic" if you must update the same record.

Security & Access Risks: When we add complexity, we must ensure we're not violating security principles. All declarative tools respect Salesforce security by default (like user profile CRUD permissions and field-level security), but some advanced scenarios might bypass that – for instance, an invocable Apex running in system context can update fields the user couldn't via UI. That's fine if intentional (you might need to), but document these elevation points as they could be abused if misconfigured. Also, external integrations: an External Service callout or Apex callout might be transmitting sensitive data to an outside service. Make sure you're complying with data protection policies (e.g., don't send customer PII to an AI service without approval). Use encryption or field masking if needed. If you leverage ChatGPT or other AI during development (like uploading data to get help), as we stressed, ensure private data is protected.

One governance idea: If your org deals with a lot of sensitive data, your company might restrict usage of certain AI features (for example, some enterprises disable file upload to ChatGPT or limit access to Operator/SORA for security). An admin should be aware of these policies. If you're championing AI usage in your Salesforce practice, work with your IT security team to define what's acceptable (maybe anonymize data before sending to ChatGPT for debugging, etc.).

Deployment & Version Control: Declarative changes, especially flows and metadata, should ideally be tracked in source control just like code. This is part of DevOps governance (likely covered in earlier chapters). For advanced setups, insist on using Salesforce's DevOps Center or a source-tracking tool so that every change is logged. Not only does this give you rollback options, it also serves as an audit trail of who changed what and why – crucial when multiple people could be tinkering with a large flow or process. It also helps with **sandboxes staying in sync** – nothing worse than a bug that was fixed in one sandbox flow but another sandbox still has the old version and reintroduces the bug later by accident. Source control and CI/CD pipelines mitigate that.

Monitoring & Alerts: Once in production, how do you know if your declarative processes are working well? Admins should set up some monitoring:

Error emails: Salesforce will send emails to the last modifier of a flow if it fails at runtime (unless that's disabled). Make sure those aren't going to a dormant inbox. Better, create a routing rule or use a system email that multiple team members check. Or use a custom solution: e.g., have a fault connector on flows that writes to a custom "Flow Error Log" object and triggers an alert. There are community packages (like "Flow Error Handling Frameworks") that you can adopt.

Performance logs: Salesforce has an "Optimizer" and also you can use Event Monitoring (with the right license) to see slow transactions or flow performance. It might be overkill for most, but if you have mission-critical flows, consider monitoring their execution time.

User feedback loops: Especially for user-facing flows like our wizard, provide a way for users to give feedback. Maybe a simple Chatter group or form where they can report "Flow was confusing at step 3" or "Option X was missing". Engaging the users ensures you catch issues that aren't outright errors but impact usability.

Feature flags: We mentioned using custom metadata as a kill switch. That's a great risk mitigation. If a new advanced process starts sending unexpected results, being able to turn it off quickly is valuable. For example, imagine our AI suggestion has a bug and is recommending wrong solutions that annoy customers. We could have a config "Enable

AI Suggestions" = true/false. Flip it to false via a metadata update – and perhaps our flow is built to skip that step if false. This avoids emergency hotfix deployment. Obviously, that kind of design needs to be built in from the start (can't add it in the moment of crisis easily), so think ahead for particularly risky steps.

Over-Engineering and Scope Creep: Advanced features are exciting, but it's possible to get carried away. One risk is adding complexity "because you can" rather than because it's needed. Always tie features back to business value. For example, using three different AI services in one flow might be cool, but does it meaningfully improve outcomes or just increase points of failure? Simplicity where possible is still king. The best architects know which fancy features **not** to use if they don't pass a cost-benefit test.

AI Specific Risks: Since this book involves ChatGPT, let's address using AI in development and in the product itself:

Development phase: Relying on AI (like ChatGPT) to generate config or code can save time but also propagate mistakes. Always review AI outputs. There's a known phenomenon of AI responses sounding confident but being incorrect (hallucinations). Double-check any guidance against official Salesforce documentation or with a proof of concept. If ChatGPT suggests a solution that seems too good or too simplistic, test it. Possibly ask it for references or reasoning – sometimes that can reveal gaps ("it might cite a feature that doesn't exist or is pilot-only).

In-product AI: If you integrate something like an OpenAI API to provide responses to users (e.g., telling a customer something), you introduce risk that the AI says something inaccurate or inappropriate. Guardrails for that could include limiting the scope of AI-generated content (maybe only drawing from your knowledge base so it doesn't go off-script), or always having a human in the loop for customer-facing communications (maybe the AI suggests a response but an agent reviews it before sending). For internal suggestions (like to support agents), the stakes are lower, but you still don't want it giving wrong info that wastes time. Monitor the quality of AI outputs regularly.

Data privacy with AI: We can't stress enough – if you feed customer data into an AI service, ensure its allowed. For example, OpenAI's terms as of 2025 allow business users to opt-out of data being used for training. Salesforce's Einstein GPT is built with data privacy in mind (data stays within Salesforce/Einstein layer supposedly), but verify with official sources. If you're in a regulated industry, run any AI usage by compliance teams. One way some solve this is by using "fake" or abstracted data in AI prompts (like encoding names or using categories instead of raw personal data).

Governance and Reviews: Treat major declarative changes with the same rigor as code changes. Have design reviews for a new complex flow – gather the team and walk through it. Maybe even do a dry run demonstration. Often a colleague can spot a potential issue ("Hey, what if two agents run this at the same time? Does that matter?"). Also plan for *governance checkpoints*: if you know a flow will need updates, maybe schedule a review of it 6 months later to see if it's grown unwieldy or if it's still serving well. Salesforce evolves, so maybe there will be a new platform feature that can simplify what you built; a periodic assessment can identify opportunities to refactor using newer, better tools.

Recovery Strategies: Despite best efforts, something might go wrong in production. Have a plan. For flows, know how to quickly deactivate them if needed (have a runbook: "go to flow X version Y, deactivate, activate alternate if exists"). For complex processes, maybe have a manual fallback. For instance, if the triage wizard fails, agents should know how to do things manually as a backup until it's fixed. Communicate outages of automation if user-facing ("We're currently experiencing an issue with the case wizard, please use the standard case form meantime"). These are operational considerations often overlooked for declarative stuff because we think "it's just clicks, what can go wrong?" – but at scale, it's part of the system and can have downtime or bugs.

Pro Tip: Leverage Sandbox "full copy" or "partial copy" environments to test not just functionality but also data-related edge cases. Advanced solutions often interact with lots of data. A flow might work fine with 10 records but behave differently with 10,000 records or unusual data combinations. A full-copy sandbox with realistic data can reveal those issues before you deploy. For example, maybe our flow runs slower in prod because it queries a big object with millions of rows – something you wouldn't see in a dev sandbox with sparse data.

Lastly, **stay educated**. The risk of not knowing is significant in a fast-moving platform. Keep up with Salesforce release notes – they might introduce a feature that obviates some custom workaround you implemented (great, you can simplify now!), or they might flag a change in behavior (like a limit increase or a deprecation that could affect you). Engage in the community: often other advanced admins share war stories of flows gone wild or lessons learned from big projects. Learn from those to avoid repeating mistakes. For instance, reading a community post about "Flow Bulkification tips" could save you from hitting a similar snag.

Managing risk in advanced declarative work is about combining the careful rigor of traditional software engineering with the agility of admin-centric development. The tools might be different (you're clicking instead of typing code), but planning, testing, monitoring, and iterating remain paramount. With sound risk management, you ensure that all the powerful solutions you build continue to run smoothly and deliver value rather than headaches.

Having navigated integration, events, metadata, invocables, limits, sophisticated use cases, and risk mitigation, you've significantly expanded your perspective on what's possible declaratively on Salesforce. In the chapter summary next, we'll recap these key insights and prepare to conclude this book by looking ahead to Salesforce's future and how you, as an architect or admin, can stay ahead of the curve.

CHAPTER SUMMARY

Advanced declarative configuration in Salesforce allows us to deliver integration and automation solutions once thought possible only with custom code. In this chapter, we explored a range of techniques that push the boundaries of "clicks, not code":

Declarative Integration: We learned how External Services and Flow HTTP Callouts enable Salesforce to connect with external APIs without Apex. By importing OpenAPI definitions or configuring HTTP steps in flows, an admin can weave outside data and actions into Salesforce processes. We also touched on Slack integrations and MuleSoft Composer, demonstrating that even multi-system workflows can often be handled with declarative tools. The key takeaway is that many integration use cases – from calling a payment service to posting a Slack alert – can be accomplished with configuration, provided we plan for security and error handling. ChatGPT can assist in crafting API schemas or troubleshooting callouts, acting as a smart companion in our integration efforts.

Platform Events for Decoupling: We examined how platform events implement a pub/sub model inside (and beyond) Salesforce, allowing processes to be decoupled for greater flexibility and reliability. Instead of chaining everything in one flow or trigger, an event-driven approach lets us trigger multiple independent flows or external listeners. This decoupling reduces conflicts and makes complex orchestrations easier to manage. Our case studies showed that a single event (like a Case update event) can fan-out to numerous actions handled separately. The reader should remember that with events, we embrace asynchronous processing and need to monitor the event bus for any bottlenecks. Nonetheless, platform events empower a modular architecture in Salesforce, aligning with modern best practices for scalability.

Custom Metadata for Scalability: We dove into custom metadata types as a means to store business rules and configuration data. By moving constants and logic out of flows and into metadata, we make our automations far more adaptable. One flow can handle many variations by looping through metadata records or looking up a value, rather than containing hardcoded branches for each scenario. This pattern – externalizing logic – is a hallmark of enterprise-grade design. It allows non-developers to update rules (by editing metadata in Setup) and promotes consistency across environments since metadata is deployable. We discussed examples like dynamic discount thresholds and feature toggles. The overarching lesson is to avoid "hardcoding in Flow" just as much as one would avoid hardcoding in Apex; use metadata to drive logic wherever feasible.

Invocable Actions and Extensibility: Recognizing that flows can't do absolutely everything, we covered how invocable Apex actions and AppExchange components extend Flow's capabilities. With small, focused bits of code (or installed packages), we can enhance flows with functions like complex calculations, external system updates, or advanced data processing. The chapter emphasized using existing solutions when available (why code an SMS sender if a verified package exists?) and writing custom invocables when needed to fill gaps. We also related how ChatGPT can help generate or review such Apex code. The big picture is that Apex and declarative can work in tandem – code becomes a reusable "power tool" in the predominantly declarative toolkit, rather than the main construction method.

Knowing the Limits – When to Use Apex: Just as crucial as knowing what declarative can do is knowing its limits. We identified scenarios where Apex or other programmatic approaches are more appropriate: extremely high-volume processing, scenarios requiring fine-grained error control or transactions, or functionality not exposed to flows (like complex regex or certain APIs). Importantly, this isn't a failure of declarative approach but a conscious architectural decision to use the right tool for the job. We stressed a "clicks-first but not clicks-only" mindset. Use flows for what they're great at – rapid development, easy maintenance, clarity – and don't shy away from invoking Apex for the heavy lifting or edge cases. This balanced approach leads to solutions that are both robust and maintainable.

Sophisticated Solutions with Minimal Code: Through the case triage wizard example, we demonstrated how all these advanced techniques can interlock to create a powerful solution that feels seamless to users. Dynamic screen flows, AI integration, event publishing, and metadata-driven branching all came together in a scenario that improves support efficiency. The example serves as inspiration for how one might architect declarative solutions that are **comprehensive** – spanning multiple features and even AI capabilities – without losing the advantages of declarative (speed of change, readability, lower technical debt). It's clear that Salesforce's platform, combined with creativity and good design, enables us to solve complex business challenges in a low-code way.

Risk Management: Finally, we highlighted that with great power comes great responsibility. Advanced declarative setups must be treated with the same diligence as complex code: careful design, thorough testing, continuous monitoring, and strong governance. We covered strategies for preventing automation chaos – from using documentation and source control, to implementing error handling and kill switches. This ensures that our elaborate flows and integrations remain assets, not liabilities. We also discussed being mindful of AI's limitations and ensuring compliance and security when using new tools like ChatGPT, Operator, or Sora in our workflow or solutions.

Throughout the chapter, we interwove the role of ChatGPT and similar AI helpers in boosting productivity and insight at each stage. From generating OpenAPI specs and Apex templates to brainstorming design and creating user prompts, AI is positioned not as a replacement for skilled professionals, but as an empowering assistant. Experienced Salesforce practitioners can leverage these AI tools to accelerate solution design, validate approaches, and even create artifacts (like diagrams or draft documentation), all while maintaining critical oversight. The future likely holds even tighter integration between Salesforce development and AI-driven support (imagine conversationally building a flow, or having an AI agent proactively suggest optimizations in your org).

In summary, Chapter 14 has armed you with advanced tactics to extend Salesforce declaratively: integrating external services, decoupling processes, scaling with metadata, and judiciously blending in Apex when needed. You've seen how to maximize the "clicks" approach without compromising on solution scope or quality. These techniques allow architects and admins to deliver innovation rapidly and with agility, which is a cornerstone of Salesforce's value proposition.

As we conclude this chapter and prepare to move into the final chapter of this book, it's a good moment to reflect on how far the Salesforce platform has come in enabling declarative development – and to look ahead at what's on the horizon. The concluding chapter will synthesize these best practices and discuss the future outlook for Salesforce and declarative config, from upcoming platform features to the evolving role of AI and automation in an admin's world. With your advanced toolkit in hand, you are not just ready to solve today's complex problems; you're also poised to adapt to tomorrow's changes and opportunities in the Salesforce ecosystem. The journey of learning and refining doesn't stop here – in fact, it's accelerating, and the next (and final) chapter will ensure you're prepared to ride that momentum forward.

15

CONCLUSION AND FUTURE OUTLOOK

You've traveled a remarkable journey from the first page of this book to this final chapter. Along the way, we've navigated the world of "clicks, not code" together—exploring how to build powerful Salesforce solutions using declarative configuration. It's been quite the ride, guided by a simple but powerful idea: pairing Salesforce's point-and-click capabilities with the intelligence of ChatGPT can transform the way you deliver value. Now that you've seen this in action, it's worth pausing to appreciate just how far you've come and how this approach has empowered your work.

For Salesforce developers, administrators, business analysts, and product owners alike, the core goal of this book has been to help you use ChatGPT to improve declarative Salesforce configuration. In practical terms, that means **bridging knowledge gaps, accelerating solution design, and refining best practices** with an AI assistant by your side. We set out with a friendly, down-to-earth tone to make advanced concepts accessible, and a practical mindset to ensure every lesson could be applied on Monday morning. The result? You've gained not only a deeper understanding of Salesforce's declarative toolkit, but also a newfound appreciation for how ChatGPT can help brainstorm, troubleshoot, and even inspire new ideas in your daily work.

In each chapter, you uncovered strategies and techniques to build scalable, maintainable Salesforce solutions without writing a line of code. You learned to think like a well-architected *Salesforce pro*—ensuring every field, automation, and page layout has a purpose—and you discovered how ChatGPT can act as a versatile partner in that process. When faced with a complex requirement, you saw how a well-crafted prompt could yield an elegant formula or Flow outline. When cleaning up an older org, ChatGPT's suggestions helped highlight redundant fields or unused automations to deprecate. As you progressed, the message became clear: armed with solid best practices and an AI helper, you can deliver high-quality configurations faster and with more confidence.

We explored **purposeful design** (making sure each configuration element serves a real need), **proactive governance** (keeping your org clean, consistent, and audited), **rigorous security** (protecting data and privacy by design), **thorough testing** (catching issues early with careful validation and user feedback), and **user-centric performance optimization** (ensuring everything you build runs efficiently and delights users). These themes, revisited throughout the book, are the pillars of declarative success. Each concept was rooted in real-world scenarios, showing that high-quality declarative work isn't abstract theory—it's practical and attainable. And at each step, we saw how following these principles in tandem with ChatGPT's assistance turns challenges into opportunities: whether it was double-checking a sharing rule configuration or brainstorming improvements to a page layout, the interplay of human expertise and AI insight proved to be a winning combination.

Through it all, ChatGPT has been a recurring sidekick in our story. From generating complex formula syntax to brainstorming Flow designs and user stories, we've seen how asking the right questions can unlock incredibly useful answers in seconds. We've uploaded snippets of configuration and received instant analysis. We've tried different prompt styles—typing out detailed instructions, speaking conversationally to ChatGPT, even feeding it documentation—to see what yields the best help. At every turn, we were reminded that while ChatGPT won't replace our own judgment or creativity, it can **augment** our skills. Think of it as a tireless junior admin or developer who's read every Salesforce manual and is always available. By leveraging ChatGPT's strengths—its vast knowledge, pattern recognition, and speed—we freed ourselves to focus more on strategy and empathy: understanding user needs, aligning with business goals, and crafting solutions that really work.

Now, as we arrive at this concluding chapter, our focus shifts to the road ahead. Salesforce is continuously evolving—and so must we. In the pages that follow, we'll **synthesize the core lessons** you've learned into a lasting mindset for the future, and then put on our futurist hats to explore what's next. We'll discuss how the role of Salesforce professionals is changing in an era of permission-set-first administration and dynamic user experiences. We'll anticipate advances in the declarative toolkit that promise to push "clicks, not code" to new heights. We'll examine the growing influence of artificial intelligence and automation (including Einstein GPT and beyond) on how we build and maintain Salesforce apps. We'll talk about cultivating a habit of continuous learning and adaptation to keep your skills sharp in Salesforce's rapid release cycle, and the importance of staying engaged with the Trailblazer community for support and inspiration. And finally, we'll tie it all back to what matters most: **delivering long-term business value** with everything you build. Consider this chapter not just a conclusion, but a springboard—an inspiring call to action to carry your newfound knowledge and confidence forward into the ever-brightening future of Salesforce.

SYNTHESIZING CORE LESSONS LEARNED

The best way to prepare for the future is to ensure we've truly absorbed the lessons of the past. So, before looking ahead, let's consolidate the core insights we've gathered throughout our exploration of declarative Salesforce best practices. By synthesizing these lessons, you'll have a crystal-clear picture of what "doing it right" looks like — a picture you can carry into every project from here on out. These aren't just bullet points to remember for a quiz; they're the guiding principles that elevate your work from good to great, ensuring that each solution you build is robust, user-friendly, and aligned with business goals.

First and foremost, this book reinforced why **"clicks, not code"** is more than a catchy motto — it's a competitive advantage. We saw how lean, configuration-driven development leads to faster delivery of features and easier changes down the road. When the sales team needs a new field or the service team asks for an automation tweak, a Salesforce admin armed with declarative tools can deliver in hours or days, not weeks. The business value here is huge: quicker time-to-market, lower development costs, and the flexibility to adapt as requirements evolve. One core lesson is that writing Apex code or building custom apps should be a last resort after exploring declarative options. Not because code is "bad," but because Salesforce's platform offers so much power out-of-the-box that leaning on those capabilities first yields simpler, more maintainable solutions. We learned to always ask: *"Can I solve this with configuration?"* — and usually, the answer was yes.

ChatGPT Prompt Example: "What are the top five benefits of using declarative (point-and-click) configuration in Salesforce instead of writing custom code?"

Another fundamental takeaway was embracing a **well-architected mindset** in everything we build. Salesforce's own guiding framework distills this into three words: **Trusted, Easy, Adaptable**. We interpreted these principles throughout our journey. *Trusted* means solutions are secure, reliable, and governable — for example, setting proper field-level security so sensitive data stays protected, and establishing naming conventions so everyone trusts what they see in the setup. *Easy* means keeping configurations as simple and user-friendly as possible — streamlining page layouts for clarity, writing validation rule error messages that actually help users fix mistakes, and generally minimizing complexity so future admins can understand what you built. *Adaptable* means designing with change in mind — using dynamic picklists and custom metadata to avoid hard-coded values, or choosing declarative tools that scale (like Record-Triggered Flows that bulk-handle records) so your org can grow without breaking. By consciously filtering design choices through **Trusted, Easy, Adaptable**, we ensured each solution wasn't just solving the problem today but was also positioned to stand the test of time.

Governance was another recurring theme — essentially, the discipline that keeps a Salesforce org from devolving into the Wild West. We learned that even the most elegant declarative solution needs the support of good processes and oversight. That means documenting your work (so the next person isn't left guessing what "Field_X_ABC" is for), establishing clear governance roles (so there's accountability for changes), and implementing change management protocols (so new automations are reviewed and tested before hitting production). We stressed the value of *proactive org management*: regularly auditing for redundant fields, unused picklist values, or outdated workflows, and cleaning them up before they become a problem. A well-governed Salesforce org is like a well-tended garden — weeds removed, plants (i.e. features) organized, everything getting the right amount of care. This not only keeps today's system healthy but also makes future projects easier, because you're building on solid, well-understood ground. The lesson: invest time in governance continuously, not just when things go wrong.

In designing any solution, we also underscored the importance of **scalability and maintainability**. Put simply, think a few steps ahead. If you're adding a custom object today, imagine how it will handle ten times the records or ten new fields in a year. We emphasized proper data modeling — picking the right relationships (master-detail vs. lookup), normalizing data where it made sense, and leveraging standard objects when possible to avoid reinventing the wheel. We also talked about keeping automation logic modular. For example, instead of one monolithic Flow that does twenty things, maybe break it into subflows or invocable actions that can be reused or updated independently. By avoiding overly tangled processes, we make maintenance so much easier. The guiding question: *"Will someone (maybe future-you) be able to understand and update this with minimal pain six months from now?"* If the answer is yes, you've likely done it right. This forward-thinking mindset ensures your configurations remain assets, not liabilities, as your org grows.

Our journey also reinforced that **security and compliance** aren't optional or "nice-to-have" — they're baked into every declarative decision. Salesforce provides a rich security model (roles, profiles, permission sets, sharing rules, field-level security, and more), and we learned how to wield those tools wisely. The mantra was *least privilege*: give users the access they need to do their jobs — no more, no less. Practically, that meant thinking about security upfront when designing objects and fields (e.g. marking a field as sensitive if it contains confidential data, and using field-level security to restrict it). We made sure record access was appropriate: open up data when collaboration is needed (say, Cases visible to a support team), but lock it down when privacy is paramount (like HR or finance records). Compliance requirements (such as GDPR or industry-specific regulations) were part of this conversation too — we noted that declarative tools like **Shield Platform Encryption** or Field Audit Trail can help address those needs with clicks. The key lesson is that a well-built declarative app doesn't sacrifice security for convenience. You can have user-friendly design *and* solid security by using Salesforce's configuration settings intelligently at every step.

We also drilled into **performance** and **user adoption**, recognizing that even a well-built feature has no value if it's slow or unused. On the performance side, we learned to be mindful of how our declarative configurations run at scale. For example, a record-triggered Flow that executes complex logic on every record save should be designed carefully to avoid hitting governor limits or slowing down data entry — perhaps by adding entry conditions or using asynchronous paths for less urgent tasks. We paid attention to page performance too: overly cluttered pages can frustrate users, so tools like Dynamic Forms and Dynamic Related Lists help show people only what they need when they need it. The takeaway is to always consider the end-user experience: Is that report taking too long to load because it's pulling millions of rows? Is that home page component helpful or just noise? Meanwhile, adoption is the ultimate test of our efforts. We discussed how involving users early (through demos or pilot groups) and incorporating their feedback leads to higher satisfaction and uptake once you go live. Every declarative tweak — whether it's a more intuitive screen flow or a clearer field label — can boost user satisfaction. And when users find Salesforce efficient and helpful, they actually use it, feeding better data into the system and proving the ROI of your work. In short, performance tuning and user-centric design aren't afterthoughts; they're integral to declarative success.

Another crucial lesson we emphasized was **testing and iteration**. Just because you built something with clicks doesn't mean you can skip testing — in fact, declarative changes can introduce unforeseen issues just like code can. We adopted the mindset of testing every Flow, validation rule, and assignment rule in a safe environment (thank you, Sandboxes!) before rolling it out. That included negative testing (does the Flow handle bad data gracefully?) and user acceptance testing (do actual end-users find the new screen intuitive?). We also stressed the importance of regression testing your automations — for example, if you add a new record-triggered Flow on an object, double-check that it's not unintentionally interfering with an existing workflow or Apex trigger. This discipline ensures that your configurations work as intended and continue to work as your org evolves. It's much better to catch an error in a sandbox or a pilot program than to have it surface in production on launch day. This rigorous approach builds trust: when users see that new features "just work," their confidence in the system (and in the team behind it) grows.

A final lesson worth underscoring is knowing the **boundaries of declarative tools** — and when to extend beyond them. Throughout the book, we championed declarative solutions, but we also acknowledged scenarios where a bit of Apex code or an AppExchange component can be the smarter choice. Perhaps you encountered a complex business rule that just can't be easily done with a formula or Flow (something involving intricate calculations or an external web service call). Or maybe you need to process an enormous volume of records in one go, which might be better handled by a batch Apex job or an ETL integration tool. The point is, being "declarative-first" doesn't mean "declarative-only." One of the best skills you've honed is the judgment to know when a requirement is pushing past what clicks can reasonably handle. In those cases, using code isn't a failure — it's using the right tool for the job, after having given declarative approaches a fair shot. What's critical is that even when you do write code, it should coexist gracefully with your declarative logic, respecting the same principles of clarity, security, and performance.

ChatGPT Prompt Example: "Create a checklist of Salesforce declarative configuration best practices I should review before deploying a new app."

Pro Tip: Consider running the Salesforce Optimizer (Salesforce's built-in org health report) periodically, and even try feeding key findings into

ChatGPT. For example, if Optimizer flags a dozen unused fields or an overly permissive profile, you could share that insight with ChatGPT and ask for strategies to address it. It's a clever way to generate improvement ideas for your org using AI.

In summary, the core lessons we've covered form a reliable playbook for any Salesforce project. If you stick to declarative-first thinking, design with foresight, govern proactively, secure diligently, optimize for performance, and test thoroughly, you'll be well ahead of the curve. Whenever you need a refresher on these principles, remember that ChatGPT is there as your quick reference — whether it's summarizing a best practice or acting as a sounding board for a design decision. With this solid foundation in place, you're prepared to tackle whatever comes next. And that's important, because Salesforce isn't static; it's a living platform, and the role of Salesforce professionals is evolving alongside it. **Next, let's explore how the landscape of Salesforce administration is changing** and what that means for you.

EMBRACING THE EVOLUTION OF SALESFORCE ADMINISTRATION

Salesforce administration is not what it was a few years ago — and that's a good thing. The platform continually evolves, and as admins (and developers, analysts, and product owners) we evolve with it. In recent years, Salesforce has been ushering in significant changes in how we manage users and design apps. Two changes stand out: a shift to "**permission-set-first**" user management (moving away from reliance on profiles), and the widespread adoption of **Dynamic Forms** for building record page layouts. These aren't just new toys to tinker with; they represent a fundamental evolution in best practices. Embracing these changes early will make your life easier and your org more future-proof. Let's break down what they mean and how you can stay ahead of the curve.

Permission-set-first administration is Salesforce's new direction for managing user access. Historically, profiles were the primary way to grant object and field permissions — every user had a profile controlling most of their access. By 2025, that model is flipping: Salesforce is heavily promoting (and even planning to enforce) the use of permission sets (and permission set groups) to handle permissions, with profiles pared down to just bare essentials (like user locale, time zone, and basic preferences). The writing on the wall is that object and field permissions will eventually live exclusively in permission sets. For admins, this means it's time to **rethink how you manage user permissions**. Instead of tweaking monolithic profiles for each role, you'll define modular permission sets that can be mixed and matched for each user's needs. It's a more flexible approach — a single user can have multiple permission sets, allowing very granular control. You can adjust someone's access by simply adding or removing a permission set rather than cloning or altering a profile.

So, how do you embrace this change? Start by auditing your current profiles. You might find, for example, that you have a "Sales User" profile and a "Sales Manager" profile that are 90% similar except for a few extra privileges for managers. In the permission-set-first world, you could create a base Sales User permission set and an add-on "Sales Manager" permission set, instead of maintaining two profiles. Salesforce has introduced tools like the User Access and Permissions Assistant (and enhancements to Permission Set Groups) to help with this transition. The key is to chip away at profiles gradually: begin assigning new permissions via permission sets rather than profiles, and migrate one team at a time to the new model. Not only does this align with Salesforce's future architecture, it also leads to cleaner security practices (since you can apply least privilege more precisely). In fact, Salesforce has even started packaging certain products with **standard permission sets** (tied to licenses) to streamline user setup, further reducing the need to edit profiles for access.

ChatGPT Prompt Example: "I've exported all object and field permissions from our profiles (see attached CSV). Can you analyze it and suggest a set of permission sets to replace these profiles?"

Be sure to follow your company's data privacy guidelines, and consider turning off ChatGPT's chat history when working with sensitive org data.

Pro Tip: If you have dozens of permission sets, organize them with Permission Set Groups for common roles or personas. For example, bundle a "Sales User Base" permission set, a "Sales Reports Access" set, and a "Sales Manager Extras" set into a single **Sales Manager** permission set group. This way, you can assign one group to managers and instantly give them all relevant permissions. It keeps your architecture modular and makes user management much easier as you scale.

Meanwhile, the way we design record pages is also evolving. Enter **Dynamic Forms.** In the past, an object's fields and sections on a record page were dictated by a single page layout (often varying by profile or record type). Dynamic Forms breaks that paradigm by letting you place fields and sections directly in the Lightning App Builder and control their visibility with conditions. As of 2025, Dynamic Forms is available for all standard objects (such as Accounts, Contacts, Opportunities, and Cases), and Salesforce is encouraging everyone to use it for more tailored UI experiences. The benefit is a more flexible, user-centric page layout. You can show or hide specific fields based on criteria like a picklist value or user role, all without needing separate page layouts for each profile. For example, a support agent and a sales rep might see different information on the same Case record page because you've made certain components conditionally visible. From an admin perspective, embracing Dynamic Forms means learning the nuances of the Lightning App Builder: how to migrate fields from old page layouts into the builder, how to use **Field Section** components, and how to set up conditional visibility rules that deliver the right information to the right people.

Adopting Dynamic Forms in practice should be done thoughtfully. You don't have to convert every page layout in your org overnight. A smart approach is to start with a high-impact page that would clearly benefit from modernization—say, the Case page that your service team finds cluttered, or the Opportunity page where different sales teams have asked for custom views. In Lightning App Builder, you can activate Dynamic Forms for that object and gradually rebuild the layout: arrange fields into logical sections, use visibility rules to hide irrelevant fields (for instance, hide "Support Plan Expiry Date" unless the customer has a support plan), and remove any components that aren't adding value. Throughout this process, gather feedback from a few end users. They might say "this new layout is cleaner" or "I still need to see X field when

Y happens" – insights you can quickly incorporate thanks to the flexibility of dynamic components. The end result is a record page that feels tailored for each user without you having to maintain multiple layouts. It's also easier to maintain in the long run: need to add a new field? Just drop it in the page once and set who should see it, rather than cloning it across several profiles' layouts.

ChatGPT Prompt Example: "What are the key steps to transition a standard Salesforce page layout to Dynamic Forms for an object like Case?"

Beyond permissions and page design, there have been other evolutions in the admin toolkit. A big one has been the consolidation of automation tools around **Flow**. Not long ago, we had Workflow Rules and Process Builder for process automation — but Salesforce has retired those in favor of the far more powerful Flow Builder. If you haven't already, it's critical to migrate any old workflow or process logic into Flows. This shift has been a learning curve for many admins, but it unlocks more advanced automation capabilities (like multi-step processes and screen wizards) all in one tool. Embracing this change means investing the time to get fluent with Flow (which we did in earlier chapters), but it ensures you're using the latest and greatest automation tech Salesforce offers.

All these changes underscore an important point: the role of a Salesforce admin is continuously evolving. Those who stick stubbornly to "the way we've always done it" — say, creating one profile after another or avoiding new features — will find themselves managing an increasingly complex and outdated org. In contrast, if you proactively adopt Salesforce's latest best practices, you'll simplify your system and stay aligned with where the platform is headed. Embracing evolution means continuously learning (sometimes unlearning and re-learning!) and not being afraid to implement changes that have long-term benefits. By updating your skills and approaches to match Salesforce's innovations, you put yourself in a position to lead your organization through transitions smoothly.

Salesforce isn't standing still, and neither should you. We've adjusted how we manage users and design pages; next, we'll look at the innovations on the horizon for the declarative tools themselves. **In the**

following section, we'll explore the emerging features and enhancements that are poised to further expand what "clicks, not code" can do.

ANTICIPATING ADVANCES IN DECLARATIVE TOOLS

As Salesforce continues to innovate, the declarative toolkit we rely on will only grow more powerful. Being an effective Salesforce professional means looking ahead and **anticipating the advances** coming down the road. We've seen how the platform has evolved (Flow taking over automation, Dynamic Forms changing UI design, etc.), and we can be sure that Salesforce's product teams have even more in store. Let's put on our futurist hat for a moment and consider some of the trends and potential enhancements that could reshape "clicks, not code" development in the near future.

One area to watch is **increased configurability** of the platform. Salesforce often takes features that once required custom code and turns them into point-and-click options. We've already witnessed this with things like external API calls (once the domain of Apex, now achievable with declarative tools like External Services and Flow HTTP callouts) and complex automations (once needing triggers, now handled by record-triggered Flows). It's a safe bet that this trend will continue. For example, admins might soon get even more powerful Flow capabilities – imagine Flows that can handle **multiple objects in one process** or new trigger types (perhaps a future where Flows can trigger after record deletion or on certain user events, closing some of the remaining gaps that still require Apex triggers). Salesforce has been steadily improving Flow Builder's usability and reach each release (enhanced debugging tools, trigger-ready flows for more scenarios, richer screen components), and we can expect more "you can do that in Flow now" moments ahead.

Alongside raw capability improvements, expect **better automation templates and guidance**. Salesforce knows that not every admin is a seasoned architect, so they've been investing in pre-built solutions and wizards. We might see expansions of tools like Flow Orchestrator (for coordinating multi-step, multi-user processes with clicks) or more out-of-the-box Flow templates for common scenarios. It wouldn't be surprising if future releases include intelligent suggestions in the Flow Builder – perhaps prompts that say "It looks like you're building an approval process; here's a recommended template." In other words, the tooling may get smarter, guiding us toward best-practice solutions even as those solutions grow more complex.

Declarative integration is another domain poised for big leaps. We covered how tools like External Services and Platform Events let us connect Salesforce to other systems without code. In the coming years, this will likely get even easier and more powerful. Salesforce might introduce more point-and-click connectors (in the spirit of MuleSoft Composer, where an admin can integrate Salesforce with, say, a SQL database or a third-party app through a guided flow). We might also see the rise of *event-driven architectures* made admin-friendly: for example, a refined interface for subscribing to events (perhaps an upgraded Pub/Sub tool) that lets admins orchestrate complex multi-system workflows through configuration. As Salesforce positions itself as a central hub in enterprise tech stacks, **low-code integration** will be a high priority. Tasks that used to require an integration specialist writing scripts could become doable with a few clicks by an admin configuring flows and connectors.

Even the user experience layer is likely to see new declarative tools. Dynamic Forms was a huge step, but Salesforce could go further. There's chatter about more drag-and-drop component creation and possibly a future where admins can design custom Lightning Web Components with low-code methods (imagine a component builder that feels like App Builder, assembling UI functionality without deep coding). While that might be a bit farther off, Salesforce has already moved in that direction with things like Dynamic Actions (showing/hiding action buttons based on criteria) and an expanding library of standard components. Keep an eye on anything that gives admins more control over the UI without code — it's a clear area of focus.

Many of these anticipated innovations boil down to one theme: **making the platform more powerful, yet easier to use.** Salesforce knows its admin community thrives when complex tasks become simpler. Think about how far we've come – building an entire approval process with email alerts once required multiple workflow rules or sneaky workarounds; now a single Approval Flow handles it cleanly. In the future, we might see declarative "health checks" or optimizers built into the platform (perhaps a tool that scans your org and suggests simplifications), or even more granular security configuration via clicks (imagine setting conditional access rules via a wizard instead of writing Apex).

Salesforce is also focusing on the admin experience around deployments and org management. The recent introduction of DevOps Center signaled an intent to simplify change deployment for declarative builders by providing source control and pipeline management in a friendly UI. We can expect continued investment here — perhaps more declarative tools for **versioning metadata**, tracking changes, or rolling back updates if something goes wrong. In the future, managing your org's lifecycle (from sandbox to production) could be as point-and-click as building a Flow, reducing the need for command-line tools or third-party release management apps.

So how do you prepare for advances that aren't here yet? Part of it is mindset: stay curious and plugged into Salesforce's roadmap. We know three releases come out each year, and with them a flood of new features. The best professionals regularly scan release notes not just for the changes happening now, but for hints of where things are headed. (The retirement of Workflow Rules, for example, was telegraphed well in advance — those paying attention had a head start migrating to Flow.) It's also wise to follow Salesforce blogs, official announcement webinars, and community chatter where upcoming features often get discussed. You might hear hints of a "Formula 2.0 builder" or an "Analytics wizard for admins" long before they become reality, giving you time to prepare.

ChatGPT Prompt Example: "What new declarative features and enhancements has Salesforce announced or hinted at for the upcoming release?"

Using ChatGPT itself can help with future-gazing. With its browsing or up-to-date knowledge, you could ask it to summarize Dreamforce announcements or to monitor the Salesforce developers' blog for you. Instead of manually digging through roadmap slides, you might have AI do some of that research (just be sure to verify rumors with official sources!). The point is to leverage tools like ChatGPT as a radar for what's coming, so you're not caught by surprise by a major platform change.

Pro Tip: Get involved in Salesforce's pre-release and pilot programs. You can sign up for a pre-release org before each major Salesforce update to experiment with new features early. And if there's a pilot (beta program) for a new declarative capability — say, a beta of a next-gen Flow feature — consider joining it. Early hands-on experience not only gives you a head start, but also allows you to provide feedback that can shape the feature before full launch. This proactive approach ensures that when new tools drop, you're already familiar and ready to capitalize on them.

By staying informed and experimenting with each release's goodies, you'll be ready to capitalize on the next wave of declarative innovations as soon as they arrive. And speaking of the next wave: few trends are as buzzworthy right now as artificial intelligence. In the next section, we'll dive into how AI and automation are intersecting with Salesforce — and how you can leverage them to work smarter.

LEVERAGING ARTIFICIAL INTELLIGENCE AND AUTOMATION

In recent years, **artificial intelligence (AI)** has moved from a buzzword to practical reality in the Salesforce ecosystem. Salesforce is infusing AI into its platform (you've probably heard of Einstein GPT by now), and savvy admins and developers are learning to leverage these capabilities to work smarter. The goal isn't to replace human expertise, but to **augment** it – automating routine tasks, surfacing insights from data, and helping users make better decisions faster. Let's explore how AI and automation are enriching declarative Salesforce work, and how you can tap into these advances.

One headline development is Salesforce's introduction of **Einstein GPT**, which brings generative AI (the technology behind ChatGPT) directly into the CRM interface. Imagine having an AI assistant in Salesforce that can draft a follow-up email to a client, summarize a lengthy case description into key bullet points, or even suggest the next best task on an opportunity — all at the click of a button. Einstein GPT is designed to do just that. For example, a sales rep can click "Generate Email" and get a proposed email to a prospect, based on context from the opportunity and past interactions. A support agent could see an AI-generated summary of a complex, multi-threaded support case, saving them time in understanding the history. In fact, some organizations are piloting features where, after a case is closed, the AI suggests a draft knowledge article capturing the resolution — turning what used to be 30 minutes of writing into a 5-minute review task. As an admin, you configure and govern these features: deciding where AI suggestions appear, which fields or records the AI is allowed to draw from, and perhaps even fine-tuning the prompt templates for your org's tone. The benefit is efficiency and consistency – routine communications and data summarization happen in seconds, letting your human team focus on more complex, value-add work (with the AI doing the boilerplate heavy-lifting).

Beyond generative text, Salesforce offers **AI-driven predictions and recommendations** that you can set up declaratively. Think of tools like Einstein Prediction Builder, Einstein Opportunity Scoring, and Next Best Action. With Prediction Builder, an admin can, say, create a model (no code needed) to score how likely a customer is to churn or how likely a lead is to convert, based on historical patterns. That prediction can surface on the record page as a field ("Churn Risk: High") for a rep to consider. (Out-of-the-box features like Opportunity Scoring do this automatically – giving each deal a 1-99 score so salespeople know at a glance which opportunities are hot or cold.) Next Best Action, on the other hand, lets you configure proactive recommendations to users – using a mix of rules and AI insight. For instance, you might build a strategy that notices when an opportunity is stagnating and, if the customer also has a high churn risk, recommends a specific retention offer to the account manager. All of this can be done with clicks: you define the criteria and content of the recommendation in a strategy flow, and Salesforce's AI can help identify which factors (like lack of recent activity or a drop in customer satisfaction) should trigger it. We're moving toward a world where the CRM doesn't just hold data, but actively interprets it and nudges users with suggestions. We may even see AI helping with data quality – for example, alerting a sales rep "This account is missing an Industry field, which similar high-value accounts usually have" or reminding an admin that "Lead Source is often blank, consider making it required." These subtle, field-level recommendations can greatly enhance data completeness over time.

AI can also supercharge your **data analysis and operations** on the platform. Salesforce has long provided analytics tools (from standard reports to Einstein Analytics/Tableau CRM) to slice and dice data, but now you have new allies. ChatGPT itself, especially with its Advanced Data Analysis capabilities, can act as a data consultant by your side. Imagine you export a spreadsheet of case closure times or lead conversion rates – you could feed it into ChatGPT and ask for trends, anomalies, or correlations, getting insights in plain language along with charts. Rather than you manually pivoting data in Excel for an hour, an AI can crunch the numbers in moments. Similarly, if you suspect a pattern (say, support cases spike on Mondays or end of quarter), you can ask an AI assistant to validate that hunch on your dataset. This kind of analysis can inform your declarative solutions: for example, if AI analysis shows spikes in support cases every Monday, you might schedule an automated email reminder to customers on Friday to head off issues, or ensure more agents are online on Mondays. In essence, an admin can have an **AI co-analyst**: you bring the data, the AI helps find the story in it, and together you implement changes in Salesforce based on those insights.

ChatGPT Prompt Example: "I've attached a CSV export of our case closure times. Can you analyze it and identify any patterns, such as specific days or weeks where our closure rate slows down?"

Automation has always been a Salesforce superpower, and AI is making it even sharper. Consider how you might combine them: you could have a Flow that invokes an AI service (via an HTTP callout) to analyze text or images and return a result used in the process. This isn't science fiction – admins today are experimenting with flows that call out to OpenAI's API or other AI services. For example, an auto-assignment flow might take an incoming customer message, send the text to an AI for sentiment analysis, and route the case to a queue based on whether the customer is upset. All of that without a single line of Apex. Salesforce is likely to bake more of these capabilities natively (we might soon see, for instance, a built-in Flow action for sentiment analysis or language translation). The line between "automation" and "AI" is blurring; you'll design a process that not only executes steps, but also makes a smart judgment call in the middle thanks to an AI helper.

Of course, leveraging AI comes with the responsibility to use it wisely. AI can sometimes be overly confident or make mistakes, and it learns from historical data which could include biases. As the admin or product owner, you'll want to put guardrails in place. That might mean requiring human approval on an AI-composed email before it's sent, or setting thresholds (e.g. only autopilot that retention offer if the confidence in the suggestion is above 80%). Salesforce provides some built-in checks – for example, you can restrict which fields an Einstein GPT feature is allowed to use, to protect sensitive info – but it's on us to monitor and refine how AI is affecting user behavior. Pilot new AI features in phases, gather feedback, and be transparent with users that these are helpers, not all-knowing oracles.

Pro Tip: Ensure your Salesforce data is clean and comprehensive before layering AI on top. Generative and predictive tools are only as good as the information they have. Experienced admins often run data quality audits (even asking ChatGPT to help spot anomalies) so that any Einstein predictions or AI-generated content are based on trustworthy data. High-quality input will yield more accurate and useful AI output.

The bottom line is that AI and advanced automation are becoming part of the declarative toolbox. Instead of writing complex code, you'll be orchestrating intelligent features with clicks – configuring AI models, deciding when and where predictions or content suggestions appear, and blending those into your flows and interfaces. By embracing these capabilities, you can dramatically streamline processes: tasks like triaging cases, populating records, or communicating with customers can happen more quickly and with less manual effort. As we continue, it's worth remembering that none of us masters these new tools in isolation. The Salesforce world is one of constant learning, and AI is just the latest chapter. In the next section, we'll discuss how to cultivate continuous learning and adaptability so you can keep pace with Salesforce's rapid changes.

CULTIVATING CONTINUOUS LEARNING AND ADAPTATION

If there's one constant in the Salesforce world, it's change. New features, updates, and even entirely new products arrive with a regular cadence (three major releases a year, plus countless smaller improvements in between). To thrive as a Salesforce professional, you need to **cultivate a habit of continuous learning and adaptation**. Rather than seeing change as a burden, the best admins and developers treat it as an opportunity – a chance to deliver more value and keep their skills sharp. But doing this requires a proactive approach. Let's talk about how you can stay current and grow alongside the platform.

First, make friends with the Salesforce release cycle. Every few months, Salesforce publishes extensive release notes detailing new features, enhancements, and sometimes deprecations. It's a lot of information (often hundreds of pages), but you don't have to read it cover to cover. Develop a strategy for each release: for example, scan the **Release Highlights** or summary blogs that pop up (sites like Salesforce Ben, or official "Release Readiness" webinars, provide great overviews). Focus on areas relevant to your org and role – if you're a Service Cloud specialist, zero in on those sections; if you manage security, watch for updates to permissions or user management. Over time, you'll get good at parsing out which updates are "cool, but not urgent" versus those you should act on immediately (like a critical update that will auto-activate next release). And if you hold Salesforce certifications, you already have a built-in reminder – the maintenance modules each release (essentially mini-exams on new features) will prompt you to learn what's changed. Even if you're not certified, it's wise to treat each release as a chance to test yourself and explore what's new.

Tools like Trailhead make staying updated more structured. With each release, Salesforce provides Trailhead modules (e.g. *Release Readiness* for admins or developers) that walk you through the top changes in an interactive way. These bite-sized lessons can quickly get you up to speed on the coolest new features, complete with hands-on challenges. Many professionals also attend webinars or local user group meet-ups that review new features – Salesforce's own "Release Readiness Live" broadcast, for example, is a fun way to hear product managers demo the highlights. These are great because you not only learn what's new, but also why it matters straight from experts.

Another key is to **learn by doing**. When a new feature catches your eye, try it out in a sandbox or Trailhead Playground. Reading about a feature is one thing; actually clicking through and using it cements your understanding. For instance, if Salesforce introduces a new Flow capability (say a "Reactive Screen Component" that dynamically updates on a Lightning page), clone a relevant flow in a sandbox and experiment with the new functionality. Not only will you learn the ins and outs, but you'll also be in a position to deploy it wisely when the time is right. It's one thing to know something exists, it's another to understand how to implement it – hands-on practice bridges that gap.

Don't underestimate the value of curiosity in everyday work, too. If a colleague mentions a tool or feature you haven't used (like "Have you tried the new DevOps Center?"), take it as a cue to investigate. Ten minutes of exploration or asking a question in the Trailblazer Community can open up new avenues. In a field as broad as Salesforce, none of us knows everything – the trick is continuously filling in knowledge gaps. Over time, these small bites of learning add up and make you the go-to person when something new arrives.

The Salesforce ecosystem is huge, and a lot of learning happens via the community. Blog posts, YouTube channels, podcasts, and community forums (like the Salesforce Stack Exchange or Reddit) often provide insights and tips that you won't find in official docs. For example, an admin might blog about "Lessons Learned from Enabling Multi-Factor Authentication" or share a demo of the new Tableau integration on YouTube. These first-hand accounts can prepare you for your own rollouts and spark ideas. Try to set aside a little time each week to peruse community content. You might follow certain hashtags (e.g. #SalesforceQA or #AwesomeAdmins on Twitter/X), subscribe to a newsletter (there are some great weekly Salesforce recap newsletters), or just bookmark a few favorite bloggers. This keeps you tapped into real-world experiences and "tribal knowledge" that formal documentation may not cover.

And remember, continuous learning isn't a solo journey – it helps to involve others. If you work on a team, consider a rotating "feature of the month" where each person learns about a new Salesforce feature and shares a quick summary or demo in a team meeting. Not only does this spread knowledge, it also builds a culture where learning is celebrated. If you're solo, you can still engage with peers through the Trailblazer Community online or at events, discussing what you've learned and asking what others are excited about. Often, someone else's perspective on a new feature (or their cautionary tale of a gotcha) can accelerate your own learning.

This is also where ChatGPT can assist in your learning journey. Instead of manually sifting through release notes or lengthy help articles, you can ask ChatGPT to summarize or explain things for you. For instance, if you're short on time you might say:

ChatGPT Prompt Example (Voice): "Explain the new Dynamic Related List feature in the latest Salesforce release in simple terms."

Imagine doing that through your phone's ChatGPT app while commuting – you get a quick, spoken rundown of the feature, almost like having a knowledgeable friend riding with you. (Yes, ChatGPT can now take voice input and talk back to you, which makes learning on-the-go much easier!) You can also have ChatGPT act as a digital tutor. For example, *"Quiz me on five key things an admin should know about Salesforce Flow best practices"* will prompt the AI to throw some Q&A at you, flashcard style. And if you come across a dense Salesforce documentation page that's hard to digest, why not paste a snippet into ChatGPT and ask for a simpler explanation or a real-world example? It's like having an on-demand mentor who can break down complex concepts whenever you need.

Pro Tip: Dedicate a small block of time in your schedule for professional development. Even one hour a week of focused learning – whether it's completing a Trailhead module, reading a few blog posts, or experimenting in a dev org – will compound over time. Consistency is key. By making learning a routine part of your job, you ensure that you're gradually, continuously improving without having to scramble to catch up.

Adaptability also means being willing to change your approach in light of new information. It's easy to fall into "we've always done it this way," but the platform's evolution often offers safer, faster, or more scalable methods. The top-notch professionals regularly re-evaluate their processes. If a new feature can replace an older workaround, they embrace it. If best practices shift (for example, moving from Profiles to Permission Sets, as we discussed earlier), they don't cling to the old way out of comfort. By staying flexible and open-minded, you'll keep your org in step with Salesforce's best thinking and avoid accumulating "technical debt" of outdated configurations.

Finally, continuous learning keeps you passionate and prevents burnout. When you actively seek out new knowledge, you maintain a growth mindset and a sense of progress. Celebrate your learning milestones – maybe you completed a tricky Superbadge on Trailhead, or you finally wrapped your head around a complex concept like External Services or Custom Metadata Types. Each skill added is not just a resume booster, but a tool you can use to deliver value.

By actively learning and staying adaptable, you'll keep your Salesforce expertise fresh and be ready to tackle whatever the platform throws your way next. This mindset not only makes you more effective in your current role, but it also future-proofs your career in an ever-changing ecosystem. Next up, we'll talk about another powerful avenue for growth and support: staying engaged with the Salesforce community.

STAYING ENGAGED WITH THE SALESFORCE COMMUNITY

One of the most rewarding aspects of the Salesforce world is its **community**. There's a famous saying: "Never underestimate the power of the Ohana" (Salesforce's term for its community family). No matter what challenge you're facing or what new skill you want to learn, chances are someone out there has insight to share. Engaging with this vibrant network of admins, developers, consultants, and users can elevate your expertise from good to great. Let's look at how staying involved with the community can enhance your continuous learning and professional growth.

First, the community is a treasure trove of knowledge and support. Online forums like the Trailblazer Community (formerly Success Community) and Salesforce Stack Exchange are filled with Q&As on just about every Salesforce topic imaginable. When you encounter a weird error in a Flow or need advice on implementing a complex requirement, a quick search often reveals that someone else has wrestled with the same thing. It's common to find not just one solution, but multiple perspectives from people in different industries or org sizes. By actively participating – asking questions when you're stuck and contributing answers when you can – you plug into a real-time brain-trust that keeps you from reinventing the wheel. Many of us have had the experience of posting a question and getting a solution from a kind stranger halfway around the world before the workday is over.

Beyond the online Q&A, Salesforce user groups (both local and virtual) are fantastic for networking and learning. These groups, often organized by city or by specialty (e.g. a Nonprofit User Group, a Marketing Cloud User Group), regularly host meetups or webinars where admins and developers share their experiences. Attending these meetups can spark ideas you never considered. For example, you might see a demo of how a company implemented a complex approval process using clicks only, or hear about a new app that boosts productivity. And it's not just the formal presentations – the casual conversations during a "happy hour" or Zoom chat can lead to tips and contacts that prove invaluable. If you're not already in a user group, check out Salesforce's Trailblazer Community Groups site and find one that fits your interests or region. There's something special about learning alongside peers who speak the same "Salesforce language" and understand the unique challenges we face.

The community is also your professional network. Engaging with others can open doors to mentorship, collaboration, and even career opportunities. For instance, if you consistently help out on forums or LinkedIn groups, you'll start to get noticed as someone knowledgeable and generous – which might lead to being nominated as a Salesforce MVP or landing that referral for a new job. But more immediately, you gain a group of friends/colleagues who you can bounce ideas off. Maybe you connect with another admin in your city and decide to co-organize a study group for the Architect cert, or you meet a reporting guru who you know you can ping when you're puzzling over a particularly tricky report requirement. These relationships can become a safety net and a springboard throughout your career.

If you're more on the shy side or feel you have nothing to contribute – trust me, everyone has something to share. Maybe you're not a Flow expert (yet), but you might be really good at organizing data or training users. The community thrives on a mix of questions and answers. Posting a thoughtful question (with details and what you've tried) is just as valuable as posting an answer, because it often triggers a great discussion that others learn from later. And when you do know the answer to someone else's question, even if it seems simple to you, taking a moment to respond earns karma (sometimes literally, in the form of community points!) and builds your confidence.

Social media and content creation are other avenues. Many in the ecosystem write blog posts or make YouTube videos to share tutorials, gotchas, and novel solutions. If you enjoy writing or presenting, consider contributing your own experiences. It might feel like "Everything's already been written about," but your perspective or the specific scenario you lived through is unique. Writing a post about how you cleaned up 1,000 unused fields might not only help someone out there with the same mess, it also deepens your own understanding of the effort you put in. Plus, you'll inevitably get feedback that furthers your learning (someone might comment with an even quicker trick, for example). Even if blogging or speaking isn't your thing, you can still boost your community presence by engaging with content – comment on posts, ask follow-up questions, share links with colleagues who might benefit. It's all part of the dialogue.

Let's not forget the big events: Dreamforce, Salesforce World Tours, and community-led "Dreamin'" events. These conferences can be energizing moments in your year. Yes, you'll learn a ton from the sessions (and even more from the unofficial conversations), but perhaps equally important, they recharge your enthusiasm. There's something about being surrounded by thousands of fellow Trailblazers, all excited about the same technology, that reminds you why you love doing this. Not everyone can attend these in person, of course, but Salesforce often streams keynotes and many sessions online – so you can partake in the knowledge-sharing remotely and join the social media chatter around it. A tip: if you do attend, go in with a plan (which sessions, which booths) but also leave room for serendipity. Some of the best insights might come from a random chat with the person next to you or a spontaneous hallway demo.

If you're looking to level up your community involvement, ChatGPT can even help brainstorm ways to do that. For example:

ChatGPT Prompt Example: "Help me come up with a 6-month plan to become more active in the Salesforce community (online and locally). I'd like to network and share knowledge, but I'm not sure where to start."

The AI might suggest something like: join and introduce yourself in an online Trailblazer Group, answer one question a week on Stack Exchange, attend a local user group meeting each month, start a blog or LinkedIn article series on a project you completed, and perhaps aim to speak at a meetup by the end of the period. Having such a plan (and yes, ChatGPT can serve as a creative brainstorming buddy here) can give you concrete steps to turn good intentions into action.

Pro Tip: Contribute in a way that feels natural to you. Not everyone needs to be a blogger or a keynote speaker – you might be great at writing detailed answers on forums, or prefer one-on-one mentoring of a colleague. What matters is the spirit of sharing. By teaching others what you know, you solidify your own knowledge and establish yourself as a leader. Many respected community figures (even future MVPs) started by consistently answering newbie questions and sharing simple tips. You don't have to be a CTA or have 10 years' experience to make an impact – you just need a willingness to help and learn.

Staying engaged with the community turns what could be a solo journey into a team effort. You'll find that as you give help, you get help; as you celebrate others' wins, they'll celebrate yours. It's a virtuous circle that makes the work more enjoyable and the tough problems more surmountable. Plus, the more people you know in the ecosystem, the more perspectives and ideas you'll be exposed to – keeping you from getting stuck in a silo. In the final section, we'll tie everything together and talk about how all these best practices and forward-thinking approaches translate into delivering long-term value for your business (and your career).

DELIVERING LONG-TERM BUSINESS VALUE

At the end of the day, all the best practices, tools, and AI assistance in the world matter only if they help the business succeed. The ultimate goal of everything we've covered – from declarative configuration and governance to continuous learning and community engagement – is to **deliver long-term value to the organization**. Salesforce isn't just a playground for cool tech; it's a platform to drive real business outcomes. Let's connect the dots between doing things the "right way" in Salesforce and seeing tangible benefits for your company (and your career).

One major theme has been **scalability and future-proofing**. By designing solutions declaratively with best practices, you ensure that as the business grows or changes, your Salesforce org can adapt with minimal friction. For example, imagine your company doubled its customer base next year – would your current processes and configurations scale smoothly? If you've followed the principles in this book, the answer should be yes. Perhaps you've designed a flexible data model and automation that can handle more volume, or you've implemented Dynamic Forms that let you quickly tweak UIs for a new sales process without a code overhaul. These investments in doing it right pay off when growth or pivots happen. The business experiences rapid time-to-market for adjustments, which can be the difference between beating a competitor to a new opportunity or scrambling while others sprint ahead.

Similarly, **governance and maintenance** best practices translate directly into reliability and trust – both crucial for business value. A well-governed Salesforce org (where redundancies are cleaned up and changes are tested) is less prone to outages or ugly data incidents. That means less downtime for your sales teams, fewer mistakes in reports to executives, and overall higher confidence in the system's data. It's hard to put a dollar value on "avoiding a crisis," but any executive who's lived through a bad deployment or a data breach can tell you: it's huge. By proactively managing your org and using tools (and even ChatGPT) to audit and document, you're preventing costly problems. The business benefits from consistent operations and compliance with regulations, avoiding fines or reputational damage – true long-term value that sometimes goes unnoticed precisely because your good practices keep issues from occurring.

Another clear business payoff is **user productivity and adoption.** When your Salesforce implementation follows declarative best practices, it tends to be more user-friendly (because you've thought about UX, performance, and training). Users actually enjoy using the system – or at least don't dread it! – and that means better data entry, more complete customer info, and more effective use of Salesforce's features. For example, if sales reps have a well-designed opportunity page with an AI-generated "next step" recommendation and an automated email template at their fingertips, they can close deals faster and spend more time selling instead of wrestling with the tool. If service agents have a console with relevant info dynamically displayed (and maybe an Einstein GPT draft response ready to go), they can resolve cases quicker, boosting customer satisfaction. These improvements show up in business metrics: higher sales conversion rates, faster case resolution times, improved CSAT scores, etc. That's direct business value coming from the marriage of good Salesforce configuration and smart use of automation/AI.

Let's consider a concrete scenario: Say your company's support team was drowning in a manual process, and customers were unhappy with response times. You, as a Salesforce expert, redesigned the process using declarative automation (Flows to route cases, email alerts for SLAs, etc.) and perhaps introduced an Einstein bot to handle common questions. You also used ChatGPT to help draft a new knowledge base and training materials for the team. The result after deployment? Case volume is handled 20% faster, and customer satisfaction scores rise by several points. By translating this into business terms – "we can handle more cases with the same staff, potentially saving $X in hiring, and improved CSAT likely leads to repeat business worth $Y" – you've made the value clear to leadership. This isn't a hypothetical; these kinds of improvements happen all the time when best practices are applied diligently. It underscores that technical excellence and business strategy go hand in hand.

Using ChatGPT and AI as your partners can also contribute to business value in less obvious ways. For one, you likely saved time (and therefore money) by accelerating solution design or troubleshooting. Maybe what would have taken you a full day to research and prototype only took a couple of hours with ChatGPT assisting – that's more time you can spend on other impactful projects (or responding to new requirements sooner). It also means your Salesforce team (even if that's just you) can deliver results faster and more consistently. Stakeholders notice when the admin team is always on top of things and can turn around requests quickly without breaking things. That builds trust in IT and encourages the business to invest more in Salesforce because they see a reliable return.

Importantly, delivering long term value isn't just about one off gains; it's about establishing a cycle of continuous improvement. Because you've embraced continuous learning, you're regularly injecting new capabilities into the org – whether it's a small tweak from a recent release or a big new feature like enabling Lightning Experience (which, recall, many resisted at first, but those who adopted early gained a productivity edge). Your engagement with the community might lead you to discover a best practice or app that saves the company money (for example, learning about a native Salesforce feature that replaces a costly third-party tool). Your adaptive mindset ensures the org doesn't stagnate; instead it evolves, often anticipating business needs rather than just reacting. Over years, this agility compounds. Companies with well-run Salesforce orgs can enter new markets or launch new initiatives in weeks instead of months, because their backbone systems are ready and their people are knowledgeable.

Pro Tip: When implementing improvements, always try to tie them to business metrics. If you automate a process, figure out roughly how much time that saves the team each week – and translate that to hours or dollars. If you clean up data quality, track the improvement (e.g., "opportunities with next steps filled in increased from 60% to 95%"). This not only demonstrates the impact of your work to executives, but also guides you to areas that yield the highest return. Measuring and communicating value is itself a best practice that ensures your Salesforce program continues to get support and investment.

In the end, your goal is to make Salesforce a driver of success, not just a system of record. By adhering to declarative best practices, harnessing tools like ChatGPT for efficiency and insight, and staying ever-curious and connected, you've positioned yourself and your org to achieve strong, sustainable results. The improvements you implement are not band-aids, but lasting enhancements that scale and flex with the business. Over time, the cumulative effect is huge – think of it as compound interest on the "investment" of good design and learning. You'll find that your Salesforce platform isn't just supporting the business, it's often shaping how the business operates and innovates.

As we wrap up this chapter and this book, reflect on how far you've come. You've reinforced a foundation of doing things the right way (for the right reasons), and you've glimpsed the future of Salesforce where AI and automation play ever-bigger roles. Armed with best practices and an AI sidekick, you are well equipped to deliver exceptional value year after year. In the summary to follow, we'll recap the major takeaways and share some final thoughts on continuing this journey – including an invitation to keep in touch with us as you forge ahead in the world of Salesforce.

SUMMARY

Bringing it all together, we've reinforced that **declarative best practices** – backed by smart use of tools and AI – are the key to unlocking Salesforce's full potential. By designing scalable, well-governed solutions and leaning on "clicks, not code" wherever possible, you create a Salesforce org that is agile, reliable, and ready for the future. We revisited core principles like purposeful design, rigorous security, proactive maintenance, and user-centric optimization, seeing how each contributes to a healthier system and happier users. We also looked ahead at the evolving landscape: new declarative features on the horizon, the growing role of AI and automation (with Einstein GPT and ChatGPT becoming invaluable sidekicks), and the importance of constantly learning and adapting. The message is clear – blending solid fundamentals with an openness to innovation positions you and your organization to thrive in the long run.

As you turn the page from this book to your real-world projects, remember that you're not alone on this journey. You have a powerful community behind you, and tools like ChatGPT by your side. Keep applying these best practices, stay curious about what's coming, and never stop refining your craft. The payoff will be seen in the business value you deliver: more efficient processes, empowered teams, and the agility to meet any new challenge with confidence. Thank you for joining us through these chapters and investing in your growth as a Salesforce Trailblazer. We're excited to see what you accomplish next – and we'd love to hear about it! Please feel free to connect with us (the authors) on LinkedIn to continue the conversation, share your successes, and keep learning together. This may be the final chapter of the book, but it's just the beginning of the next phase of your Salesforce journey – one where declarative excellence and AI partnership will take you to new heights. Good luck, and happy Trailblazing!

APPENDIX: THE BOOK, SUMMARIZED

THE BOOK IN ONE SENTENCE

This book explains how experienced practitioners can use ChatGPT as a disciplined thinking partner to design, configure, and govern Salesforce solutions with greater clarity, speed, and confidence, without surrendering professional judgment or accountability.

THE BOOK IN ONE PARAGRAPH

Modern Salesforce configuration is no longer a mechanical exercise in enabling features but a series of architectural and organizational decisions with long-term consequences. This book shows how ChatGPT can be integrated into that work as a cognitive amplifier, helping practitioners reason through requirements, data models, security, automation, integrations, and change management more effectively. Rather than treating AI as an answer engine, the book emphasizes framing, critique, iteration, and context, teaching readers how to ask better questions, validate responses, and translate insights into durable configurations. The result is a practical approach to using ChatGPT that strengthens judgment, improves communication, reduces rework, and supports sustainable Salesforce delivery in complex enterprise environments.

THE TEN MOST IMPORTANT THINGS YOU WILL LEARN FROM THIS BOOK

1. **How to treat ChatGPT as a thinking partner, not an authority.** You will learn how to use AI to explore options and challenge assumptions while keeping final decisions firmly in human hands.

2. **How to provide the right context so AI answers are actually useful.** The book shows how role, industry, constraints, scale, and governance dramatically change the quality of ChatGPT's output.

3. **How to reason through Salesforce architecture, not just configuration.** Readers learn to use ChatGPT to evaluate tradeoffs across data models, automation, security, and integrations.

4. **How to identify and avoid confident but incorrect AI output.** You will learn practical techniques for pressure-testing and validating AI-generated recommendations.

5. **How to accelerate requirements analysis and design conversations.** The book demonstrates how ChatGPT can help clarify vague inputs and surface hidden assumptions early.

6. **How to improve documentation without creating busywork.** You will see how to use AI to explain intent, decisions, and impacts in clear, durable language.

7. **How to align business language with Salesforce implementation realities.** The book shows how ChatGPT can translate between stakeholder goals and platform constraints.

8. **How to reduce rework by thinking further ahead.** Readers learn how to use AI to anticipate downstream impacts before configuration choices are locked in.

9. **How to use AI safely in regulated or high-governance environments.** The book addresses data boundaries, validation, and responsible usage patterns.

10. **How to evolve your role as AI becomes part of everyday delivery.** You will gain a framework for remaining valuable as tools become more powerful.

TEN QUOTABLE IDEAS FROM THE BOOK

1. **"ChatGPT does not replace judgment; it exposes whether judgment is present."** This captures the idea that AI amplifies the quality of the thinking it is given.

2. **"Configuration decisions are architecture decisions, whether we admit it or not."** The quote reinforces that even declarative choices shape long-term system behavior.

3. **"Good prompts are a form of design documentation."** This highlights how clear framing reflects clear thinking.

4. **"Speed without understanding is just faster rework."** A reminder that AI acceleration must be paired with rigor.

5. **"If you cannot explain a configuration choice, you do not fully understand it."** This emphasizes the role of articulation in sound system design.

6. **"AI is most dangerous when it sounds confident and you stop asking why."** A warning against uncritical acceptance of fluent answers.

7. **"The value of ChatGPT lies less in answers and more in alternatives."** This reframes AI as a tool for exploration rather than finality.

8. **"Salesforce works best when decisions are made intentionally, not incrementally."** This speaks to the cumulative cost of ad hoc configuration.

9. **"Using AI responsibly means knowing when not to use it."** A reminder that discretion is part of professional maturity.

10. **"The future Salesforce practitioner is defined by thinking, not clicking."** This captures the shift from tool operation to system reasoning.

ABOUT THE AUTHORS

Anirudh Ganesh Bharadwaj brings a rare blend of financial rigor, analytical depth, and hands-on consulting experience to the world of Salesforce configuration and AI-powered delivery. With a background that spans accounting, financial modeling, business analysis, and technology consulting, he approaches Salesforce work with a precision and clarity shaped by years of working with complex data, high-stakes operations, and cross-functional teams. Before stepping into Salesforce consulting, Anirudh built his career across roles where accuracy, structure, and insight weren't optional—they were the job. He has led financial forecasting efforts, analyzed multimillion-dollar variances, supported ERP transitions involving millions of transactions, and developed valuation models that helped teams make smarter, faster decisions. His work has uncovered fraud risks, improved operational efficiency, and delivered measurable performance gains across diverse industries and organizational environments. In parallel, Anirudh has been a dedicated educator and mentor, teaching financial modeling, valuation, and business fundamentals to students and professionals. His workshops and one-on-one sessions helped learners boost exam scores, sharpen analytical thinking, and approach business problems with confidence and structure. Today, he brings that same discipline to Salesforce—combining analytical strength with an enthusiasm for emerging technology. He's part of a new generation of consultants who see AI not as a distant trend but as a working tool that's already reshaping how configuration, analysis, and delivery get done. Curious, driven, and grounded, Anirudh represents the future of AI-first Salesforce delivery: a consultant who pairs business acumen with modern technical instincts and a commitment to helping organizations work smarter in an accelerating world.

Andy Forbes is a seasoned Salesforce expert who has spent his career guiding organizations through the fast-changing intersection of business, technology, and human behavior. He has led large-scale Salesforce programs, advised global enterprises on digital transformation, and helped teams adopt new tools and new ways of working as cloud platforms, automation, and now AI reshape what "getting work done" actually means. With deep experience across configuration, architecture, and delivery, Andy specializes in translating complex ideas into clear, practical guidance that real teams can use. His current work focuses on how generative AI tools can elevate the everyday work of admins, consultants, and business leaders by accelerating configuration, sharpening decision-making, and enabling faster, more reliable Salesforce delivery. Andy writes with a belief that the future of enterprise technology belongs to those who embrace AI early, experiment often, and stay curious as the pace of change accelerates. This book reflects that philosophy: practical, forward-looking, and grounded in the realities of building and deploying Salesforce at scale. He lives in Virginia with his family, writes about the future of AI and organizational change, and is always happy to connect with readers (via LinkedIn) who want to discuss the next chapter of AI-first enterprise delivery.

Made in the USA
Monee, IL
05 November 2026

67914380R00325